GENERAL MOTORS | BUICK/OLDSMOBILE/PONTIAC FWD 1985-05 REPAIR MANUAL

CHILTON

Covers U.S. and Canadian models of

Buick LeSabre, Electra and Park Avenue - 1985 thru 2005

Pontiac Bonneville - 1985 thru 2005

Oldsmobile Eighty Eight, Delta 88, Royale, Ninety Eight, LLS and Regency - 1985 thru 2002

Front Wheel Drive

Does not cover diesel engine and related information, supercharger information, rear-wheel drive models or V8 models

by Christine L. Sheeky, S.A.E. and Mike Stubblefield

CHILTON *Automotive Books*

PUBLISHED BY **HAYNES NORTH AMERICA. Inc.**

Manufactured in USA
©1998, 2006 Haynes North America, Inc.
ISBN-13: 978-1-56392-627-3
ISBN-10: 1-56392-627-X
Library of Congress Control Number: 2006934953

Haynes Publishing Group
Sparkford Nr Yeovil
Somerset BA22 7JJ England

Haynes North America, Inc
861 Lawrence Drive
Newbury Park
California 91320 USA

ABCD

9J1

Contents

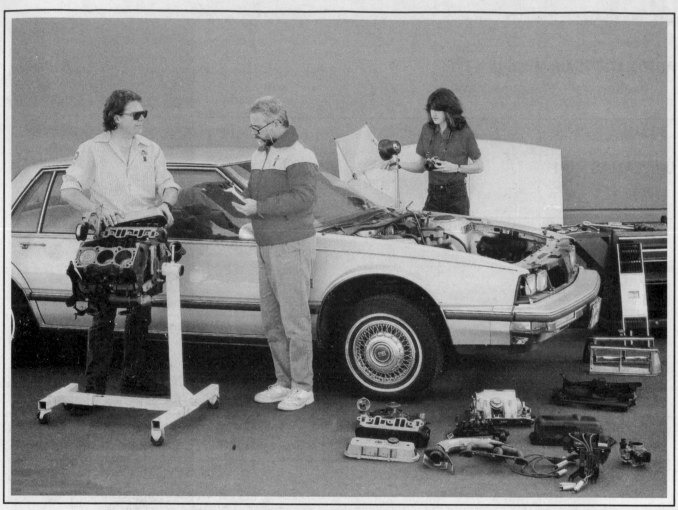

Mechanic, author and photographer with Oldsmobile Delta 88

ACKNOWLEDGEMENTS

Wiring diagrams provided exclusively for the publisher by Valley Forge Technical Information Services. Technical writers who contributed to this project include Jon La Course, Jeff Killingsworth and Ken Freund.

While every attempt is made to ensure that the information in this manual is correct, no liability can be accepted by the authors or publishers for loss, damage or injury caused by any errors in, or omissions from, the information given.

About this manual

ITS PURPOSE

The purpose of this manual is to help you get the best value from your vehicle. It can do so in several ways. It can help you decide what work must be done, even if you choose to have it done by a dealer service department or a repair shop; it provides information and procedures for routine maintenance and servicing; and it offers diagnostic and repair procedures to follow when trouble occurs.

We hope you use the manual to tackle the work yourself. For many simpler jobs, doing it yourself may be quicker than arranging an appointment to get the vehicle into a shop and making the trips to leave it and pick it up. More importantly, a lot of money can be saved by avoiding the expense the shop must pass on to you to cover its labor and overhead costs. An added benefit is the sense of satisfaction and accomplishment that you feel after doing the job yourself.

USING THE MANUAL

The manual is divided into Chapters. Each Chapter is divided into numbered Sections. Each Section consists of consecutively numbered paragraphs.

At the beginning of each numbered Section you will be referred to any illustrations which apply to the procedures in that Section. The reference numbers used in illustration captions pinpoint the pertinent Section and the Step within that Section. That is, illustration 3.2 means the illustration refers to Section 3 and Step (or paragraph) 2 within that Section.

Procedures, once described in the text, are not normally repeated. When it's necessary to refer to another Chapter, the reference will be given as Chapter and Section number. Cross references given without use of the word "Chapter" apply to Sections and/or paragraphs in the same Chapter. For example, "see Section 8" means in the same Chapter.

References to the left or right side of the vehicle assume you are sitting in the driver's seat, facing forward.

Even though we have prepared this manual with extreme care, neither the publisher nor the author can accept responsibility for any errors in, or omissions from, the information given.

➡**NOTE**

A *Note* provides information necessary to properly complete a procedure or information which will make the procedure easier to understand.

CAUTION

A *Caution* provides a special procedure or special steps which must be taken while completing the procedure where the Caution is found. Not heeding a Caution can result in damage to the assembly being worked on.

WARNING

A *Warning* provides a special procedure or special steps which must be taken while completing the procedure where the Warning is found. Not heeding a Warning can result in personal injury.

Introduction to the Buick, Oldsmobile and Pontiac full-size, front-wheel drive models

The full-size General Motors models covered by this manual are front engine/front-wheel drive "C" and "H" body vehicles only. Most are four-door sedans, although some two-door models are available.

All models are powered by a transversely mounted V6 engine which drives the front wheels through an automatic transaxle and independent driveaxles.

Independent suspension, featuring coil springs and struts or shock absorbers, is used at all four wheels. The rack and pinion steering unit is mounted behind the engine.

The brakes are disc at the front and drums or discs at the rear, with power assist standard.

Vehicle Identification Numbers

Modifications are a continuing and unpublicized part of vehicle manufacturing. Since spare parts manuals and lists are compiled on a numerical basis, the individual vehicle numbers are essential to correctly identify the component required.

VEHICLE IDENTIFICATION NUMBER (VIN)

This very important identification number is stamped on a plate attached to the left side of the dashboard and is visible through the driver's side of the windshield (see illustration). The VIN also appears on the Vehicle Certificate of Title and Registration. It contains informa-

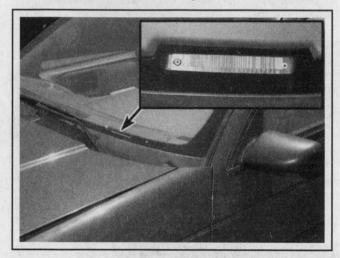

The vehicle Identification Number (VIN) is on a plate attached to the top of the dashboard on the driver's side of the vehicle - it can be seen through the windshield

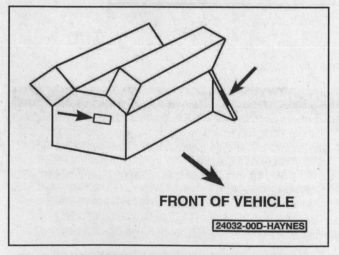

FRONT OF VEHICLE

24032-00D-HAYNES

The engine identification number is in one of two places, at the right end of the block adjacent to the water pump, or on the front side of the block near the starter - on later models, the numbers are in both locations

tion such as where and when the vehicle was manufactured, the model year and the body style.

VEHICLE CERTIFICATION PLATE

The Vehicle Certification Plate (VC label) is affixed to the rear of the left front door. The plate contains the name of the manufacturer, the month and year of production, the Gross Vehicle Weight Rating (GVWR) and the certification statement.

BODY IDENTIFICATION PLATE

The body identification plate is located in the engine compartment on the upper surface of the radiator support. Like the VIN, it contains valuable information concerning the production of the vehicle, as well as information on the options with which it is equipped. This plate is especially useful for matching the color and type of paint for repair work.

ENGINE IDENTIFICATION NUMBER

The engine ID number is located on a pad at the drivebelt (right) end of the engine block, adjacent to the water pump or on the front surface of the block at the transaxle (left) end, adjacent to the starter (see illustration).

SERVICE PARTS IDENTIFICATION LABEL

This label is located inside the trunk (see illustration). It lists the VIN number, wheelbase, paint number, options and other information specific to your vehicle. Always refer to this label when ordering parts.

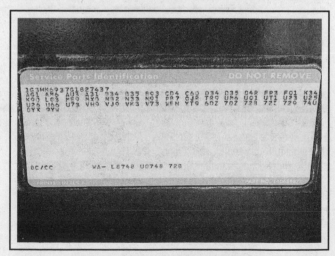

The service parts identification label is located on the inside of the trunk lid or under the spare tire (2005 Bonneville models)

TRANSAXLE IDENTIFICATION NUMBER

The transaxle identification number is located on the right rear side of the transaxle (see illustration).

VEHICLE EMISSIONS CONTROL INFORMATION LABEL

The Vehicle Emissions Control Information label is under the hood, often attached to the left shock tower (see Chapter 6 for more information and an illustration of the label).

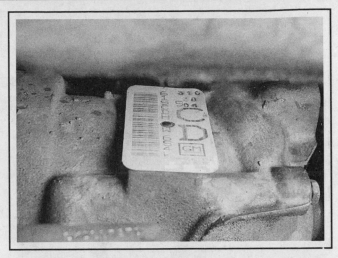

The transaxle identification number is on the right rear side of the transaxle

Buying parts

Replacement parts are available from many sources, which generally fall into one of two categories - authorized dealer parts departments and independent retail auto parts stores. Our advice concerning these parts is as follows:

Retail auto parts stores: Good auto parts stores will stock frequently needed components which wear out relatively fast, such as clutch components, exhaust systems, brake parts, tune-up parts, etc. These stores often supply new or reconditioned parts on an exchange basis, which can save a considerable amount of money. Discount auto parts stores are often very good places to buy materials and parts needed for general vehicle maintenance such as oil, grease, filters, spark plugs, belts, touch-up paint, bulbs, etc. They also usually sell tools and general accessories, have convenient hours, charge lower prices and can often be found not far from home.

Authorized dealer parts department: This is the best source for parts which are unique to the vehicle and not generally available elsewhere (such as major engine parts, transmission parts, trim pieces, etc.).

Warranty information: If the vehicle is still covered under warranty, be sure that any replacement parts purchased - regardless of the source - do not invalidate the warranty!

To be sure of obtaining the correct parts, have engine and chassis numbers available and, if possible, take the old parts along for positive identification.

Maintenance techniques, tools and working facilities

MAINTENANCE TECHNIQUES

There are a number of techniques involved in maintenance and repair that will be referred to throughout this manual. Application of these techniques will enable the home mechanic to be more efficient, better organized and capable of performing the various tasks properly, which will ensure that the repair job is thorough and complete.

Fasteners

Fasteners are nuts, bolts, studs and screws used to hold two or more parts together. There are a few things to keep in mind when work- ing with fasteners. Almost all of them use a locking device of some type, either a lockwasher, locknut, locking tab or thread adhesive. All threaded fasteners should be clean and straight, with undamaged threads and undamaged corners on the hex head where the wrench fits. Develop the habit of replacing all damaged nuts and bolts with new ones. Special locknuts with nylon or fiber inserts can only be used once. If they are removed, they lose their locking ability and must be replaced with new ones.

Rusted nuts and bolts should be treated with a penetrating fluid to ease removal and prevent breakage. Some mechanics use turpentine in a spout-type oil can, which works quite well. After applying the rust penetrant, let it work for a few minutes before trying to loosen the nut

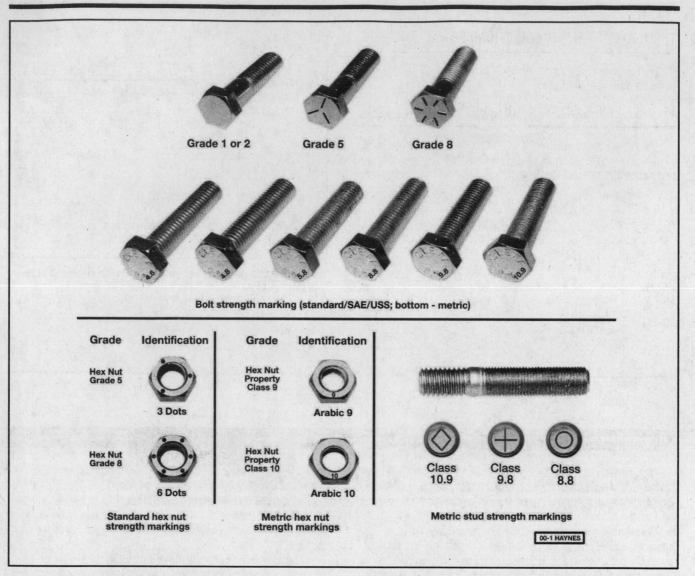

Grade 1 or 2   Grade 5   Grade 8

Bolt strength marking (standard/SAE/USS; bottom - metric)

Grade	Identification
Hex Nut Grade 5	3 Dots
Hex Nut Grade 8	6 Dots

Standard hex nut strength markings

Grade	Identification
Hex Nut Property Class 9	Arabic 9
Hex Nut Property Class 10	Arabic 10

Metric hex nut strength markings

Class 10.9   Class 9.8   Class 8.8

Metric stud strength markings

00-1 HAYNES

or bolt. Badly rusted fasteners may have to be chiseled or sawed off or removed with a special nut breaker, available at tool stores.

If a bolt or stud breaks off in an assembly, it can be drilled and removed with a special tool commonly available for this purpose. Most automotive machine shops can perform this task, as well as other repair procedures, such as the repair of threaded holes that have been stripped out.

Flat washers and lockwashers, when removed from an assembly, should always be replaced exactly as removed. Replace any damaged washers with new ones. Never use a lockwasher on any soft metal surface (such as aluminum), thin sheet metal or plastic.

Fastener sizes

For a number of reasons, automobile manufacturers are making wider and wider use of metric fasteners. Therefore, it is important to be able to tell the difference between standard (sometimes called U.S. or SAE) and metric hardware, since they cannot be interchanged.

All bolts, whether standard or metric, are sized according to diameter, thread pitch and length. For example, a standard 1/2 - 13 x 1 bolt is 1/2 inch in diameter, has 13 threads per inch and is 1 inch long. An M12 - 1.75 x 25 metric bolt is 12 mm in diameter, has a thread pitch of 1.75 mm (the distance between threads) and is 25 mm long. The two bolts are nearly identical, and easily confused, but they are not interchangeable.

In addition to the differences in diameter, thread pitch and length, metric and standard bolts can also be distinguished by examining the bolt heads. To begin with, the distance across the flats on a standard bolt head is measured in inches, while the same dimension on a metric bolt is sized in millimeters (the same is true for nuts). As a result, a standard wrench should not be used on a metric bolt and a metric wrench should not be used on a standard bolt. Also, most standard bolts have slashes radiating out from the center of the head to denote the grade or strength of the bolt, which is an indication of the amount of torque that can be applied to it. The greater the number of slashes, the greater the strength of the bolt. Grades 0 through 5 are commonly used on automobiles. Metric bolts have a property class (grade) number, rather than a slash, molded into their heads to indicate bolt strength. In this case, the higher the number, the stronger the bolt. Property class numbers 8.8, 9.8 and 10.9 are commonly used on automobiles.

Strength markings can also be used to distinguish standard hex nuts from metric hex nuts. Many standard nuts have dots stamped into one side, while metric nuts are marked with a number. The greater the number of dots, or the higher the number, the greater the strength of the nut.

Metric studs are also marked on their ends according to property class (grade). Larger studs are numbered (the same as metric bolts), while smaller studs carry a geometric code to denote grade.

Metric thread sizes	Ft-lbs	Nm
M-6	6 to 9	9 to 12
M-8	14 to 21	19 to 28
M-10	28 to 40	38 to 54
M-12	50 to 71	68 to 96
M-14	80 to 140	109 to 154
Pipe thread sizes		
1/8	5 to 8	7 to 10
1/4	12 to 18	17 to 24
3/8	22 to 33	30 to 44
1/2	25 to 35	34 to 47
U.S. thread sizes		
1/4 - 20	6 to 9	9 to 12
5/16 - 18	12 to 18	17 to 24
5/16 - 24	14 to 20	19 to 27
3/8 - 16	22 to 32	30 to 43
3/8 - 24	27 to 38	37 to 51
7/16 - 14	40 to 55	55 to 74
7/16 - 20	40 to 60	55 to 81
1/2 - 13	55 to 80	75 to 108

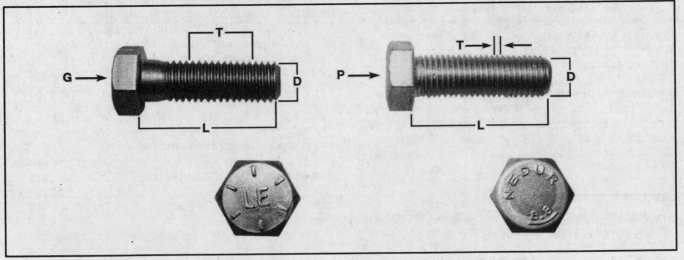

Standard (SAE and USS) bolt dimensions/grade marks

G Grade marks (bolt strength)
L Length (in inches)
T Thread pitch (number of threads per inch)
D Nominal diameter (in inches)

Metric bolt dimensions/grade marks

P Property class (bolt strength)
L Length (in millimeters)
T Thread pitch (distance between threads in millimeters)
D Diameter

It should be noted that many fasteners, especially Grades 0 through 2, have no distinguishing marks on them. When such is the case, the only way to determine whether it is standard or metric is to measure the thread pitch or compare it to a known fastener of the same size.

Standard fasteners are often referred to as SAE, as opposed to metric. However, it should be noted that SAE technically refers to a non-metric fine thread fastener only. Coarse thread non-metric fasteners are referred to as USS sizes.

Since fasteners of the same size (both standard and metric) may have different strength ratings, be sure to reinstall any bolts, studs or nuts removed from your vehicle in their original locations. Also, when replacing a fastener with a new one, make sure that the new one has a strength rating equal to or greater than the original.

Tightening sequences and procedures

Most threaded fasteners should be tightened to a specific torque value (torque is the twisting force applied to a threaded component such as a nut or bolt). Overtightening the fastener can weaken it and cause it to break, while undertightening can cause it to eventually come loose. Bolts, screws and studs, depending on the material they are made of and their thread diameters, have specific torque values, many of which are noted in the Specifications at the end of each Chapter. Be sure to follow the torque recommendations closely. For fasteners not assigned a specific torque, a general torque value chart is presented here as a guide. These torque values are for dry (unlubricated) fasteners threaded into steel or cast iron (not aluminum). As was previously mentioned, the size and grade of a fastener determine the amount of torque that can

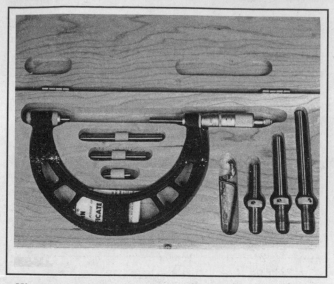

Micrometer set

Dial indicator set

safely be applied to it. The figures listed here are approximate for Grade 2 and Grade 3 fasteners. Higher grades can tolerate higher torque values.

Fasteners laid out in a pattern, such as cylinder head bolts, oil pan bolts, differential cover bolts, etc., must be loosened or tightened in sequence to avoid warping the component. This sequence will normally be shown in the appropriate Chapter. If a specific pattern is not given, the following procedures can be used to prevent warping.

Initially, the bolts or nuts should be assembled finger-tight only. Next, they should be tightened one full turn each, in a criss-cross or diagonal pattern. After each one has been tightened one full turn, return to the first one and tighten them all one-half turn, following the same pattern. Finally, tighten each of them one-quarter turn at a time until each fastener has been tightened to the proper torque. To loosen and remove the fasteners, the procedure would be reversed.

Component disassembly

Component disassembly should be done with care and purpose to help ensure that the parts go back together properly. Always keep track of the sequence in which parts are removed. Make note of special characteristics or marks on parts that can be installed more than one way, such as a grooved thrust washer on a shaft. It is a good idea to lay the disassembled parts out on a clean surface in the order that they were removed. It may also be helpful to make sketches or take instant photos of components before removal.

When removing fasteners from a component, keep track of their locations. Sometimes threading a bolt back in a part, or putting the washers and nut back on a stud, can prevent mix-ups later. If nuts and bolts cannot be returned to their original locations, they should be kept in a compartmented box or a series of small boxes. A cupcake or muffin tin is ideal for this purpose, since each cavity can hold the bolts and nuts from a particular area (i.e. oil pan bolts, valve cover bolts, engine mount bolts, etc.). A pan of this type is especially helpful when working on assemblies with very small parts, such as the carburetor, alternator, valve train or interior dash and trim pieces. The cavities can be marked with paint or tape to identify the contents.

Whenever wiring looms, harnesses or connectors are separated, it is a good idea to identify the two halves with numbered pieces of masking tape so they can be easily reconnected.

Gasket sealing surfaces

Throughout any vehicle, gaskets are used to seal the mating surfaces between two parts and keep lubricants, fluids, vacuum or pressure contained in an assembly.

Many times these gaskets are coated with a liquid or paste-type gasket sealing compound before assembly. Age, heat and pressure can sometimes cause the two parts to stick together so tightly that they are very difficult to separate. Often, the assembly can be loosened by striking it with a soft-face hammer near the mating surfaces. A regular hammer can be used if a block of wood is placed between the hammer and the part. Do not hammer on cast parts or parts that could be easily damaged. With any particularly stubborn part, always recheck to make sure that every fastener has been removed.

Avoid using a screwdriver or bar to pry apart an assembly, as they can easily mar the gasket sealing surfaces of the parts, which must remain smooth. If prying is absolutely necessary, use an old broom handle, but keep in mind that extra clean up will be necessary if the wood splinters.

After the parts are separated, the old gasket must be carefully scraped off and the gasket surfaces cleaned. Stubborn gasket material can be soaked with rust penetrant or treated with a special chemical to soften it so it can be easily scraped off.

✳✳ CAUTION:

Never use gasket removal solutions or caustic chemicals on plastic or other composite components.

A scraper can be fashioned from a piece of copper tubing by flattening and sharpening one end. Copper is recommended because it is usually softer than the surfaces to be scraped, which reduces the chance of gouging the part. Some gaskets can be removed with a wire brush, but regardless of the method used, the mating surfaces must be left clean and smooth. If for some reason the gasket surface is gouged, then a gasket sealer thick enough to fill scratches will have to be used during reassembly of the components. For most applications, a non-drying (or semi-drying) gasket sealer should be used.

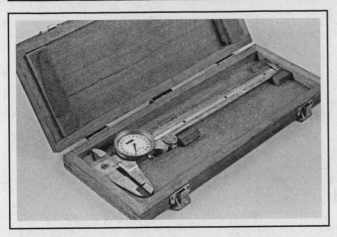

Dial caliper

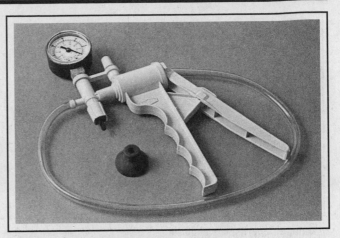

Hand-operated vacuum pump

Timing light

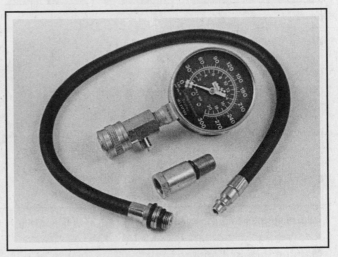

Compression gauge with spark plug hole adapter

Hose removal tips

⁂ WARNING:

If the vehicle is equipped with air conditioning, do not disconnect any of the A/C hoses without first having the system depressurized by a dealer service department or a service station.

Hose removal precautions closely parallel gasket removal precautions. Avoid scratching or gouging the surface that the hose mates against or the connection may leak. This is especially true for radiator hoses. Because of various chemical reactions, the rubber in hoses can bond itself to the metal spigot that the hose fits over. To remove a hose, first loosen the hose clamps that secure it to the spigot. Then, with slip-joint pliers, grab the hose at the clamp and rotate it around the spigot. Work it back and forth until it is completely free, then pull it off. Silicone or other lubricants will ease removal if they can be applied between the hose and the outside of the spigot. Apply the same lubricant to the inside of the hose and the outside of the spigot to simplify installation.

As a last resort (and if the hose is to be replaced with a new one anyway), the rubber can be slit with a knife and the hose peeled from the spigot. If this must be done, be careful that the metal connection is not damaged.

If a hose clamp is broken or damaged, do not reuse it. Wire-type clamps usually weaken with age, so it is a good idea to replace them with screw-type clamps whenever a hose is removed.

TOOLS

A selection of good tools is a basic requirement for anyone who plans to maintain and repair his or her own vehicle. For the owner who has few tools, the initial investment might seem high, but when compared to the spiraling costs of professional auto maintenance and repair, it is a wise one.

To help the owner decide which tools are needed to perform the tasks detailed in this manual, the following tool lists are offered: *Maintenance and minor repair, Repair/overhaul and Special.*

The newcomer to practical mechanics should start off with the *maintenance and minor repair* tool kit, which is adequate for the simpler jobs performed on a vehicle. Then, as confidence and experience grow, the owner can tackle more difficult tasks, buying additional tools as they are needed. Eventually the basic kit will be expanded into the *repair and overhaul* tool set. Over a period of time, the experienced do-it-yourselfer will assemble a tool set complete enough for most repair and overhaul procedures and will add tools from the special category when it is felt that the expense is justified by the frequency of use.

Damper/steering wheel puller

General purpose puller

Hydraulic lifter removal tool

Valve spring compressor

Valve spring compressor

Ridge reamer

Maintenance and minor repair tool kit

The tools in this list should be considered the minimum required for performance of routine maintenance, servicing and minor repair work. We recommend the purchase of combination wrenches (box-end and open-end combined in one wrench). While more expensive than open end wrenches, they offer the advantages of both types of wrench.

Combination wrench set (1/4-inch to 1 inch or 6 mm to 19 mm)
Adjustable wrench, 8 inch
Spark plug wrench with rubber insert
Spark plug gap adjusting tool
Feeler gauge set
Brake bleeder wrench
Standard screwdriver (5/16-inch x 6 inch)
Phillips screwdriver (No. 2 x 6 inch)
Combination pliers - 6 inch
Hacksaw and assortment of blades
Tire pressure gauge
Grease gun
Oil can
Fine emery cloth
Wire brush
Battery post and cable cleaning tool
Oil filter wrench
Funnel (medium size)
Safety goggles
Jackstands (2)
Drain pan

➡ **Note: If basic tune-ups are going to be part of routine maintenance, it will be necessary to purchase a good quality stroboscopic timing light and combination tachometer/dwell meter. Although they are included in the list of special tools, it is mentioned here because they are absolutely necessary for tuning most vehicles properly.**

Repair and overhaul tool set

These tools are essential for anyone who plans to perform major repairs and are in addition to those in the maintenance and minor repair tool kit. Included is a comprehensive set of sockets which, though expensive, are invaluable because of their versatility, especially when various extensions and drives are available. We recommend the 1/2-inch drive over the 3/8-inch drive. Although the larger drive is bulky and more expensive, it has the capacity of accepting a very wide range of large sockets. Ideally, however, the mechanic should have a 3/8-inch drive set and a 1/2-inch drive set.

Socket set(s)
Reversible ratchet
Extension - 10 inch
Universal joint
Torque wrench (same size drive as sockets)
Ball peen hammer - 8 ounce
Soft-face hammer (plastic/rubber)
Standard screwdriver (1/4-inch x 6 inch)
Standard screwdriver (stubby - 5/16-inch)
Phillips screwdriver (No. 3 x 8 inch)
Phillips screwdriver (stubby - No. 2)
Pliers - vise grip

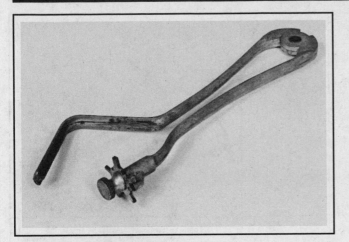

Piston ring groove cleaning tool

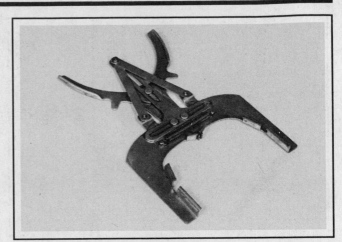

Ring removal/installation tool

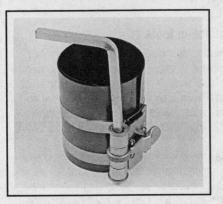

Ring compressor

Cylinder hone

Brake hold-down spring tool

Pliers - lineman's
Pliers - needle nose
Pliers - snap-ring (internal and external)
Cold chisel - 1/2-inch
Scribe
Scraper (made from flattened copper tubing)
Centerpunch
Pin punches (1/16, 1/8, 3/16-inch)
Steel rule/straightedge - 12 inch
Allen wrench set (1/8 to 3/8-inch or 4 mm to 10 mm)
A selection of files
Wire brush (large)
Jackstands (second set)
Jack (scissor or hydraulic type)

➡**Note: Another tool which is often useful is an electric drill with a chuck capacity of 3/8-inch and a set of good quality drill bits.**

Special tools

The tools in this list include those which are not used regularly, are expensive to buy, or which need to be used in accordance with their manufacturer's instructions. Unless these tools will be used frequently, it is not very economical to purchase many of them. A consideration would be to split the cost and use between yourself and a friend or friends. In addition, most of these tools can be obtained from a tool rental shop on a temporary basis.

This list primarily contains only those tools and instruments widely available to the public, and not those special tools produced by the vehicle manufacturer for distribution to dealer service depart-ments. Occasionally, references to the manufacturer's special tools are included in the text of this manual. Generally, an alternative method of doing the job without the special tool is offered. However, sometimes there is no alternative to their use. Where this is the case, and the tool cannot be purchased or borrowed, the work should be turned over to the dealer service department or an automotive repair shop.

Valve spring compressor
Piston ring groove cleaning tool
Piston ring compressor
Piston ring installation tool
Cylinder compression gauge
Cylinder ridge reamer
Cylinder surfacing hone
Cylinder bore gauge
Micrometers and/or dial calipers
Hydraulic lifter removal tool
Balljoint separator
Universal-type puller
Impact screwdriver
Dial indicator set
Stroboscopic timing light (inductive pick-up)
Hand operated vacuum/pressure pump
Tachometer/dwell meter
Universal electrical multimeter
Cable hoist
Brake spring removal and installation tools
Floor jack

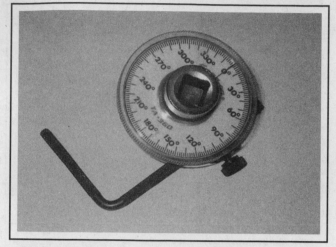

Torque angle gauge

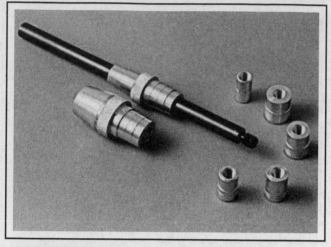

Clutch plate alignment tool

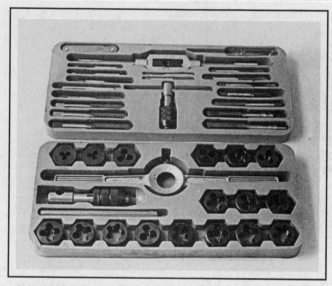

Tap and die set

Buying tools

For the do-it-yourselfer who is just starting to get involved in vehicle maintenance and repair, there are a number of options available when purchasing tools. If maintenance and minor repair is the extent of the work to be done, the purchase of individual tools is satisfactory. If, on the other hand, extensive work is planned, it would be a good idea to purchase a modest tool set from one of the large retail chain stores. A set can usually be bought at a substantial savings over the individual tool prices, and they often come with a tool box. As additional tools are needed, add-on sets, individual tools and a larger tool box can be purchased to expand the tool selection. Building a tool set gradually allows the cost of the tools to be spread over a longer period of time and gives the mechanic the freedom to choose only those tools that will actually be used.

Tool stores will often be the only source of some of the special tools that are needed, but regardless of where tools are bought, try to avoid cheap ones, especially when buying screwdrivers and sockets, because they won't last very long. The expense involved in replacing cheap tools will eventually be greater than the initial cost of quality tools.

Care and maintenance of tools

Good tools are expensive, so it makes sense to treat them with respect. Keep them clean and in usable condition and store them properly when not in use. Always wipe off any dirt, grease or metal chips before putting them away. Never leave tools lying around in the work area. Upon completion of a job, always check closely under the hood for tools that may have been left there so they won't get lost during a test drive.

Some tools, such as screwdrivers, pliers, wrenches and sockets, can be hung on a panel mounted on the garage or workshop wall, while others should be kept in a tool box or tray. Measuring instruments, gauges, meters, etc. must be carefully stored where they cannot be damaged by weather or impact from other tools.

When tools are used with care and stored properly, they will last a very long time. Even with the best of care, though, tools will wear out if used frequently. When a tool is damaged or worn out, replace it. Subsequent jobs will be safer and more enjoyable if you do.

HOW TO REPAIR DAMAGED THREADS

Sometimes, the internal threads of a nut or bolt hole can become stripped, usually from overtightening. Stripping threads is an all-too-common occurrence, especially when working with aluminum parts, because aluminum is so soft that it easily strips out.

Usually, external or internal threads are only partially stripped. After they've been cleaned up with a tap or die, they'll still work. Sometimes, however, threads are badly damaged. When this happens, you've got three choices:

1) *Drill and tap the hole to the next suitable oversize and install a larger diameter bolt, screw or stud.*
2) *Drill and tap the hole to accept a threaded plug, then drill and tap the plug to the original screw size. You can also buy a plug already threaded to the original size. Then you simply drill a hole to the specified size, then run the threaded plug into the hole with a bolt and jam nut. Once the plug is fully seated, remove the jam nut and bolt.*
3) *The third method uses a patented thread repair kit like Heli-Coil or Slimsert. These easy-to-use kits are designed to repair damaged threads in straight-through holes and blind holes. Both are available as kits which can handle a variety of sizes and thread*

patterns. Drill the hole, then tap it with the special included tap. Install the Heli-Coil and the hole is back to its original diameter and thread pitch.

Regardless of which method you use, be sure to proceed calmly and carefully. A little impatience or carelessness during one of these relatively simple procedures can ruin your whole day's work and cost you a bundle if you wreck an expensive part.

WORKING FACILITIES

Not to be overlooked when discussing tools is the workshop. If anything more than routine maintenance is to be carried out, some sort of suitable work area is essential.

It is understood, and appreciated, that many home mechanics do not have a good workshop or garage available, and end up removing an engine or doing major repairs outside. It is recommended, however, that the overhaul or repair be completed under the cover of a roof.

A clean, flat workbench or table of comfortable working height is an absolute necessity. The workbench should be equipped with a vise that has a jaw opening of at least four inches.

As mentioned previously, some clean, dry storage space is also required for tools, as well as the lubricants, fluids, cleaning solvents, etc. which soon become necessary.

Sometimes waste oil and fluids, drained from the engine or cooling system during normal maintenance or repairs, present a disposal problem. To avoid pouring them on the ground or into a sewage system, pour the used fluids into large containers, seal them with caps and take them to an authorized disposal site or recycling center. Plastic jugs, such as old antifreeze containers, are ideal for this purpose.

Always keep a supply of old newspapers and clean rags available. Old towels are excellent for mopping up spills. Many mechanics use rolls of paper towels for most work because they are readily available and disposable. To help keep the area under the vehicle clean, a large cardboard box can be cut open and flattened to protect the garage or shop floor.

Whenever working over a painted surface, such as when leaning over a fender to service something under the hood, always cover it with an old blanket or bedspread to protect the finish. Vinyl covered pads, made especially for this purpose, are available at auto parts stores.

Booster battery (jump) starting

Observe these precautions when using a booster battery to start a vehicle:

a) *Before connecting the booster battery, make sure the ignition switch is in the Off position.*
b) *Turn off the lights, heater and other electrical loads.*
c) *Your eyes should be shielded. Safety goggles are a good idea.*
d) *Make sure the booster battery is the same voltage as the dead one in the vehicle.*
e) *The two vehicles MUST NOT TOUCH each other!*
f) *Make sure the transaxle is in Neutral (manual) or Park (automatic).*
g) *If the booster battery is not a maintenance-free type, remove the vent caps and lay a cloth over the vent holes.*

Connect the red jumper cable to the positive (+) terminals of each battery (see illustration).

Connect one end of the black jumper cable to the negative (-) terminal of the booster battery. The other end of this cable should be connected to a good ground on the vehicle to be started, such as a bolt or bracket on the body.

Start the engine using the booster battery, then, with the engine running at idle speed, disconnect the jumper cables in the reverse order of connection.

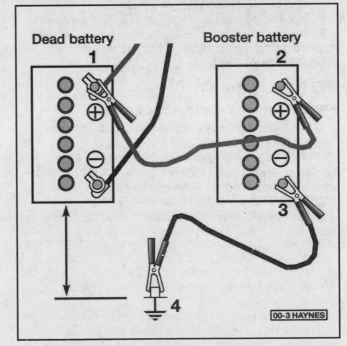

Make the booster battery cable connections in the numerical order shown (note that the negative cable of the booster battery is NOT attached to the negative terminal of the dead battery)

Anti-theft audio systems

GENERAL INFORMATION

1   Some models are equipped with an anti-theft audio system which includes an anti-theft feature that will render the stereo inoperative if stolen. If the power source to the stereo is disconnected with the anti-theft feature activated, the stereo must be "unlocked" using the appropriate secret code to become operative again. Even if the power is immediately reconnected, the stereo will not function. If your vehicle is equipped with this anti-theft system, do not disconnect the battery, remove the stereo or disconnect related components unless you have disabled the anti-theft feature first. In order to deactivate this feature you must know the secret code.

2   Refer to your owner's manual for more complete information on your particular audio system and it's anti-theft feature. If you loose or forget your code, contact your local dealer service department.

DISABLING THE ANTI-THEFT FEATURE

Delco LOC II system

➡**Note: This system uses a six digit code.**

3   With the ignition ON and the stereo OFF, press the stereo's 1 and 4 buttons at the same time for five seconds. The display will show SEC, indicating the unit is in the secure mode (anti-theft feature activated).

4   Press the SET button. The display will show "000."

5   Press the SEEK right or SEEK left arrows as required to display the first digit of the secret code.

6   Press the SCAN button to make the next two digits of your code appear. On 1992 and 1993 models, rotate the TUNE knob to display your code.

7   Press the BAND knob, "000" will be displayed.

8   Repeat steps 5 and 6 to enter the last three digits of your code.

9   Press the BAND knob. If the display shows "—" the anti-theft feature has been deactivated. If SEC is displayed, the code entered was incorrect and the anti-theft feature is still enabled.

THEFTLOCK system

➡**Note: This system uses a three or four digit code.**

10   In order to successfully deactivate the anti-theft feature, pause no more than fifteen seconds between each step.

11   With the ignition ON and the stereo OFF, press the stereo's 1 and 4 buttons at the same time until the display shows SEC, indicating the unit is in the secure mode (anti-theft feature activated).

12   Press the MN button. The display will show "000."

13   Press the MN button again to display the LAST two digits of your code.

14   Next, press the HR button to display the FIRST one or two digits of your code appear.

15   Press the AM-FM button. If the display shows "—" the anti-theft feature has been deactivated. If SEC is displayed, the code entered was incorrect and the anti-theft feature is still enabled.

UNLOCKING THE STEREO AFTER A POWER LOSS

Delco LOC II system

16   If the anti-theft feature was not deactivated before a power interruption occurred, when power is restored to the stereo LOC will appear on the display. In order to successfully unlock the anti-theft feature, pause no more than fifteen seconds between each step.

17   With the ignition ON and the stereo OFF, press the SET button. The display will show "000."

18   Press the SEEK right or SEEK left arrows as required to display the first digit of the secret code.

19   Press the SCAN button to make the next two digits of your code appear. On 1992 and 1993 models, rotate the TUNE knob to display your code.

20   Press the BAND knob, "000" will be displayed.

21   Repeat steps 18 and 19 to enter the last three digits of your code.

22   Press the BAND knob. If the display shows the time (hours and minutes) the anti-theft feature has been successfully deactivated and the stereo will function. If SEC is displayed, the code entered was incorrect and the anti-theft feature is still enabled.

THEFTLOCK system

23   If the anti-theft feature was not deactivated before a power interruption occurred, when power is restored to the stereo LOC will appear on the display. In order to successfully unlock the anti-theft feature, pause no more than fifteen seconds between each step.

24   With the ignition ON and the stereo OFF, press the MN button. The display will show "000."

25   Press the MN button again to display the LAST two digits of your code.

26   Next, press the HR button to display the FIRST one or two digits of your code appear.

27   Press the AM-FM button. If the display shows "SEC" the anti-theft feature has been successfully deactivated and the stereo will function. If you enter the wrong code eight times, the display will show "INOP." If this occurs, the ignition must remain in the ON position for one hour before you can try to unlock the stereo again. Then on the next try, you will only get three chances to unlock the system before the display shows "INOP."

Jacking and towing

JACKING

❈❈ WARNING:

The jack supplied with the vehicle should only be used for raising the vehicle when changing a tire or placing jackstands under the frame. Never work under the vehicle or start the engine while the jack is being used as the only means of support.

The vehicle must be on a level surface with the wheels blocked and the transaxle in Park. Apply the parking brake if the front of the vehicle must be raised. Make sure no one is in the vehicle as it's being raised with the jack.

Remove the jack, lug nut wrench and spare tire (if needed) from the vehicle. If a tire is being replaced, use the lug wrench to remove the wheel cover.

❈❈ WARNING:

Wheel covers may have sharp edges - be very careful not to cut yourself.

Loosen the lug nuts one-half turn, but leave them in place until the tire is raised off the ground. Position the jack under the vehicle at the indicated jacking point. There's a front and rear jacking point on each side of the vehicle (see illustrations). On later models, the proper location is marked with the word "Jack."

Turn the jack handle clockwise until the tire clears the ground.

Remove the lug nuts, pull the tire off and replace it with the spare. Replace the lug nuts with the beveled edges facing in and tighten them snugly. Don't attempt to tighten them completely until the vehicle is lowered or it could slip off the jack.

Turn the jack handle counterclockwise to lower the vehicle. Remove the jack and tighten the lug nuts in a criss-cross pattern. If possible, tighten the nuts with a torque wrench (see Chapter 1 for the torque figures). If you don't have access to a torque wrench, have the nuts checked by a service station or repair shop as soon as possible.

Stow the tire, jack and wrench and unblock the wheels.

TOWING

As a general rule, these vehicles should be towed with the front (drive) wheels off the ground. You may tow the vehicle with the front wheels on the ground for distances up to 500 miles provided speed does not exceed 55 mph. These vehicles should not be towed with all four wheels on the ground.

Be sure to release the parking brake. If the vehicle is being towed with the front wheels on the ground, place the transaxle in Neutral. Also, the ignition key must be in the ACC position, since the steering lock mechanism isn't strong enough to hold the front wheels straight while towing.

Equipment specifically designed for towing should be used. It must be attached to the main structural members of the vehicle, not the bumpers or brackets.

Safety is a major consideration when towing and all applicable state and local laws must be obeyed. A safety chain must be used at all times. Remember that power steering and brakes won't work with the engine off.

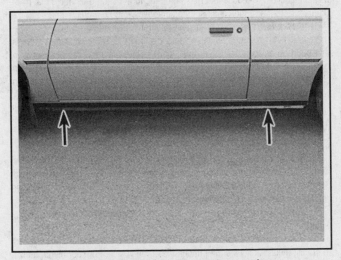

Location of the rocker panel flange notches used for jack placement

Make sure the head of the jack securely engages with the rocker panel flange

Automotive chemicals and lubricants

A number of automotive chemicals and lubricants are available for use during vehicle maintenance and repair. They include a wide variety of products ranging from cleaning solvents and degreasers to lubricants and protective sprays for rubber, plastic and vinyl.

CLEANERS

Carburetor cleaner and choke cleaner is a strong solvent for gum, varnish and carbon. Most carburetor cleaners leave a dry-type lubricant film which will not harden or gum up. Because of this film it is not recommended for use on electrical components.

Brake system cleaner is used to remove brake dust, grease and brake fluid from the brake system, where clean surfaces are absolutely necessary. It leaves no residue and often eliminates brake squeal caused by contaminants.

Electrical cleaner removes oxidation, corrosion and carbon deposits from electrical contacts, restoring full current flow. It can also be used to clean spark plugs, carburetor jets, voltage regulators and other parts where an oil-free surface is desired.

Demoisturants remove water and moisture from electrical components such as alternators, voltage regulators, electrical connectors and fuse blocks. They are non-conductive and non-corrosive.

Degreasers are heavy-duty solvents used to remove grease from the outside of the engine and from chassis components. They can be sprayed or brushed on and, depending on the type, are rinsed off either with water or solvent.

LUBRICANTS

Motor oil is the lubricant formulated for use in engines. It normally contains a wide variety of additives to prevent corrosion and reduce foaming and wear. Motor oil comes in various weights (viscosity ratings) from 0 to 50. The recommended weight of the oil depends on the season, temperature and the demands on the engine. Light oil is used in cold climates and under light load conditions. Heavy oil is used in hot climates and where high loads are encountered. Multi-viscosity oils are designed to have characteristics of both light and heavy oils and are available in a number of weights from 5W-20 to 20W-50.

Gear oil is designed to be used in differentials, manual transmissions and other areas where high-temperature lubrication is required.

Chassis and wheel bearing grease is a heavy grease used where increased loads and friction are encountered, such as for wheel bearings, balljoints, tie-rod ends and universal joints.

High-temperature wheel bearing grease is designed to withstand the extreme temperatures encountered by wheel bearings in disc brake equipped vehicles. It usually contains molybdenum disulfide (moly), which is a dry-type lubricant.

White grease is a heavy grease for metal-to-metal applications where water is a problem. White grease stays soft under both low and high temperatures (usually from -100 to +190-degrees F), and will not wash off or dilute in the presence of water.

Assembly lube is a special extreme pressure lubricant, usually containing moly, used to lubricate high-load parts (such as main and rod bearings and cam lobes) for initial start-up of a new engine. The assembly lube lubricates the parts without being squeezed out or washed away until the engine oiling system begins to function.

Silicone lubricants are used to protect rubber, plastic, vinyl and nylon parts.

Graphite lubricants are used where oils cannot be used due to contamination problems, such as in locks. The dry graphite will lubricate metal parts while remaining uncontaminated by dirt, water, oil or acids. It is electrically conductive and will not foul electrical contacts in locks such as the ignition switch.

Moly penetrants loosen and lubricate frozen, rusted and corroded fasteners and prevent future rusting or freezing.

Heat-sink grease is a special electrically non-conductive grease that is used for mounting electronic ignition modules where it is essential that heat is transferred away from the module.

SEALANTS

RTV sealant is one of the most widely used gasket compounds. Made from silicone, RTV is air curing, it seals, bonds, waterproofs, fills surface irregularities, remains flexible, doesn't shrink, is relatively easy to remove, and is used as a supplementary sealer with almost all low and medium temperature gaskets.

Anaerobic sealant is much like RTV in that it can be used either to seal gaskets or to form gaskets by itself. It remains flexible, is solvent resistant and fills surface imperfections. The difference between an anaerobic sealant and an RTV-type sealant is in the curing. RTV cures when exposed to air, while an anaerobic sealant cures only in the absence of air. This means that an anaerobic sealant cures only after the assembly of parts, sealing them together.

Thread and pipe sealant is used for sealing hydraulic and pneumatic fittings and vacuum lines. It is usually made from a Teflon compound, and comes in a spray, a paint-on liquid and as a wrap-around tape.

CHEMICALS

Anti-seize compound prevents seizing, galling, cold welding, rust and corrosion in fasteners. High-temperature anti-seize, usually made with copper and graphite lubricants, is used for exhaust system and exhaust manifold bolts.

Anaerobic locking compounds are used to keep fasteners from vibrating or working loose and cure only after installation, in the absence of air. Medium strength locking compound is used for small nuts, bolts and screws that may be removed later. High-strength locking compound is for large nuts, bolts and studs which aren't removed on a regular basis.

Oil additives range from viscosity index improvers to chemical treatments that claim to reduce internal engine friction. It should be noted that most oil manufacturers caution against using additives with their oils.

Gas additives perform several functions, depending on their chemical makeup. They usually contain solvents that help dissolve gum and varnish that build up on carburetor, fuel injection and intake parts. They also serve to break down carbon deposits that form on the inside surfaces of the combustion chambers. Some additives contain upper cylinder lubricants for valves and piston rings, and others contain chemicals to remove condensation from the gas tank.

MISCELLANEOUS

Brake fluid is specially formulated hydraulic fluid that can withstand the heat and pressure encountered in brake systems. Care must be taken so this fluid does not come in contact with painted surfaces or plastics. An opened container should always be resealed to prevent contamination by water or dirt.

Weatherstrip adhesive is used to bond weatherstripping around doors, windows and trunk lids. It is sometimes used to attach trim pieces.

Undercoating is a petroleum-based, tar-like substance that is designed to protect metal surfaces on the underside of the vehicle from corrosion. It also acts as a sound-deadening agent by insulating the bottom of the vehicle.

Waxes and polishes are used to help protect painted and plated surfaces from the weather. Different types of paint may require the use of different types of wax and polish. Some polishes utilize a chemical or abrasive cleaner to help remove the top layer of oxidized (dull) paint on older vehicles. In recent years many non-wax polishes that contain a wide variety of chemicals such as polymers and silicones have been introduced. These non-wax polishes are usually easier to apply and last longer than conventional waxes and polishes.

CONVERSION FACTORS

LENGTH (distance)

Inches (in)	X	25.4	= Millimeters (mm)	X	0.0394 = Inches (in)
Feet (ft)	X	0.305	= Meters (m)	X	3.281 = Feet (ft)
Miles	X	1.609	= Kilometers (km)	X	0.621 = Miles

VOLUME (capacity)

Cubic inches (cu in; in^3)	X	16.387	= Cubic centimeters (cc; cm^3)	X	0.061 = Cubic inches (cu in; in^3)
Imperial pints (Imp pt)	X	0.568	= Liters (l)	X	1.76 = Imperial pints (Imp pt)
Imperial quarts (Imp qt)	X	1.137	= Liters (l)	X	0.88 = Imperial quarts (Imp qt)
Imperial quarts (Imp qt)	X	1.201	= US quarts (US qt)	X	0.833 = Imperial quarts (Imp qt)
US quarts (US qt)	X	0.946	= Liters (l)	X	1.057 = US quarts (US qt)
Imperial gallons (Imp gal)	X	4.546	= Liters (l)	X	0.22 = Imperial gallons (Imp gal)
Imperial gallons (Imp gal)	X	1.201	= US gallons (US gal)	X	0.833 = Imperial gallons (Imp gal)
US gallons (US gal)	X	3.785	= Liters (l)	X	0.264 = US gallons (US gal)

MASS (weight)

Ounces (oz)	X	28.35	= Grams (g)	X	0.035 = Ounces (oz)
Pounds (lb)	X	0.454	= Kilograms (kg)	X	2.205 = Pounds (lb)

FORCE

Ounces-force (ozf; oz)	X	0.278	= Newtons (N)	X	3.6 = Ounces-force (ozf; oz)
Pounds-force (lbf; lb)	X	4.448	= Newtons (N)	X	0.225 = Pounds-force (lbf; lb)
Newtons (N)	X	0.1	= Kilograms-force (kgf; kg)	X	9.81 = Newtons (N)

PRESSURE

Pounds-force per square inch (psi; lbf/in^2; lb/in^2)	X	0.070	= Kilograms-force per square centimeter (kgf/cm^2; kg/cm^2)	X	14.223 = Pounds-force per square inch (psi; lbf/in^2; lb/in^2)
Pounds-force per square inch (psi; lbf/in^2; lb/in^2)	X	0.068	= Atmospheres (atm)	X	14.696 = Pounds-force per square inch (psi; lbf/in^2; lb/in^2)
Pounds-force per square inch (psi; lbf/in^2; lb/in^2)	X	0.069	= Bars	X	14.5 = Pounds-force per square inch (psi; lbf/in^2; lb/in^2)
Pounds-force per square inch (psi; lbf/in^2; lb/in^2)	X	6.895	= Kilopascals (kPa)	X	0.145 = Pounds-force per square inch (psi; lbf/in^2; lb/in^2)
Kilopascals (kPa)	X	0.01	= Kilograms-force per square centimeter (kgf/cm^2; kg/cm^2)	X	98.1 = Kilopascals (kPa)

TORQUE (moment of force)

Pounds-force inches (lbf in; lb in)	X	1.152	= Kilograms-force centimeter (kgf cm; kg cm)	X	0.868 = Pounds-force inches (lbf in; lb in)
Pounds-force inches (lbf in; lb in)	X	0.113	= Newton meters (Nm)	X	8.85 = Pounds-force inches (lbf in; lb in)
Pounds-force inches (lbf in; lb in)	X	0.083	= Pounds-force feet (lbf ft; lb ft)	X	12 = Pounds-force inches (lbf in; lb in)
Pounds-force feet (lbf ft; lb ft)	X	0.138	= Kilograms-force meters (kgf m; kg m)	X	7.233 = Pounds-force feet (lbf ft; lb ft)
Pounds-force feet (lbf ft; lb ft)	X	1.356	= Newton meters (Nm)	X	0.738 = Pounds-force feet (lbf ft; lb ft)
Newton meters (Nm)	X	0.102	= Kilograms-force meters (kgf m; kg m)	X	9.804 = Newton meters (Nm)

VACUUM

Inches mercury (in. Hg)	X	3.377	= Kilopascals (kPa)	X	0.2961 = Inches mercury
Inches mercury (in. Hg)	X	25.4	= Millimeters mercury (mm Hg)	X	0.0394 = Inches mercury

POWER

Horsepower (hp)	X	745.7	= Watts (W)	X	0.0013 = Horsepower (hp)

VELOCITY (speed)

Miles per hour (miles/hr; mph)	X	1.609	= Kilometers per hour (km/hr; kph)	X	0.621 = Miles per hour (miles/hr; mph)

FUEL CONSUMPTION *

Miles per gallon, Imperial (mpg)	X	0.354	= Kilometers per liter (km/l)	X	2.825 = Miles per gallon, Imperial (mpg)
Miles per gallon, US (mpg)	X	0.425	= Kilometers per liter (km/l)	X	2.352 = Miles per gallon, US (mpg)

TEMPERATURE

Degrees Fahrenheit = (°C x 1.8) + 32   Degrees Celsius (Degrees Centigrade; °C) = (°F - 32) x 0.56

It is common practice to convert from miles per gallon (mpg) to liters/100 kilometers (l/100km), where mpg (Imperial) x l/100 km = 282 and mpg (US) x l/100 km = 235

FRACTION/DECIMAL/MILLIMETER EQUIVALENTS

DECIMALS to MILLIMETERS

Decimal	mm	Decimal	mm
0.001	0.0254	0.500	12.7000
0.002	0.0508	0.510	12.9540
0.003	0.0762	0.520	13.2080
0.004	0.1016	0.530	13.4620
0.005	0.1270	0.540	13.7160
0.006	0.1524	0.550	13.9700
0.007	0.1778	0.560	14.2240
0.008	0.2032	0.570	14.4780
0.009	0.2286	0.580	14.7320
		0.590	14.9860
0.010	0.2540		
0.020	0.5080		
0.030	0.7620		
0.040	1.0160	0.600	15.2400
0.050	1.2700	0.610	15.4940
0.060	1.5240	0.620	15.7480
0.070	1.7780	0.630	16.0020
0.080	2.0320	0.640	16.2560
0.090	2.2860	0.650	16.5100
		0.660	16.7640
0.100	2.5400	0.670	17.0180
0.110	2.7940	0.680	17.2720
0.120	3.0480	0.690	17.5260
0.130	3.3020		
0.140	3.5560		
0.150	3.8100		
0.160	4.0640	0.700	17.7800
0.170	4.3180	0.710	18.0340
0.180	4.5720	0.720	18.2880
0.190	4.8260	0.730	18.5420
		0.740	18.7960
0.200	5.0800	0.750	19.0500
0.210	5.3340	0.760	19.3040
0.220	5.5880	0.770	19.5580
0.230	5.8420	0.780	19.8120
0.240	6.0960	0.790	20.0660
0.250	6.3500		
0.260	6.6040		
0.270	6.8580	0.800	20.3200
0.280	7.1120	0.810	20.5740
0.290	7.3660	0.820	21.8280
		0.830	21.0820
0.300	7.6200	0.840	21.3360
0.310	7.8740	0.850	21.5900
0.320	8.1280	0.860	21.8440
0.330	8.3820	0.870	22.0980
0.340	8.6360	0.880	22.3520
0.350	8.8900	0.890	22.6060
0.360	9.1440		
0.370	9.3980		
0.380	9.6520		
0.390	9.9060		
		0.900	22.8600
0.400	10.1600	0.910	23.1140
0.410	10.4140	0.920	23.3680
0.420	10.6680	0.930	23.6220
0.430	10.9220	0.940	23.8760
0.440	11.1760	0.950	24.1300
0.450	11.4300	0.960	24.3840
0.460	11.6840	0.970	24.6380
0.470	11.9380	0.980	24.8920
0.480	12.1920	0.990	25.1460
0.490	12.4460	1.000	25.4000

FRACTIONS to DECIMALS to MILLIMETERS

Fraction	Decimal	mm	Fraction	Decimal	mm
1/64	0.0156	0.3969	33/64	0.5156	13.0969
1/32	0.0312	0.7938	17/32	0.5312	13.4938
3/64	0.0469	1.1906	35/64	0.5469	13.8906
1/16	0.0625	1.5875	9/16	0.5625	14.2875
5/64	0.0781	1.9844	37/64	0.5781	14.6844
3/32	0.0938	2.3812	19/32	0.5938	15.0812
7/64	0.1094	2.7781	39/64	0.6094	15.4781
1/8	0.1250	3.1750	5/8	0.6250	15.8750
9/64	0.1406	3.5719	41/64	0.6406	16.2719
5/32	0.1562	3.9688	21/32	0.6562	16.6688
11/64	0.1719	4.3656	43/64	0.6719	17.0656
3/16	0.1875	4.7625	11/16	0.6875	17.4625
13/64	0.2031	5.1594	45/64	0.7031	17.8594
7/32	0.2188	5.5562	23/32	0.7188	18.2562
15/64	0.2344	5.9531	47/64	0.7344	18.6531
1/4	0.2500	6.3500	3/4	0.7500	19.0500
17/64	0.2656	6.7469	49/64	0.7656	19.4469
9/32	0.2812	7.1438	25/32	0.7812	19.8438
19/64	0.2969	7.5406	51/64	0.7969	20.2406
5/16	0.3125	7.9375	13/16	0.8125	20.6375
21/64	0.3281	8.3344	53/64	0.8281	21.0344
11/32	0.3438	8.7312	27/32	0.8438	21.4312
23/64	0.3594	9.1281	55/64	0.8594	21.8281
3/8	0.3750	9.5250	7/8	0.8750	22.2250
25/64	0.3906	9.9219	57/64	0.8906	22.6219
13/32	0.4062	10.3188	29/32	0.9062	23.0188
27/64	0.4219	10.7156	59/64	0.9219	23.4156
7/16	0.4375	11.1125	15/16	0.9375	23.8125
29/64	0.4531	11.5094	61/64	0.9531	24.2094
15/32	0.4688	11.9062	31/32	0.9688	24.6062
31/64	0.4844	12.3031	63/64	0.9844	25.0031
1/2	0.5000	12.7000	1	1.0000	25.4000

Safety first!

Regardless of how enthusiastic you may be about getting on with the job at hand, take the time to ensure that your safety is not jeopardized. A moment's lack of attention can result in an accident, as can failure to observe certain simple safety precautions. The possibility of an accident will always exist, and the following points should not be considered a comprehensive list of all dangers. Rather, they are intended to make you aware of the risks and to encourage a safety conscious approach to all work you carry out on your vehicle.

ESSENTIAL DOS AND DON'TS

DON'T rely on a jack when working under the vehicle. Always use approved jackstands to support the weight of the vehicle and place them under the recommended lift or support points.

DON'T attempt to loosen extremely tight fasteners (i.e. wheel lug nuts) while the vehicle is on a jack - it may fall.

DON'T start the engine without first making sure that the transmission is in Neutral (or Park where applicable) and the parking brake is set.

DON'T remove the radiator cap from a hot cooling system - let it cool or cover it with a cloth and release the pressure gradually.

DON'T attempt to drain the engine oil until you are sure it has cooled to the point that it will not burn you.

DON'T touch any part of the engine or exhaust system until it has cooled sufficiently to avoid burns.

DON'T siphon toxic liquids such as gasoline, antifreeze and brake fluid by mouth, or allow them to remain on your skin.

DON'T inhale brake lining dust - it is potentially hazardous (see *Asbestos* below).

DON'T allow spilled oil or grease to remain on the floor - wipe it up before someone slips on it.

DON'T use loose fitting wrenches or other tools which may slip and cause injury.

DON'T push on wrenches when loosening or tightening nuts or bolts. Always try to pull the wrench toward you. If the situation calls for pushing the wrench away, push with an open hand to avoid scraped knuckles if the wrench should slip.

DON'T attempt to lift a heavy component alone - get someone to help you.

DON'T rush or take unsafe shortcuts to finish a job.

DON'T allow children or animals in or around the vehicle while you are working on it.

DO wear eye protection when using power tools such as a drill, sander, bench grinder, etc. and when working under a vehicle.

DO keep loose clothing and long hair well out of the way of moving parts.

DO make sure that any hoist used has a safe working load rating adequate for the job.

DO get someone to check on you periodically when working alone on a vehicle.

DO carry out work in a logical sequence and make sure that everything is correctly assembled and tightened.

DO keep chemicals and fluids tightly capped and out of the reach of children and pets.

DO remember that your vehicle's safety affects that of yourself and others. If in doubt on any point, get professional advice.

ASBESTOS

Certain friction, insulating, sealing, and other products - such as brake linings, brake bands, clutch linings, torque converters, gaskets, etc. - may contain asbestos. Extreme care must be taken to avoid inhalation of dust from such products, since it is hazardous to health. If in doubt, assume that they do contain asbestos.

FIRE

Remember at all times that gasoline is highly flammable. Never smoke or have any kind of open flame around when working on a vehicle. But the risk does not end there. A spark caused by an electrical short circuit, by two metal surfaces contacting each other, or even by static electricity built up in your body under certain conditions, can ignite gasoline vapors, which in a confined space are highly explosive. Do not, under any circumstances, use gasoline for cleaning parts. Use an approved safety solvent.

Always disconnect the battery ground (-) cable at the battery before working on any part of the fuel system or electrical system. Never risk spilling fuel on a hot engine or exhaust component. It is strongly recommended that a fire extinguisher suitable for use on fuel and electrical fires be kept handy in the garage or workshop at all times. Never try to extinguish a fuel or electrical fire with water.

FUMES

Certain fumes are highly toxic and can quickly cause unconsciousness and even death if inhaled to any extent. Gasoline vapor falls into this category, as do the vapors from some cleaning solvents. Any draining or pouring of such volatile fluids should be done in a well ventilated area.

When using cleaning fluids and solvents, read the instructions on the container carefully. Never use materials from unmarked containers.

Never run the engine in an enclosed space, such as a garage. Exhaust fumes contain carbon monoxide, which is extremely poisonous. If you need to run the engine, always do so in the open air, or at least have the rear of the vehicle outside the work area.

If you are fortunate enough to have the use of an inspection pit, never drain or pour gasoline and never run the engine while the vehicle is over the pit. The fumes, being heavier than air, will concentrate in the pit with possibly lethal results.

THE BATTERY

Never create a spark or allow a bare light bulb near a battery. They normally give off a certain amount of hydrogen gas, which is highly explosive.

Always disconnect the battery ground (-) cable at the battery before working on the fuel or electrical systems.

If possible, loosen the filler caps or cover when charging the battery from an external source (this does not apply to sealed or maintenance-free batteries). Do not charge at an excessive rate or the battery may burst.

Take care when adding water to a non maintenance-free battery and when carrying a battery. The electrolyte, even when diluted, is very corrosive and should not be allowed to contact clothing or skin.

Always wear eye protection when cleaning the battery to prevent the caustic deposits from entering your eyes.

HOUSEHOLD CURRENT

When using an electric power tool, inspection light, etc., which operates on household current, always make sure that the tool is correctly connected to its plug and that, where necessary, it is properly grounded. Do not use such items in damp conditions and, again, do not create a spark or apply excessive heat in the vicinity of fuel or fuel vapor.

SECONDARY IGNITION SYSTEM VOLTAGE

A severe electric shock can result from touching certain parts of the ignition system (such as the spark plug wires) when the engine is running or being cranked, particularly if components are damp or the insulation is defective. In the case of an electronic ignition system, the secondary system voltage is much higher and could prove fatal.

Troubleshooting

CONTENTS

This section provides an easy reference guide to the more common problems which may occur during the operation of your vehicle. Various symptoms and their possible causes are grouped under headings denoting components or systems, such as *Engine, Cooling system*, etc. They also refer to the Chapter and/or Section that deals with the problem.

Remember that successful troubleshooting isn't a mysterious "black art" practiced only by professional mechanics. It's simply the result of knowledge combined with an intelligent, systematic approach to a problem. Always use a process of elimination, starting with the simplest solution and working through to the most complex - and never over-

look the obvious. Anyone can run the gas tank dry or leave the lights on overnight, so don't assume that you're exempt from such oversights.

Finally, always establish a clear idea why a problem has occurred and take steps to ensure that it doesn't happen again. If the electrical system fails because of a poor connection, check all other connections in the system to make sure they don't fail as well. If a particular fuse continues to blow, find out why - don't just go on replacing fuses. Remember, failure of a small component can often be indicative of potential failure or incorrect functioning of a more important component or system.

ENGINE AND PERFORMANCE

1 Engine will not rotate when attempting to start

1 Battery terminal connections loose or corroded (Chapter 1).
2 Battery discharged or faulty (Chapter 1).
3 Automatic transaxle not completely engaged in Park (Chapter 7).
4 Broken, loose or disconnected wiring in the starting circuit (Chapters 5 and 12).
5 Starter motor pinion jammed in flywheel ring gear (Chapter 5).
6 Starter solenoid faulty (Chapter 5).
7 Starter motor faulty (Chapter 5).
8 Ignition switch faulty (Chapter 12).
9 Starter pinion or driveplate teeth worn or broken (Chapter 5).

2 Engine rotates but will not start

1 Fuel tank empty.
2 Battery discharged (engine rotates slowly) (Chapter 5).
3 Battery terminal connections loose or corroded (Chapter 1).
4 Leaking fuel injector(s), fuel pump, pressure regulator, etc. (Chapter 4).
5 Fuel not reaching fuel injection system (Chapter 4).
6 Ignition components damp or damaged (Chapter 5).
7 Worn, faulty or incorrectly gapped spark plugs (Chapter 1).
8 Broken, loose or disconnected wiring in the starting circuit (Chapter 5).
9 Damaged or failed crankshaft or camshaft position sensor (Chapter 6).
10 Broken, loose or disconnected wires at the ignition coil(s) or faulty coil(s) (Chapter 5).

3 Engine hard to start when cold

1 Battery discharged or low (Chapter 1).
2 Fuel system malfunctioning (Chapter 4).
3 Injector(s) leaking (Chapter 4).

4 Engine hard to start when hot

1 Air filter clogged (Chapter 1).
2 Fuel not reaching the fuel injection system (Chapter 4).
3 Corroded battery connections, especially ground (Chapter 1).

5 Starter motor noisy or excessively rough in engagement

1 Pinion or driveplate gear teeth worn or broken (Chapter 5).
2 Starter motor mounting bolts loose or missing (Chapter 5).

6 Engine starts but stops immediately

1 Loose or faulty electrical connections at coil pack or alternator (Chapter 5).
2 Insufficient fuel reaching the fuel injectors (Chapter 4).
3 Vacuum leak at the gasket between the intake manifold/plenum and throttle body (Chapters 1 and 4).
4 Restricted exhaust system (most likely the catalytic converter) (Chapters 4 and 6).

7 Oil puddle under engine

1 Oil pan gasket and/or oil pan drain bolt seal leaking (Chapters 1 and 2).
2 Oil pressure sending unit leaking (Chapter 2).
3 Rocker arm cover gaskets leaking (Chapter 2).
4 Engine oil seals leaking (Chapter 2).
5 Timing cover sealant or sealing flange leaking (Chapter 2).

8 Engine lopes while idling or idles erratically

1 Vacuum leakage (Chapter 4).
2 Leaking EGR valve or plugged PCV valve (Chapters 1 and 6).
3 Air filter clogged (Chapter 1).
4 Fuel pump not delivering sufficient fuel to the fuel injection system (Chapter 4).
5 Leaking head gasket (Chapter 2).
6 Timing chain and/or gears worn (Chapter 2).
7 Camshaft lobes worn (Chapter 2).

9 Engine misses at idle speed

1 Spark plugs worn or not gapped properly (Chapter 1).
2 Faulty spark plug wires (Chapter 1).
3 Vacuum leaks (Chapters 1 and 4).
4 Incorrect ignition timing (Chapter 5).
5 Uneven or low compression (Chapter 2).

10 Engine misses throughout driving speed range

1 Fuel filter clogged and/or impurities in the fuel system (Chapters 1 and 4).
2 Low fuel output at the injector (Chapter 4).
3 Faulty or incorrectly gapped spark plugs (Chapter 1).
4 Incorrect ignition timing (Chapter 5).
5 Leaking spark plug wires (Chapter 1).
6 Faulty emission system components (Chapter 6).
7 Low or uneven cylinder compression pressures (Chapter 2).
8 Weak or faulty ignition system (Chapter 5).
9 Vacuum leak in fuel injection system, intake manifold or vacuum hoses (Chapter 4).

11 Engine stumbles on acceleration

1 Spark plugs fouled (Chapter 1).
2 Fuel injection system needs adjustment or repair (Chapter 4).
3 Fuel filter clogged (Chapter 1).
4 Incorrect ignition timing (Chapter 5).
5 Intake manifold air leak (Chapter 4).

12 Engine surges while holding accelerator steady

1 Intake air leak (Chapter 4).
2 Fuel pump faulty (Chapter 4).
3 Loose fuel injector harness connections (Chapter 4).
4 Defective ECM (Chapter 6).

13 Engine stalls

1 Idle speed incorrect (Chapters 1 and 4).

2 Fuel filter clogged and/or water and impurities in the fuel system (Chapters 1 and 4).

3 Ignition components damp or damaged (Chapter 5).

4 Faulty emissions system components (Chapter 6).

5 Faulty or incorrectly gapped spark plugs (Chapter 1).

6 Faulty spark plug wires (Chapter 1).

7 Vacuum leak in the fuel injection system, intake manifold or vacuum hoses (Chapter 4).

14 Engine lacks power

1 Incorrect ignition timing (Chapter 5).

2 Faulty or incorrectly gapped spark plugs (Chapter 1).

3 Fuel injection system malfunctioning (Chapter 4).

4 Faulty coil(s) (Chapter 5).

5 Brakes binding (Chapter 1).

6 Automatic transaxle fluid level incorrect (Chapter 1).

7 Fuel filter clogged and/or impurities in the fuel system (Chapter 1).

8 Emission control system not functioning properly (Chapter 6).

9 Low or uneven cylinder compression pressures (Chapter 2).

10 Restricted exhaust system (most likely the catalytic converter (Chapters 4 and 6).

15 Engine backfires

1 Emissions system not functioning properly (Chapter 6).

2 Ignition timing incorrect (Chapter 5).

3 Faulty secondary ignition system (Chapter 5).

4 Fuel injection system malfunctioning (Chapter 4).

5 Vacuum leak at fuel injectors, intake manifold or vacuum hoses (Chapter 4).

6 Valves sticking (Chapter 2).

16 Pinging or knocking engine sounds during acceleration or uphill

1 Incorrect grade of fuel.

2 Ignition timing incorrect (Chapter 5).

3 Fuel injection system malfunctioning Chapter 4).

4 Improper or damaged spark plugs or wires (Chapter 1).

5 Worn or damaged ignition components (Chapter 5).

6 Faulty emissions system (Chapter 6).

7 Vacuum leak (Chapter 4).

17 Engine runs with oil pressure light on

1 Low oil level (Chapter 1).

2 Idle rpm below specification (Chapter 1).

3 Short in wiring circuit (Chapter 12).

4 Faulty oil pressure sender (Chapter 2).

5 Oil viscosity too low or oil diluted.

6 Worn engine bearings and/or oil pump (Chapter 2).

18 Engine diesels (continues to run) after switching off

1 Idle speed too high (Chapters 1 and 4).

2 Excessive engine operating temperature (Chapter 3).

3 Hot spots and/or carbon deposits in the combustion chamber(s).

ENGINE ELECTRICAL SYSTEM

19 Battery will not hold a charge

1 Alternator drivebelt defective or not adjusted properly (Chapter 1).

2 Battery terminals loose or corroded (Chapter 1).

3 Alternator not charging properly (Chapter 5).

4 Loose, broken or faulty wiring in the charging circuit (Chapter 5).

5 Short in vehicle wiring (Chapters 5 and 12).

6 Internally defective battery (Chapters 1 and 5).

20 Voltage warning light fails to go out

1 Faulty alternator or charging circuit (Chapter 5).

2 Alternator drivebelt defective or out of adjustment (Chapter 1).

3 Alternator voltage regulator inoperative (Chapter 5).

21 Voltage warning light fails to come on when key is turned on

1 Warning light bulb defective (Chapter 12).

2 Fault in the printed circuit, dash wiring or bulb holder (Chapter 12).

FUEL SYSTEM

22 Excessive fuel consumption

1 Dirty or clogged air filter element (Chapter 1).

2 Incorrectly set ignition timing (Chapter 5).

3 Emissions system not functioning properly (Chapter 6).

4 Fuel injection system malfunctioning (Chapter 4).

5 Low tire pressure or incorrect tire size (Chapter 1).

23 Fuel leakage and/or fuel odor

1 Leak in a fuel feed or vent line (Chapter 4).

2 Tank overfilled.

3 Evaporative emissions control canister filter clogged (Chapters 1 and 6).

4 Fuel injector internal parts excessively worn (Chapter 4).

COOLING SYSTEM

24 Overheating

1 Insufficient coolant in system (Chapter 1).

2 Water pump drivebelt defective or out of adjustment (Chapter 1).

3 Radiator core blocked or grille restricted (Chapter 3).

4 Thermostat faulty (Chapter 3).

5 Electric cooling fan blades broken or cracked (Chapter 3).

6 Radiator cap not maintaining proper pressure (Chapter 3).

7 Ignition timing incorrect (Chapter 5).

25   Overcooling

Incorrect (opening temperature too low) or faulty thermostat (Chapter 3).

26   External coolant leakage

1   Deteriorated/damaged hoses or loose clamps (Chapters 1 and 3).
2   Water pump seal defective (Chapters 1 and 3).
3   Leakage from radiator core or header tank (Chapter 3).
4   Engine drain or water jacket core plugs leaking (Chapter 2).

27   Internal coolant leakage

1   Leaking cylinder head gasket (Chapter 2).
2   Cracked cylinder bore or cylinder head (Chapter 2).

28   Coolant loss

1   Too much coolant in system (Chapter 1).
2   Coolant boiling away because of overheating (Chapter 3).
3   Internal or external leakage (Chapter 3).
4   Faulty radiator cap (Chapter 3).

29   Poor coolant circulation

1   Inoperative water pump (Chapter 3).
2   Restriction in cooling system (Chapters 1 and 3).
3   Water pump drivebelt defective or out of adjustment (Chapter 1).
4   Thermostat sticking (Chapter 3).

AUTOMATIC TRANSAXLE

➡**Note: Due to the complexity of the automatic transaxle, it's difficult for the home mechanic to properly diagnose and service this component. For problems other than the following, the vehicle should be taken to a dealer service department or a transmission shop.**

30   Fluid leakage

1   Automatic transmission fluid is a deep red color. Fluid leaks should not be confused with engine oil, which can easily be blown by airflow to the transaxle.
2   To pinpoint a leak, first remove all built-up dirt and grime from the transaxle housing with degreasing agents and/or steam cleaning. Drive the vehicle at low speeds so air flow will not blow the leak far from its source. Raise the vehicle and determine where the leak is coming from. Common areas of leakage are:
   a) Pan (Chapters 1 and 7)
   b) Filler pipe (Chapter 7)
   c) Fluid cooler lines (Chapter 7)
   d) Speedometer gear or sensor (Chapter 7)

31   Transaxle fluid brown or has a burned smell

Transaxle overheated. Change fluid (Chapter 1).

32   General shift mechanism problems

1   Chapter 7 deals with checking and adjusting the shift linkage on automatic transaxles. Common problems which may be attributed to a poorly adjusted linkage are:
   a) Engine starting in gears other than Park or Neutral.
   b) Indicator on shifter pointing to a gear other than the one actually being used.
   c) Vehicle moves when in Park.
2   Refer to Chapter 7 for the shift linkage adjustment procedure.

33   Transaxle will not downshift with accelerator pedal pressed to the floor

Throttle valve (TV) cable out of adjustment (Chapter 7).

34   Engine will start in gears other than Park or Neutral

Starter safety switch malfunctioning (Chapter 7).

35   Transaxle slips, shifts roughly, is noisy or has no drive in forward or reverse gears

There are many probable causes for the above problems, but the home mechanic should be concerned with only one possibility - fluid level. Before taking the vehicle to a repair shop, check the level and condition of the fluid as described in Chapter 1.
Correct the fluid level as necessary or change the fluid and filter if needed. If the problem persists, have a professional diagnose the probable cause.

DRIVEAXLES

36   Clicking noise in turns

Worn or damaged outer CV joint. Check for cut or damaged boots (Chapter 1). Repair as necessary (Chapter 8).

37   Knock or clunk when accelerating after coasting

Worn or damaged outer CV joint. Check for cut or damaged boots (Chapter 1). Repair as necessary (Chapter 8).

38   Shudder or vibration during acceleration

1   Excessive inner CV joint angle. Check and correct as necessary (Chapter 8).
2   Worn or damaged CV joints. Repair or replace as necessary (Chapter 8).
3   Sticking inner joint assembly. Correct or replace as necessary (Chapter 8).

BRAKES

➡**Note: Before assuming that a brake problem exists, make sure:**

    a) The tires are in good condition and properly inflated (Chapter 1).
    b) The front end alignment is correct (Chapter 10).
    c) The vehicle isn't loaded with weight in an unequal manner.

39   Vehicle pulls to one side during braking

1   Incorrect tire pressures (Chapter 1).
2   Front end out of line (have the front end aligned).
3   Unmatched tires on same axle.
4   Restricted brake lines or hoses (Chapter 9).
5   Malfunctioning brake assembly (Chapter 9).
6   Loose suspension parts (Chapter 10).
7   Loose brake calipers (Chapter 9).

40   Noise (high-pitched squeal when the brakes are applied)

    Front disc brake pads worn out. The noise comes from the wear sensor rubbing against the disc. Replace pads with new ones immediately (Chapter 9).

41   Brake roughness or chatter (pedal pulsates)

1   Excessive front brake disc lateral runout (Chapter 9).
2   Parallelism not within specifications (Chapter 9).
3   Uneven pad wear caused by caliper not sliding due to improper clearance or dirt (Chapter 9).
4   Defective brake disc (Chapter 9).
5   Rear brake drum out-of-round.

42   Excessive pedal effort required to stop vehicle

1   Malfunctioning power brake booster (Chapter 9).
2   Partial system failure (Chapter 9).
3   Excessively worn pads or shoes (Chapter 9).
4   One or more caliper pistons or wheel cylinders seized or sticking (Chapter 9).
5   Brake pads or shoes contaminated with oil or grease (Chapter 9).
6   New pads or shoes installed and not yet seated. It will take a while for the new material to seat.

43   Excessive brake pedal travel

1   Partial brake system failure (Chapter 9).
2   Insufficient fluid in master cylinder (Chapters 1 and 9).
3   Air trapped in system (Chapters 1 and 9).

44   Dragging brakes

1   Master cylinder pistons not returning correctly (Chapter 9).
2   Restricted brakes lines or hoses (Chapters 1 and 9).
3   Incorrect parking brake adjustment (Chapter 9).

45   Grabbing or uneven braking action

1   Malfunction of proportioner valves (Chapter 9).
2   Malfunction of power brake booster unit (Chapter 9).
3   Binding brake pedal mechanism (Chapter 9).

46   Brake pedal feels spongy when depressed

1   Air in hydraulic lines (Chapter 9).
2   Master cylinder mounting bolts loose (Chapter 9).
3   Master cylinder defective (Chapter 9).

47   Brake pedal travels to the floor with little resistance

    Little or no fluid in the master cylinder reservoir caused by leaking caliper or wheel cylinder pistons, loose, damaged or disconnected brake lines (Chapter 9).

48   Parking brake does not hold

    Parking brake linkage improperly adjusted (Chapter 9).

SUSPENSION AND STEERING SYSTEMS

➡**Note: Before attempting to diagnose the suspension and steering systems, perform the following preliminary checks:**

    a) Check the tire pressures and look for uneven wear.
    b) Check the steering universal joints or coupling from the column to the steering gear for loose fasteners and wear.
    c) Check the front and rear suspension and the steering gear assembly for loose and damaged parts.
    d) Look for out-of-round or out-of-balance tires, bent rims and loose and/or rough wheel bearings.

49   Vehicle pulls to one side

1   Mismatched or uneven tires (Chapter 10).
2   Broken or sagging springs (Chapter 10).
3   Wheel alignment incorrect (Chapter 10).
4   Front brakes dragging (Chapter 9).

50   Abnormal or excessive tire wear

1   Front wheel alignment incorrect (Chapter 10).
2   Sagging or broken springs (Chapter 10).
3   Tire out-of-balance (Chapter 10).
4   Worn shock absorber (Chapter 10).
5   Overloaded vehicle.
6   Tires not rotated regularly.

51   Wheel makes a "thumping" noise

1   Blister or bump on tire (Chapter 1).
2   Improper shock absorber action (Chapter 10).

52 Shimmy, shake or vibration

1 Tire or wheel out-of-balance or out-of-round (Chapter 10).
2 Loose or worn wheel bearings (Chapter 10).
3 Worn tie-rod ends (Chapter 10).
4 Worn balljoints (Chapter 10).
5 Excessive wheel runout (Chapter 10).
6 Blister or bump on tire (Chapter 1).

53 Hard steering

1 Lack of lubrication at balljoints, tie-rod ends and steering gear assembly (Chapter 10).
2 Front wheel alignment incorrect (Chapter 10).
3 Low tire pressure (Chapter 1).

54 Steering wheel does not return to center position correctly

1 Lack of lubrication at balljoints and tie-rod ends (Chapter 10).
2 Binding in steering column (Chapter 10).
3 Defective rack-and-pinion assembly (Chapter 10).
4 Front wheel alignment problem (Chapter 10).

55 Abnormal noise at the front end

1 Lack of lubrication at balljoints and tie-rod ends (Chapter 1).
2 Loose upper strut mount (Chapter 10).
3 Worn tie-rod ends (Chapter 10).
4 Loose stabilizer bar (Chapter 10).
5 Loose wheel lug nuts (Chapter 1).
6 Loose suspension bolts (Chapter 10).

56 Wander or poor steering stability

1 Mismatched or uneven tires (Chapter 10).
2 Lack of lubrication at balljoints or tie-rod ends (Chapters 1 and 10).
3 Worn shock absorbers (Chapter 10).
4 Loose stabilizer bar (Chapter 10).
5 Broken or sagging springs (Chapter 10).
6 Front wheel alignment incorrect (Chapter 10).
7 Worn steering gear clamp bushings (Chapter 10).

57 Erratic steering when braking

1 Wheel bearings worn (Chapters 8 and 10).
2 Broken or sagging springs (Chapter 10).
3 Leaking wheel cylinder or caliper (Chapter 9).
4 Warped rotors or brake drums (Chapter 9).
5 Worn steering gear clamp bushings (Chapter 10).

58 Excessive pitching and/or rolling around corners or during braking

1 Loose stabilizer bar (Chapter 10).
2 Worn shock absorbers or mounts (Chapter 10).
3 Broken or sagging springs (Chapter 10).
4 Overloaded vehicle.

59 Suspension bottoms

1 Overloaded vehicle.
2 Worn shock absorbers (Chapter 10).
3 Incorrect, broken or sagging springs (Chapter 10).

60 Cupped tires

1 Front wheel alignment incorrect (Chapter 10).
2 Worn shock absorbers (Chapter 10).
3 Wheel bearings worn (Chapters 8 and 10).
4 Excessive tire or wheel runout (Chapter 10).
5 Worn balljoints (Chapter 10).

61 Excessive tire wear on outside edge

1 Inflation pressures incorrect (Chapter 1).
2 Excessive speed in turns.
3 Wheel alignment incorrect (excessive toe-in or positive camber). Have professionally aligned.
4 Suspension arm bent or twisted (Chapter 10).

62 Excessive tire wear on inside edge

1 Inflation pressures incorrect (Chapter 1).
2 Wheel alignment incorrect (toe-out or excessive negative camber). Have professionally aligned.
3 Loose or damaged steering components (Chapter 10).

63 Tire tread worn in one place

1 Tires out-of-balance.
2 Damaged or buckled wheel. Inspect and replace if necessary.
3 Defective tire (Chapter 1).

64 Excessive play or looseness in steering system

1 Wheel bearings worn (Chapter 10).
2 Tie-rod end loose or worn (Chapter 10).
3 Steering gear loose (Chapter 10).

65 Rattling or clicking noise in rack and pinion

Steering gear clamps loose (Chapter 10).

Notes

Section

1

TUNE-UP AND ROUTINE MAINTENANCE

1   Introduction

➡**Note: On models equipped with the Delco Loc II audio system, be sure the lockout feature is turned off before performing any procedure which requires disconnecting the battery.**

This Chapter is designed to help home mechanics maintain their vehicles with the goals of maximum performance, economy, safety and reliability in mind.

Included is a master maintenance schedule, followed by procedures dealing specifically with each item on the schedule. Visual checks, adjustments, component replacement and other helpful items are included. Refer to the accompanying illustrations of the engine compartment and the underside of the vehicle for the locations of various components.

Servicing your vehicle in accordance with the mileage/time maintenance schedule and the step-by-step procedures will result in a planned maintenance program that should produce a long and reliable service life. Keep in mind that it's a comprehensive plan, so maintaining some items but not others at the specified intervals will not produce the same results.

As you service your vehicle, you'll discover that many of the procedures can - and should - be grouped together because of the nature of the particular procedure you're performing or because of the close proximity of two otherwise unrelated components to one another.

For example, if the vehicle is raised, you should inspect the exhaust, suspension, steering and fuel systems while you're under the vehicle. When you're rotating the tires, it makes good sense to check the brakes since the wheels are already removed. Finally, let's suppose you have to borrow or rent a torque wrench. Even if you only need it to tighten the spark plugs, you might as well check the torque of as many critical fasteners as time allows.

The first step in this maintenance program is to prepare yourself before the actual work begins. Read through all the procedures you're planning to do, then gather up all the parts and tools needed. If it looks like you might run into problems during a particular job, seek advice from a mechanic or an experienced do-it-yourselfer.

Typical engine compartment component layout (earlier model shown)

1	Power steering fluid dipstick	5	EGR valve	9	Battery
2	Coolant reservoir	6	Engine oil filter cap	10	Windshield washer fluid reservoir
3	Alternator	7	Engine oil dipstick	11	Brake fluid reservoir
4	Radiator cap	8	Air cleaner assembly	12	Automatic transaxle fluid dipstick

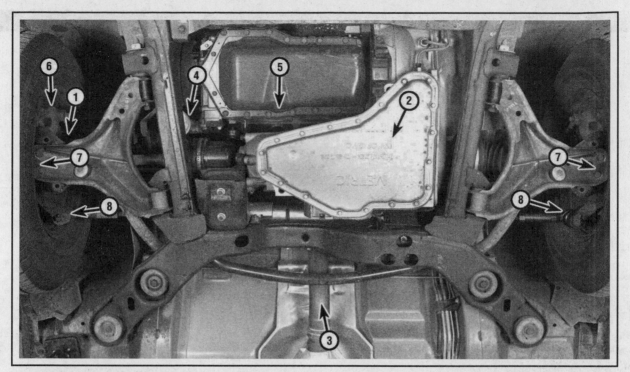

Typical front underside components

1	CV joint boot	3	Exhaust pipe	5	Engine oil drain plug	7	Balljoint
2	Automatic transaxle	4	Engine oil filter	6	Disc brake caliper	8	Tie-rod end

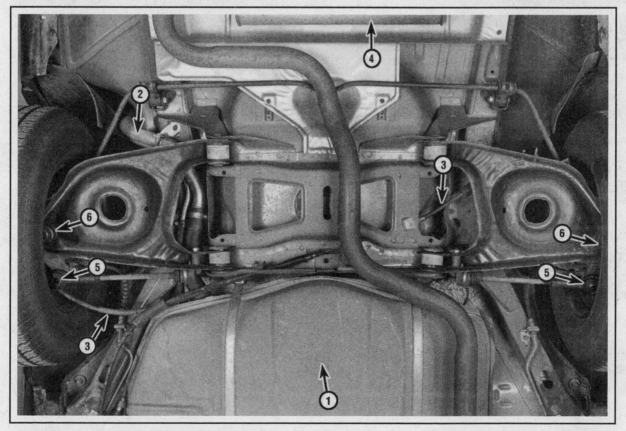

Typical rear underside component layout

1	Fuel tank	3	Parking brake cable	5	Toe-link end lube fitting
2	Fuel tank filler hose and pipe	4	Muffler	6	Balljoint lube fitting

2 General Motors Full-size FWD car Maintenance Schedule

The following maintenance intervals are based on the assumption that the vehicle owner will be doing the maintenance or service work, as opposed to having a dealer service department do the work. Although the time/mileage intervals are loosely based on factory recommendations, most have been shortened to ensure, for example, that such items as lubricants and fluids are checked/changed at intervals that promote maximum engine/driveline service life. Also, subject to the preference of the individual owner interested in keeping his or her vehicle in peak condition at all times, and with the vehicle's ultimate resale in mind, many of the maintenance procedures may be performed more often than recommended in the following schedule. We encourage such owner initiative.

When the vehicle is new it should be serviced initially by a factory authorized dealer service department to protect the factory warranty. In many cases the initial maintenance check is done at no cost to the owner (check with your dealer service department for more information).

EVERY 250 MILES OR WEEKLY, WHICHEVER COMES FIRST

Check the engine oil level (Section 4)
Check the engine coolant level (Section 4)
Check the windshield washer fluid level (Section 4)
Check the brake fluid level (Section 4)
Check the tires and tire pressures (Section 5)

EVERY 3000 MILES OR 3 MONTHS, WHICHEVER COMES FIRST

All items listed above plus:
Check the automatic transaxle fluid level (Section 6)*
Check the power steering fluid level (Section 7)*
Check the cooling system (Section 9)
Change the engine oil and filter (Section 12)*

EVERY 7500 MILES OR 6 MONTHS, WHICHEVER COMES FIRST

Check and service the battery (Section 8)
Inspect and replace, if necessary, all underhood hoses (Section 10)
Inspect and replace, if necessary, the windshield wiper blades (Section 11)
Rotate the tires (Section 17).
Inspect and replace the air filter if necessary (Section 20)*

EVERY 15,000 MILES OR 12 MONTHS, WHICHEVER COMES FIRST

All items listed above plus:
Lubricate the chassis components (Section 13)
Check the driveaxle boots (Section 14)
Inspect the suspension and steering components (Section 15)
Inspect the exhaust system (Section 16)
Check the brakes (Section 18)*
Inspect the fuel system (Section 19)

Inspect the air filter (Section 20)
Check the engine drivebelts (Section 21)
Check the seat belts (Section 22)
Check the neutral start switch (Section 23)
Check the seatback latch (Section 24)
Check the spare tire and jack (Section 25)

EVERY 30,000 MILES OR 24 MONTHS, WHICHEVER COMES FIRST

All items listed above plus:
Replace the air filter (Section 20)
Replace the fuel filter (Section 26)
Change the automatic transaxle fluid (Section 27)**
Service the cooling system (drain, flush and refill) (green-colored ethylene glycol anti-freeze only) (Section 28)
Inspect and replace, if necessary, the PCV valve (Section 29)
Inspect the evaporative emissions control system (Section 30)
Check the EGR system (Section 31)
Replace the spark plugs (conventional [non-platinum] spark plugs) (Section 32)
Inspect the spark plug wires (Section 33)
Inspect the distributor cap and rotor (1985 models only) (Section 34)
Check and adjust, if necessary, the ignition timing (1985 models only) (Section 35)
Replace the cabin air filter (later models, Section 36)

EVERY 60,000 MILES OR 48 MONTHS, WHICHEVER COMES FIRST

Check and replace if necessary the oxygen sensor (Chapter 6)

EVERY 100,000 MILES OR 5 YEARS, WHICHEVER COMES FIRST

Service the cooling system (drain, flush and refill) (orange-colored "DEX-COOL" silicate-free coolant only) (Section 28)
Replace the spark plugs (platinum-tipped spark plugs) (Section 32)

* This item is affected by "severe" operating conditions as described below. If the vehicle is operated under severe conditions, perform all maintenance indicated with an asterisk (*) at 3000 mile/3 month intervals. Severe conditions are indicated if the vehicle is operated mainly . . .

In dusty areas
While towing a trailer
When allowed to idle for extended periods and/or at low speeds when outside temperatures remain below freezing and most trips are less than four miles long

** If operated under one or more of the following conditions, change the automatic transaxle fluid every 15,000 miles.

In heavy city traffic where the outside temperature regularly reaches 90-degrees F or higher
In hilly or mountainous terrain
Frequent trailer pulling

3 Tune-up general information

The term tune-up is used in this manual to represent a combination of individual operations rather than one specific procedure.

If, from the time the vehicle is new, the routine maintenance schedule is followed closely and frequent checks are made of fluid levels and high wear items, as suggested throughout this manual, the engine will be kept in relatively good running condition and the need for additional work will be minimized.

More likely than not, however, there will be times when the engine is running poorly due to lack of regular maintenance. This is even more likely if a used vehicle, which has not received regular and frequent maintenance checks, is purchased. In such cases, an engine tune-up will be needed outside of the regular routine maintenance intervals.

The first step in any tune-up or diagnostic procedure to help correct a poor running engine is a cylinder compression check. A compression check (see Chapter 2, Part B) will help determine the condition of internal engine components and should be used as a guide for tune-up and repair procedures. If, for instance, a compression check indicates serious internal engine wear, a conventional tune-up won't improve the performance of the engine and would be a waste of time and money. Because of its importance, the compression check should be done by someone with the right equipment and the knowledge to use it properly.

The following procedures are those most often needed to bring a generally poor running engine back into a proper state of tune.

MINOR TUNE-UP

Check all engine related fluids (Section 4)
Clean, inspect and test the battery (Section 8)
Check the cooling system (Section 9)
Check all underhood hoses (Section 10)
Check the air filter (Section 20)
Check the drivebelt (Section 21)
Check the PCV valve (Section 29)
Replace the spark plugs (Section 32)
Inspect the spark plug wires (Section 33)
Inspect the distributor cap and rotor (1985 models only)
   (Section 34)
Check and adjust, if necessary, the ignition timing
   (1985 models only) (Section 35)

MAJOR TUNE-UP

All items listed under Minor tune-up plus . . .

Check the fuel system (Section 19)
Replace the air filter (Section 20)
Check the EGR system (Section 31)
Check the ignition system (Section 33 and Chapter 5)
Check the charging system (Chapter 5)
Replace the spark plug wires (Section 33)
Replace the distributor cap and rotor (1985 models only)
   (Section 34)

4 Fluid level checks

➡Note: The following are fluid level checks to be done on a 250 mile or weekly basis. Additional fluid level checks can be found in specific maintenance procedures which follow. Regardless of intervals, be alert to fluid leaks under the vehicle which would indicate a problem to be corrected immediately.

1 Fluids are an essential part of the lubrication, cooling, brake and windshield washer systems. Because the fluids gradually become depleted and/or contaminated during normal operation of the vehicle, they must be periodically replenished. See *Recommended lubricants and fluids* at the end of this Chapter before adding fluid to any of the following components.

➡Note: The vehicle must be on level ground when fluid levels are checked.

ENGINE OIL

▶ Refer to illustrations 4.2 and 4.4

2 The engine oil level is checked with a dipstick (see illustration). The dipstick extends through a metal tube down into the oil pan.

3 The oil level should be checked before the vehicle has been driven, or about 15 minutes after the engine has been shut off. If the oil is checked immediately after driving the vehicle, some of the oil will remain in the upper part of the engine, resulting in an inaccurate reading on the dipstick.

4 Pull the dipstick from the tube and wipe all the oil from the end with a clean rag or paper towel. Insert the clean dipstick all the way

4.2 On most models, the engine oil dipstick is clearly marked "ENGINE OIL" (arrow), as is the oil filler cap, which threads into the rocker arm cover (arrow) - on later models the filler cap is at the right side of the valve cover

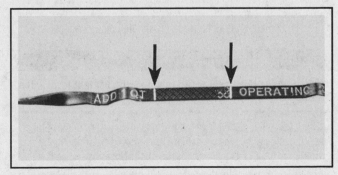

4.4 The oil level should be in the cross-hatched area - if it's below the ADD line, add enough oil to bring the level into the cross-hatched area

back into the tube and pull it out again. Note the oil at the end of the dipstick. Add oil as necessary to keep the level above the ADD mark in the cross-hatched area of the dipstick (see illustration).

5   Do not overfill the engine by adding too much oil since this may result in oil fouled spark plugs, oil leaks or oil seal failures.

6   Oil is added to the engine after removing a twist off cap located on the rocker arm cover (see illustration 4.2). An oil can spout or funnel may help to reduce spills.

7   Checking the oil level is an important preventive maintenance step. A consistently low oil level indicates oil leakage through damaged seals, defective gaskets or past worn rings or valve guides. If the oil looks milky in color or has water droplets in it, the cylinder head gasket may be blown or the head or block may be cracked. The engine should be checked immediately. The condition of the oil should also be checked. Whenever you check the oil level, slide your thumb and index finger up the dipstick before wiping off the oil. If you see small dirt or metal particles clinging to the dipstick, the oil should be changed (Section 12).

ENGINE COOLANT

▶ **Refer to illustration 4.9**

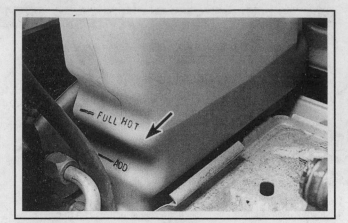

4.9 The coolant level must be maintained between the FULL HOT and ADD marks on the reservoir

> **※ WARNING:**
>
> **Do not allow antifreeze to come in contact with your skin or painted surfaces of the vehicle. Flush contaminated areas immediately with plenty of water. Do not store new coolant or leave old coolant lying around where it's accessible to children or pets - they're attracted by its sweet smell. Ingestion of even a small amount of coolant can be fatal! Wipe up garage floor and drip pan coolant spills immediately. Keep antifreeze containers covered and repair leaks in the cooling system immediately.**

> **※ CAUTION:**
>
> **Never mix green-colored ethylene glycol anti-freeze and orange-colored "DEX-COOL" silicate-free coolant because doing so will destroy the efficiency of the "DEX-COOL" coolant which is designed to last for 100,000 miles or five years.**

8   All vehicles covered by this manual are equipped with a pressurized coolant recovery system. A plastic coolant reservoir located in the right front corner of the engine compartment (in the right rear corner of the engine compartment on later models) is connected by a hose to the radiator filler neck. If the engine overheats, coolant escapes through a valve in the radiator cap and travels through the hose into the reservoir. As the engine cools, the coolant is automatically drawn back into the cooling system to maintain the correct level.

9   The coolant level in the reservoir should be checked regularly.

> **※ WARNING:**
>
> **Do not remove the radiator cap to check the coolant level when the engine is warm.**

The level in the reservoir varies with the temperature of the engine. When the engine is cold, the coolant level should be at or slightly above the FULL COLD mark on the reservoir. Once the engine has warmed up, the level should be at or near the FULL HOT mark (see illustration). If it isn't, allow the engine to cool, then unscrew the cap from the reservoir and add a 50/50 mixture of ethylene glycol based green colored antifreeze or orange colored "DEX-COOL" and water (see

Caution above). The coolant and windshield washer reservoirs are similar-looking, so be sure to add the correct fluids; the caps are clearly marked.

10 Drive the vehicle and recheck the coolant level. If only a small amount of coolant is required to bring the system up to the proper level, water can be used. However, repeated additions of water will dilute the antifreeze and water solution. In order to maintain the proper ratio of antifreeze and water, always top up the coolant level with the correct mixture. An empty plastic milk jug or bleach bottle makes an excellent container for mixing coolant. Do not use rust inhibitors or additives.

11 If the coolant level drops consistently, there may be a leak in the system. Inspect the radiator, hoses, filler cap, drain plugs and water pump (see Section 9). If no leaks are noted, have the radiator cap pressure tested by a service station.

12 If you have to remove the radiator cap, wait until the engine has cooled completely, then wrap a thick cloth around the cap and turn it to the first stop. If coolant or steam escapes, let the engine cool down longer, then remove the cap.

13 Check the condition of the coolant as well. It should be relatively clear. If it is brown or rust colored, the system should be drained, flushed and refilled. Even if the coolant appears to be normal, the corrosion inhibitors wear out, so it must be replaced at the specified intervals.

WINDSHIELD WASHER FLUID

▶ **Refer to illustration 4.14**

14 Fluid for the windshield washer system is located in a plastic reservoir on the right side of the engine compartment (see illustration). On 2002 and later models, the reservoir is located close to the radiator, and only the filler cap is visible from above. In milder climates, plain water can be used in the reservoir, but it should be kept no more than two-thirds full to allow for expansion if the water freezes. In colder climates, use windshield washer system antifreeze, available at any auto parts store, to lower the freezing point of the fluid. Mix the antifreeze with water in accordance with the manufacturer's directions on the container.

> **※ CAUTION:**
>
> **Do not use cooling system antifreeze - it will damage the vehicle's paint.**

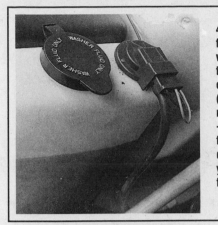

4.14 The reservoir for the windshield washer is located on the right side of the engine compartment (fluid is added after removing the top - how often you use the washers will dictate how often you need to check the reservoir)

15 To help prevent icing in cold weather, warm the windshield with the defroster before using the washer.

BATTERY ELECTROLYTE

16 All vehicles covered by this manual are equipped with a battery which is permanently sealed (except for vent holes) and has no filler caps. Water does not have to be added to these batteries at any time.

BRAKE FLUID

▶ **Refer to illustration 4.18**

17 The brake master cylinder is mounted on the front of the power booster unit in the engine compartment.

18 On anti-lock brake reservoirs, unscrew the cap and make sure the fluid level is even with the bottom of the filler neck slot. On translucent white plastic reservoirs, the fluid inside is readily visible (see illustration). If a low level is indicated, be sure to clean the reservoir cap, to prevent contamination of the brake system, before removing it.

19 When adding fluid, pour it carefully into the reservoir to avoid spilling it on surrounding painted surfaces. Be sure the specified fluid is used, since mixing different types of brake fluid can cause damage to the system. See *Recommended lubricants and fluids* at the end of this Chapter or your owner's manual.

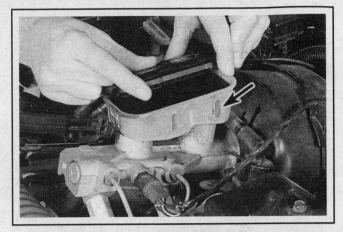

4.18 The fluid level inside the white plastic brake reservoir is easily checked through the inspection windows (when adding fluid, grasp the tabs and rotate the cover up as shown)

❋❋ WARNING:

Brake fluid can harm your eyes and damage painted surfaces, so use extreme caution when handling or pouring it. Do not use brake fluid that has been standing open or is more than one year old. Brake fluid absorbs moisture from the air. Excess moisture can cause a dangerous loss of braking effectiveness.

20 At this time the fluid and master cylinder can be inspected for contamination. The system should be drained and refilled if deposits, dirt particles or contamination are seen in the fluid.

21 After filling the reservoir to the proper level, make sure the cap is on tight to prevent fluid leakage.

22 The brake fluid level in the master cylinder will drop slightly as the pads and the brake shoes at each wheel wear down during normal operation. If the master cylinder requires repeated replenishing to keep it at the proper level, this is an indication of leakage in the brake system, which should be corrected immediately. Check all brake lines and connections (see Section 18 for more information).

23 If, when checking the master cylinder fluid level, you discover one or both reservoirs empty or nearly empty, the brake system should be bled (see Chapter 9) and the cause of the fluid loss found.

5 Tire and tire pressure checks

▶ **Refer to illustrations 5.2, 5.3, 5.4a, 5.4b and 5.8**

1 Periodic inspection of the tires may spare you the inconvenience of being stranded with a flat tire. It can also provide you with vital information regarding possible problems in the steering and suspension systems before major damage occurs.

2 The original tires on this vehicle are equipped with 1/2-inch side bands that appear when tread depth reaches 1/16-inch, but they don't appear until the tires are worn out. Tread wear can be monitored with a simple, inexpensive device known as a tread depth indicator (see illustration).

3 Note any abnormal tread wear (see illustration). Tread pattern

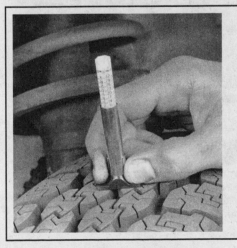

5.2 Use a tire tread depth indicator to monitor tire wear - they are available at auto parts stores and service stations and cost very little

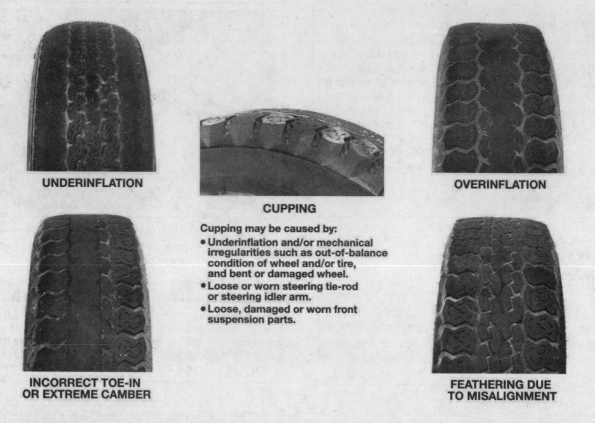

UNDERINFLATION

CUPPING

OVERINFLATION

Cupping may be caused by:

- Underinflation and/or mechanical irregularities such as out-of-balance condition of wheel and/or tire, and bent or damaged wheel.
- Loose or worn steering tie-rod or steering idler arm.
- Loose, damaged or worn front suspension parts.

INCORRECT TOE-IN OR EXTREME CAMBER

FEATHERING DUE TO MISALIGNMENT

5.3 This chart will help you determine the condition of the tires, the probable cause(s) of abnormal wear and the corrective action necessary

irregularities such as cupping, flat spots and more wear on one side than the other are indications of front end alignment and/or balance problems. If any of these conditions are noted, take the vehicle to a tire shop or service station to correct the problem.

4 Look closely for cuts, punctures and embedded nails or tacks. Sometimes a tire will hold air pressure for a short time or leak down very slowly after a nail has embedded itself in the tread. If a slow leak persists, check the valve stem core to make sure it's tight (see

illustration). Examine the tread for an object that may have embedded itself in the tire or for a "plug" that may have begun to leak (radial tire punctures are repaired with a plug that's installed in a puncture). If a puncture is suspected, it can be easily verified by spraying a solution of soapy water onto the suspected area (see illustration). The soapy solution will bubble if there's a leak. Unless the puncture is unusually large, a tire shop or service station can usually repair the tire.

5 Carefully inspect the inner sidewall of each tire for evidence of

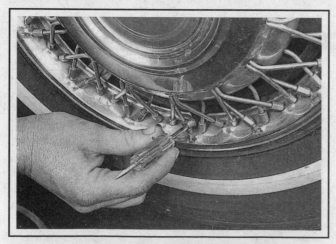

5.4a If a tire loses air on a steady basis, check the valve core first to make sure it's snug (special inexpensive wrenches are commonly available at auto parts stores)

5.4b If the valve core is tight, raise the corner of the vehicle with the low tire and spray a soapy water solution onto the tread as the tire is turned slowly - leaks will cause small bubbles to appear

brake fluid. If you see any, inspect the brakes immediately.

6 Correct air pressure adds miles to the lifespan of the tires, improves mileage and enhances overall ride quality. Tire pressure cannot be accurately estimated by looking at a tire, especially if it's a radial. A tire pressure gauge is essential. Keep an accurate gauge in the vehicle. The pressure gauges attached to the nozzles of air hoses at gas stations are often inaccurate.

7 Always check tire pressure when the tires are cold. Cold, in this case, means the vehicle has not been driven over a mile in the three hours preceding a tire pressure check. A pressure rise of four to eight pounds is not uncommon once the tires are warm.

8 Unscrew the valve cap protruding from the wheel or hubcap and push the gauge firmly onto the valve stem (see illustration). Note the reading on the gauge and compare the figure to the recommended tire pressure shown on the label attached to the rear edge of the driver's door. Be sure to reinstall the valve cap to keep dirt and moisture out of the valve stem mechanism. Check all four tires and, if necessary, add enough air to bring them up to the recommended pressure.

9 Don't forget to keep the spare tire inflated to the specified pressure (refer to your owner's manual or the tire sidewall).

5.8 To extend the life of the tires, check the air pressure at least once a week with an accurate gauge (don't forget the spare!)

6   Automatic transaxle fluid level check

▶ **Refer to illustrations 6.3 and 6.6**

1 The automatic transaxle fluid level should be carefully maintained. Low fluid level can lead to slipping or loss of drive, while overfilling can cause foaming and loss of fluid.

2 With the parking brake set, start the engine, then move the shift lever through all the gear ranges, ending in Park. The fluid level must be checked with the vehicle level and the engine running at idle.

➡**Note: Incorrect fluid level readings will result if the vehicle has just been driven at high speeds for an extended period, in hot weather in city traffic, or if it has been pulling a trailer. If any of these conditions apply, wait until the fluid has cooled (about 30 minutes).**

3 With the transaxle at normal operating temperature, remove the dipstick from the filler tube. The dipstick is located at the rear of the

engine compartment (see illustration).

4 Carefully touch the fluid at the end of the dipstick to determine if the fluid is cool, warm or hot. Wipe the fluid from the dipstick with a clean rag and push it back into the filler tube until the cap seats.

5 Pull the dipstick out again and note the fluid level.

6 If the fluid felt cool, the level should be about 1/8-to-3/8 inch below the "ADD 1 PT" mark (see illustration). If it felt warm, the level should be close to the "ADD 1 PT" mark. If the fluid was hot, the level should be within the cross-hatched area. If additional fluid is required, pour it directly into the tube using a funnel. It takes about one pint to raise the level from the ADD mark to the upper edge of the cross-hatched area with a hot transaxle, so add the fluid a little at a time and keep checking the level until it's correct.

7 The condition of the fluid should also be checked along with the level. If the fluid at the end of the dipstick is a dark reddish-brown color, or if the fluid has a burned smell, the fluid should be changed. If you're in doubt about the condition of the fluid, purchase some new fluid and compare the two for color and smell.

6.3 The automatic transaxle fluid dipstick is clearly marked ("TRANS FLUID") and is located at the rear of the engine compartment

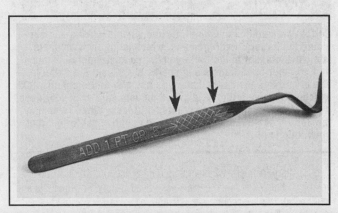

6.6 The automatic transaxle fluid level must be maintained within the cross-hatched area of the dipstick

7   Power steering fluid level check

▶ **Refer to illustrations 7.2 and 7.6**

1   The power steering system relies on fluid which may, over a period of time, require replenishing.

2   The fluid reservoir for the power steering pump is located behind the radiator near the front (drivebelt end) of the engine (see illustration). On 2003 and later models, it's located at the rear of the engine compartment on the passenger side.

3   For the check, the front wheels should be pointed straight ahead and the engine should be off.

4   Use a clean rag to wipe off the reservoir cap and the area around the cap. This will help prevent any foreign matter from entering the reservoir during the check.

5   Twist off the cap and check the temperature of the fluid at the end of the dipstick with your finger.

6   Wipe off the fluid with a clean rag, reinsert it, then withdraw it and read the fluid level. The level should be at the HOT mark if the fluid was hot to the touch (see illustration). It should be at the COLD mark if the fluid was cool to the touch. Note that on some models the marks (FULL HOT and COLD) are on opposite sides of the dipstick. At no time should the fluid level drop below the ADD mark.

7   If additional fluid is required, pour the specified type directly into the reservoir, using a funnel to prevent spills.

8   If the reservoir requires frequent fluid additions, all power steering hoses, hose connections, the power steering pump and the rack and pinion assembly should be carefully checked for leaks.

7.2   The power steering fluid reservoir is located near the front (drivebelt end) of the engine; turn the cap clockwise for removal

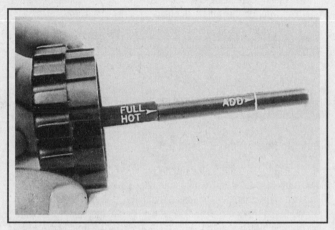

7.6   The marks on the power steering fluid dipstick indicate the safe fluid level range

8   Battery check and maintenance

▶ **Refer to illustration 8.1**

❋❋ WARNING:

Certain precautions must be followed when checking and servicing the battery. Hydrogen gas, which is highly flammable, is always present in the battery cells, so keep lighted tobacco and all other open flames and sparks away from the battery. The electrolyte inside the battery is actually dilute sulfuric acid, which will cause injury if splashed on your skin or in your eyes. It will also ruin clothes and painted surfaces. When removing the battery cables, always detach the negative cable first and hook it up last!

1   Battery maintenance is an important procedure which will help ensure you aren't stranded because of a dead battery. Several tools are required for this procedure (see illustration).

2   A sealed battery is standard equipment on all vehicles covered by this manual. Although this type of battery has many advantages over the older, capped cell type, and never requires the addition of water, it should still be routinely maintained according to the procedure which follows.

3   The battery is located on the right side of the engine compartment.

➡ **Note: On 2000 and later models, the battery is located under the rear seat cushion.**

The exterior of the battery should be inspected periodically for damage such as a cracked case or cover.

4   Check the tightness of the battery cable bolts to ensure good electrical connections and check the entire length of each cable for cracks and frayed conductors.

5   If corrosion (visible as white, fluffy deposits) is evident, remove the cables from the terminals, clean them with a battery brush and reinstall the cables. Corrosion can be kept to a minimum by using special treated fiber washers available at auto parts stores or by apply-

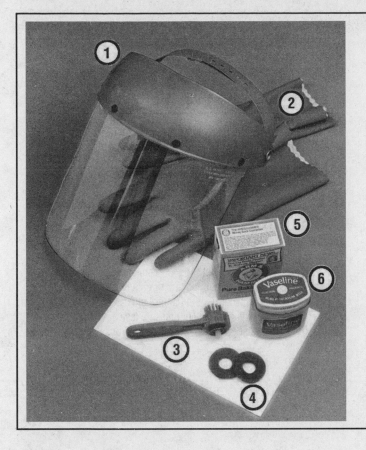

8.1 Tools and materials required for battery maintenance

1 *Face shield/safety goggles* - When removing corrosion with a brush, the acidic particles can easily fly up into your eyes
2 *Rubber gloves* - Another safety item to consider when servicing the battery - remember that's acid inside the battery!
3 *Battery terminal/cable cleaner* - This wire brush cleaning tool will remove all traces of corrosion from the battery posts and cable clamps
4 *Treated felt washers* - Placing one of these on each post, directly under the cable clamps, will help prevent corrosion
5 *Baking soda* - A solution of baking soda and water can be used to neutralize corrosion
6 *Petroleum jelly* - A layer of this on the battery posts will help prevent corrosion

ing a layer of petroleum jelly to the terminals and cables after they are assembled.

6 Make sure that the battery tray is in good condition and the hold-down clamp bolt is tight. If the battery is removed from the tray, make sure no parts remain in the bottom of the tray when the battery is reinstalled. When reinstalling the hold-down clamp bolt, do not overtighten it.

7 Corrosion on the hold-down components, battery case and surrounding areas can be removed with a solution of water and baking soda. Thoroughly rinse all cleaned areas with plain water.

8 Any metal parts of the vehicle damaged by corrosion should be covered with a zinc-based primer then painted.

9 Further information on the battery, charging and jump starting can be found in Chapter 5 and at the front of this manual.

9 Cooling system check

▶ **Refer to illustration 9.4**

※ CAUTION:

Never mix green-colored ethylene glycol anti-freeze and orange-colored "DEX-COOL" silicate-free coolant because doing so will destroy the efficiency of the "DEX-COOL" coolant which is designed to last for 100,000 miles or five years.

1 Many major engine failures can be attributed to a faulty cooling system. The cooling system also cools the transaxle fluid and plays an important role in prolonging transaxle life.

2 The cooling system should be checked with the engine cold. Do this before the vehicle is driven for the day or after the engine has been shut off for at least three hours.

3 Remove the radiator cap by turning it to the left until it reaches a stop. If you hear any hissing sounds (indicating there is still pressure in the system), wait until it stops. Now press down on the cap with the palm of your hand and continue turning to the left until the cap can be removed. Thoroughly clean the cap, inside and out, with clean water. Also clean the filler neck on the radiator. All traces of corrosion should be removed. The coolant inside the radiator should be relatively transparent. If it is rust colored, the system should be drained and refilled (see Section 28). If the coolant level is not up to the top, add additional antifreeze/coolant mixture (see Section 4).

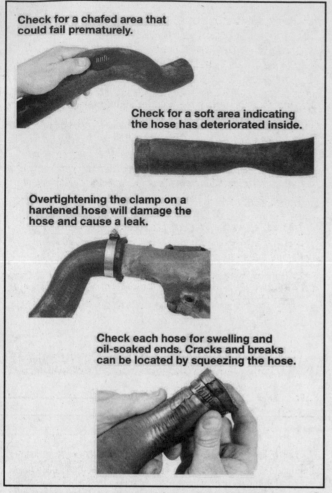

Check for a chafed area that could fail prematurely.

Check for a soft area indicating the hose has deteriorated inside.

Overtightening the clamp on a hardened hose will damage the hose and cause a leak.

Check each hose for swelling and oil-soaked ends. Cracks and breaks can be located by squeezing the hose.

9.4 Hoses, like drivebelts, have a habit of failing at the worst possible time - to prevent the inconvenience of a blown radiator or heater hose, inspect them carefully as shown here

4 Carefully check the large upper and lower radiator hoses along with any smaller diameter heater hoses which run from the engine to the firewall. Inspect each hose along its entire length, replacing any hose which is cracked, swollen or shows signs of deterioration. Cracks may become more apparent if the hose is squeezed (see illustration).

5 Make sure all hose connections are tight. A leak in the cooling system will usually show up as white or rust colored deposits on the areas adjoining the leak. If wire-type clamps are used at the ends of the hoses, it may be wise to replace them with more secure screw-type clamps.

6 Use compressed air or a soft brush to remove bugs, leaves, etc. from the front of the radiator or air conditioning condenser. Be careful not to damage the delicate cooling fins or cut yourself on them.

7 Every other inspection, or at the first indication of cooling system problems, have the cap and system pressure tested. If you don't have a pressure tester, most gas stations and repair shops will do this for a minimal charge.

10 Underhood hose check and replacement

GENERAL

❊❊ CAUTION:

Replacement of air conditioning hoses must be left to a dealer service department or air conditioning shop that has the equipment to depressurize the system safely. Never remove air conditioning components or hoses until the system has been depressurized.

1 High temperatures under the hood can cause the deterioration of the rubber and plastic hoses used for engine, accessory and emission systems operation. Periodic inspection should be made for cracks, loose clamps, material hardening and leaks. Information specific to the cooling system hoses can be found in Section 9.

2 Some, but not all, hoses are secured to the fittings with clamps. Where clamps are used, check to be sure they haven't lost their tension, allowing the hose to leak. If clamps aren't used, make sure the hose hasn't expanded and/or hardened where it slips over the fitting, allowing it to leak.

VACUUM HOSES

3 It's quite common for vacuum hoses, especially those in the emissions system, to be color coded or identified by colored stripes molded into each hose. Various systems require hoses with different wall thicknesses, collapse resistance and temperature resistance. When replacing hoses, be sure the new ones are made of the same material.

4 Often the only effective way to check a hose is to remove it completely from the vehicle. If more than one hose is removed, be sure to label the hoses and fittings to ensure correct installation.

5 When checking vacuum hoses, be sure to include any plastic T-fittings in the check. Inspect the fittings for cracks and the hose where it fits over the fitting for distortion, which could cause leakage.

6 A small piece of vacuum hose (1/4-inch inside diameter) can be used as a stethoscope to detect vacuum leaks. Hold one end of the hose to your ear and probe around vacuum hoses and fittings, listening for the "hissing" sound characteristic of a vacuum leak.

When probing with the vacuum hose stethoscope, be careful not to allow your body or the hose to come into contact with moving engine components such as the drivebelt, cooling fan, etc.

FUEL HOSE

There are certain precautions which must be taken when inspecting or servicing fuel system components. Work in a well ventilated area and do not allow open flames (cigarettes, appliance pilot lights, etc.) or bare light bulbs near the work area. Mop up any spills immediately and do not store fuel soaked rags where they could ignite. The fuel system is under pressure, so if any fuel lines must be disconnected, the pressure in the system must be relieved first (see Chapter 4 for more information).

7 Check all rubber fuel lines for deterioration and chafing. Check especially for cracks in areas where the hose bends and just before fittings, such as where a hose attaches to the fuel filter and fuel injection unit.

8 High quality fuel line, usually identified by the word Fluroelastomer printed on the hose, should be used for fuel line replacement. Never, under any circumstances, use unreinforced vacuum line, clear plastic tubing or water hose for fuel lines.

9 Spring-type clamps are commonly used on fuel lines. These clamps often lose their tension over a period of time, and can be "sprung" during the removal process. As a result spring-type clamps be replaced with screw-type clamps whenever a hose is replaced.

METAL LINES

10 Sections of steel tubing often used for fuel line between the fuel pump and fuel injection unit. Check carefully for cracks, kinks and flat spots in the line.

11 If a section of metal fuel line must be replaced, only seamless steel tubing should be used, since copper and aluminum tubing do not have the strength necessary to withstand normal engine vibration.

12 Check the metal brake lines where they enter the master cylinder and brake proportioning unit (if used) for cracks in the lines and loose fittings. Any sign of brake fluid leakage calls for an immediate thorough inspection of the brake system.

11 Windshield wiper blade inspection and replacement

▶ **Refer to illustrations 11.5a, 11.5b, 11.5c and 11.7**

1 The windshield wiper and blade assembly should be inspected periodically for damage, loose components and cracked or worn blade elements.

2 Road film can build up on the wiper blades and affect their efficiency, so they should be washed regularly with a mild detergent solution.

3 The action of the wiping mechanism can loosen the bolts, nuts and fasteners, so they should be checked and tightened, as necessary, at the same time the wiper blades are checked.

4 If the wiper blade elements (sometimes called inserts) are cracked, worn or warped, they should be replaced with new ones.

5 On most models, it will be necessary to remove the wiper blade assembly from the wiper arm by inserting a small screwdriver into the opening and gently prying on the spring while pulling on the blade

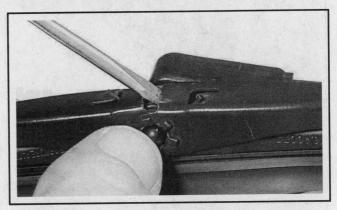

11.5a On most models, use a small screwdriver to gently pry on the spring at the center of the wiper arm . . .

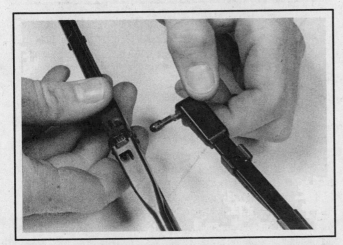

11.5b . . . while pulling the blade assembly away from the arm

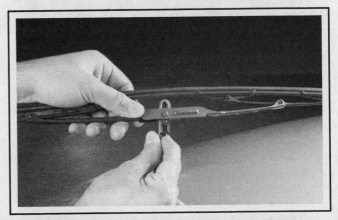

11.5c On other models it will be necessary to press on the release tab, then push the blade assembly down and out of the hook in the arm

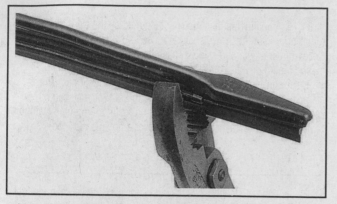

11.7 The rubber element is retained in the blade by small clips - the metal backing of the rubber element can be compressed at one end with pliers, allowing the element to slide out of the clips

to release it (see illustrations). On other models, it will be necessary to push in on the button of the wiper blade assembly clip and remove the blade assembly from the wiper arm. Carefully withdraw the blade assembly from the wiper arm

6   With the blade removed from the vehicle, you can remove the rubber element from the blade.

7   Using pliers, pinch the metal backing of the element (see illustration), then slide the element out of the blade assembly.

8   Compare the new element with the old for length, design, etc.

9   Slide the new element into place. It will automatically lock at the correct location.

10 Reinstall the blade assembly on the arm, wet the windshield glass and test for proper operation.

12 Engine oil and filter change

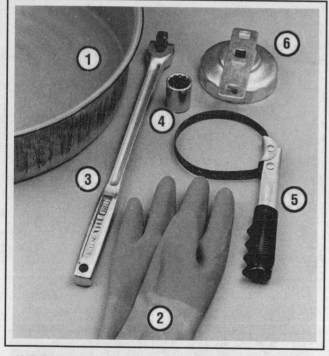

12.3 These tools are required when changing the engine oil and filter

1   **Drain pan** - It should be fairly shallow in depth, but wide to prevent spills

2   **Rubber gloves** - When removing the drain plug and filter, you will get oil on your hands (the gloves will prevent burns)

3   **Breaker bar** - Sometimes the oil drain plug is tight, and a long breaker bar is needed to loosen it

4   **Socket** - To be used with the breaker bar or a ratchet (must be the correct size to fit the drain plug - six-point preferred)

5   **Filter wrench** - This is a metal band-type wrench, which requires clearance around the filter to be effective

6   **Filter wrench** - This type fits on the bottom of the filter and can be turned with a ratchet or breaker bar (different-size wrenches are available for different types of filters)

▶ **Refer to illustrations 12.3, 12.9, 12.14 and 12.18**

1   Frequent oil changes are the most important preventive maintenance procedures that can be done by the home mechanic. As engine oil ages, it becomes diluted and contaminated, which leads to premature engine wear.

2   Although some sources recommend oil filter changes every other oil change, we feel that the minimal cost of an oil filter and the relative ease with which it is installed dictate that a new filter be used every time the oil is changed.

3   Gather together all necessary tools and materials before beginning the procedure (see illustration).

4   In addition, you should have plenty of clean rags and newspapers handy to mop up any spills. Access to the underside of the vehicle is greatly improved if the vehicle can be lifted on a hoist, driven onto ramps or supported by jackstands.

✳✳ WARNING:

Do not work under a vehicle which is supported only by a bumper, hydraulic or scissors-type jack.

5   If this is your first oil change, get under the vehicle and familiarize yourself with the locations of the oil drain plug and the oil filter. The engine and exhaust components will be warm during the actual work, so note how they are situated to avoid touching them when working under the vehicle.

6   Warm the engine to normal operating temperature. If the new oil or any tools are needed, use this warm-up time to gather everything necessary for the job. The correct type of oil for your application can be found in *Recommended lubricants and fluids* at the end of this Chapter.

7   With the engine oil warm (warm engine oil will drain better and more built-up sludge will be removed with the oil), raise and support the vehicle. Make sure it's safely supported.

8   Move all necessary tools, rags and newspapers under the vehicle. Position the drain pan under the drain plug. Keep in mind that the oil will initially flow from the pan with some force, so place the pan accordingly.

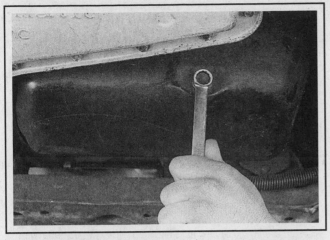

12.9 The engine oil drain plug is located at the rear of the oil pan - it's usually very tight, so use a box-end wrench to avoid rounding off the hex

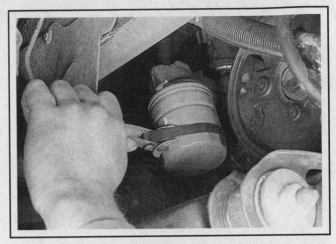

12.14 The oil filter is usually on very tight as well and will require a special wrench for removal - DO NOT use the wrench to tighten the new filter!

9   Being careful not to touch any of the hot exhaust components, remove the drain plug at the bottom of the oil pan (see illustration). Depending on how hot the oil is, you may want to wear gloves while unscrewing the plug the final few turns.

10 Allow the old oil to drain into the pan. It may be necessary to move the pan farther under the engine as the oil flow slows to a trickle.

11 After all the oil has drained, wipe off the drain plug with a clean rag. Small metal particles may cling to the plug which would immediately contaminate the new oil.

12 Clean the area around the drain plug opening and reinstall the plug. Tighten the plug securely with the wrench. If a torque wrench is available, use it to tighten the plug.

13 Move the drain pan into position under the oil filter.

14 Use the filter wrench to loosen the oil filter (see illustration). Chain or metal band filter wrenches may distort the filter canister, but this is of no concern as the filter will be discarded anyway.

15 Completely unscrew the old filter. Be careful - it's full of oil. Empty the oil inside the filter into the drain pan.

16 Compare the old filter with the new one to make sure they are the same type.

17 Use a clean rag to remove all oil, dirt and sludge from the area where the oil filter mounts on the engine. Check the old filter to make sure the rubber gasket isn't stuck to the engine. If the gasket is stuck to the engine (use a flashlight if necessary), remove it.

18 Apply a light coat of oil to the rubber gasket on the new oil filter (see illustration). Open a container of oil and partially fill the oil filter with fresh oil. Oil pressure will not build in the engine until the oil pump has filled the filter with oil, so partially filling it at this time will reduce the amount of time the engine runs with no oil pressure.

19 Attach the new filter to the engine, following the tightening directions printed on the filter canister or box. Most filter manufacturers recommend against using a filter wrench due to the possibility of over-tightening and damage to the seal.

20 Remove all tools, rags, etc. from under the vehicle, being careful not to spill the oil in the drain pan, then lower the vehicle.

21 Move to the engine compartment and locate the oil filler cap.

22 Pour the fresh oil through the filler opening. A funnel can be used.

23 Pour three quarts of fresh oil into the engine. Wait a few minutes to allow the oil to drain into the pan, then check the level on the dipstick (see Section 4 if necessary). If the oil level is above the ADD mark, start the engine and allow the new oil to circulate.

24 Run the engine for only about a minute and then shut it off. Immediately look under the vehicle and check for leaks at the oil pan drain plug and around the oil filter. If either is leaking, tighten with a bit more force.

25 With the new oil circulated and the filter now completely full, recheck the level on the dipstick and add more oil as necessary.

26 During the first few trips after an oil change, make it a point to check frequently for leaks and proper oil level.

27 The old oil drained from the engine cannot be reused in its present state and should be disposed of. Oil reclamation centers, auto repair shops and gas stations will normally accept the oil, which can be refined and used again. After the oil has cooled it can be drained into a container (capped plastic jugs, topped bottles, milk cartons, etc.) for transport to a disposal site.

12.18 Lubricate the oil filter gasket with clean engine oil before installing the filter on the engine

13 Chassis lubrication

▶ **Refer to illustrations 13.1 and 13.6**

1 Refer to *Recommended lubricants and fluids* at the end of this Chapter to obtain the necessary lubricants. You'll also need a grease gun (see illustration). Occasionally plugs will be installed rather than grease fittings. If so, grease fittings will have to be purchased and installed.

2 Look under the vehicle and see if grease fittings or plugs are installed in the balljoints and tie-rod ends. If there are plugs, remove them and buy grease fittings, which will thread into the component. A dealer or auto parts store will be able to supply the correct fittings. Straight, as well as angled, fittings are available.

3 For easier access under the vehicle, raise it with a jack and place jackstands under the frame. Make sure it's securely supported by the stands. If the wheels are being removed at this interval for rotation or brake inspection, loosen the lug nuts slightly while the vehicle is still on the ground.

4 Before beginning, force a little grease out of the nozzle to remove any dirt from the end of the gun. Wipe the nozzle clean with a rag.

5 With the grease gun and plenty of clean rags, crawl under the vehicle and begin lubricating the components (see the under-vehicle photos at the beginning of this Chapter).

6 Wipe off the grease fitting and push the nozzle firmly over it (see illustration). Squeeze the trigger on the grease gun to force grease into the component. The balljoints and tie-rod ends should be lubricated until each rubber seal is firm to the touch. Do not pump too much grease into the fitting or it could rupture the seal. If the grease escapes around the grease gun nozzle, the fitting is clogged or the nozzle isn't completely seated on the fitting. Resecure the gun nozzle to the fitting and try again. If necessary, replace the fitting with a new one.

7 Wipe the excess grease off the components and the grease fitting. Repeat the procedure for the remaining fittings.

8 Lubricate the shift linkage with a little multi-purpose grease. While you are under the vehicle, clean and lubricate the parking brake cable along with the cable guides and levers. This can be done by smearing some of the chassis grease onto the cable and its related parts with your fingers.

9 Open the hood and smear a little chassis grease on the hood latch mechanism. Have an assistant pull the hood release lever from inside the vehicle as you lubricate the cable at the latch.

10 Lubricate all the hinges (door, hood, etc.) with engine oil.

11 The key lock cylinders can be lubricated with spray-on graphite or silicone lubricant, which is available at auto parts stores.

✳✳ CAUTION:

The manufacturer doesn't recommend using oil in black plastic lock cylinders - it could damage them by washing out the factory-applied lubricant.

12 Lubricate the door weatherstripping with silicone spray. This will reduce chafing and retard wear.

13.1 Materials required for chassis and body lubrication

1 **Engine oil** - *Light engine oil in a can like this can be used for door and hood hinges*
2 **Graphite spray** - *Used to lubricate lock cylinders*
3 **Grease** - *Grease, in a variety of types and weights, is available for use in a grease gun. Check the Specifications for your requirements*
4 **Grease gun** - *A common grease gun, shown here with a detachable hose and nozzle, is needed for chassis lubrication. After use, clean it thoroughly*

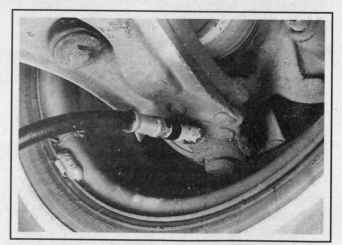

13.6 After cleaning the grease fitting, push the gun nozzle firmly into place and pump the grease into the component (usually about two pumps will be sufficient)

14 Driveaxle boot check

14.2 Push on the driveaxle boots to check for cracks

▶ **Refer to illustration 14.2**

1 The driveaxle boots are very important because they prevent dirt, water and foreign material from entering and damaging the constant velocity (CV) joints.

2 Inspect the boots for tears and cracks as well as loose clamps (see illustration). If there is any evidence of cracks or leaking lubricant, they must be replaced as described in Chapter 8.

5 Suspension and steering check

1 Raise the front of the vehicle periodically and visually check the suspension and steering components for wear.

2 Be alert for excessive play in the steering wheel before the front wheels react, excessive sway around corners, body movement over rough roads and binding at some point as the steering wheel is turned. If you notice any of the above symptoms, the steering and suspension systems should be checked.

3 Support the vehicle on jackstands placed under the frame rails. Because of the work to be done, make sure the vehicle cannot fall off the stands.

4 Check the front wheel hub nuts and make sure they are securely locked in place.

5 Working under the vehicle, check for loose bolts, broken or disconnected parts and deteriorated rubber bushings on all suspension and steering components. Look for grease or fluid leaking from the steering assembly. Check the power steering hoses and connections for leaks.

6 Have an assistant turn the steering wheel from side-to-side and check the steering components for free movement, chafing and binding. If the steering doesn't react with the movement of the steering wheel, try to determine where the slack is located.

6 Exhaust system check

1 With the engine cold (at least three hours after the vehicle has been driven), check the complete exhaust system from the engine to the end of the tailpipe. Ideally, the inspection should be done with the vehicle on a hoist to permit unrestricted access. If a hoist is not available, raise the vehicle and support it securely on jackstands.

2 Check the exhaust pipes and connections for evidence of leaks, severe corrosion and damage. Make sure that all brackets and hangers are in good condition and tight.

3 At the same time, inspect the underside of the body for holes, corrosion, open seams, etc. which may allow exhaust gases to enter the interior. Seal all body openings with silicone or body putty.

4 Rattles and other noises can often be traced to the exhaust system, especially the mounts and hangers. Try to move the pipes, muffler and catalytic converter. If the components can come in contact with the body or suspension parts, secure the exhaust system with new mounts.

5 Check the running condition of the engine by inspecting inside the end of the tailpipe. The exhaust deposits here are an indication of engine state-of-tune. If the pipe is black and sooty or coated with white deposits, the engine is in need of a tune-up, including a thorough fuel system inspection and adjustment.

17 Tire rotation

▶ **Refer to illustration 17.2**

1   The tires should be rotated at the specified intervals and whenever uneven wear is noticed.

2   Front wheel drive vehicles require a special tire rotation pattern (see illustration).

3   Refer to the information in *Jacking and towing* at the front of this manual for the proper procedures to follow when raising the vehicle and changing a tire. If the brakes are going to be checked, don't apply the parking brake as stated. Make sure the tires are blocked to prevent the vehicle from rolling as it's raised.

4   The entire vehicle should be raised at the same time. This can be done on a hoist or by jacking up each corner and then lowering the vehicle onto jackstands placed under the frame rails. Always use four jackstands and make sure the vehicle is safely supported.

5   After rotation, check and adjust the tire pressures as necessary and be sure to check the lug nut tightness.

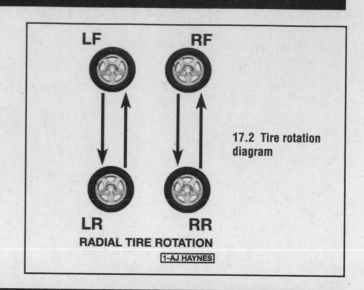

17.2 Tire rotation diagram

18 Brake check

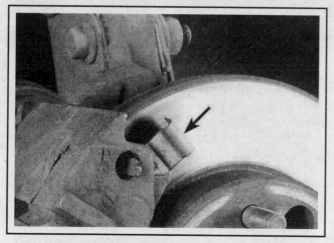

18.3 The brake pad wear indicator (arrow) will contact the disc and make a squealing noise when the pad is worn

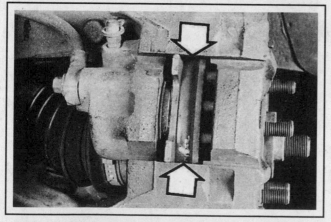

18.5 Look through the opening in the front of the caliper to check the brake pads (arrows) - the pad lining, which rubs against the disc, can also be inspected by looking through each end of the caliper

➡**Note: For detailed photographs of the brake system, refer to Chapter 9.**

❄❄ **WARNING:**

Brake system dust is hazardous to your health. DO NOT blow it out with compressed air or inhale it. An approved filtering mask should be worn whenever working on the brakes. DO NOT use gasoline or solvents to remove the dust. Use brake system cleaner or denatured alcohol only.

1   In addition to the specified intervals, the brakes should be inspected every time the wheels are removed or whenever a defect is suspected. Loosen the wheel lug nuts, then raise the vehicle and place it securely on jackstands. Remove the wheels (see *Jacking and towing* at the front of this manual, if necessary).

DISC BRAKES

▶ **Refer to illustrations 18.3 and 18.5**

2   Front disc brakes are standard equipment on all vehicles covered by this manual, while rear disc brakes are optional equipment on 1997 and later models. Extensive rotor damage can occur if the pads are not replaced when needed.

3   The disc brake pads have built-in wear indicators which make a high-pitched squealing or cricket-like warning sound when the pads are worn (see illustration).

❄❄ **CAUTION:**

Expensive rotor damage can result if the pads are not replaced soon after the wear indicators start squealing.

4   The disc brake calipers, which contain the pads, are visible once the wheels have been removed. There is an outer pad and an inner pad in each caliper. All pads should be inspected.

5   Each caliper has an opening to inspect the pads (see illustration).

18.11 Knock out the access hole cover in the brake drum, push the self-adjusting lever aside and turn the star wheel

18.12 Measure the depth of the rivet hole (arrow), to determine the thickness of remaining brake shoe material

If the pad material has worn to about 1/8-inch thick or less, the pads should be replaced.

6   If you're unsure about the exact thickness of the remaining lining material, remove the pads for further inspection or replacement (refer to Chapter 9).

7   Before installing the wheels, check for leakage and/or damage at the brake hoses and connections. Replace the hose or fittings as necessary, referring to Chapter 9.

8   Check the condition of the brake rotor. Look for score marks, deep scratches and overheated areas (they will appear blue or discolored). If damage or wear is noted, the rotor can be removed and resurfaced by an automotive machine shop or replaced with a new one. Refer to Chapter 9 for more detailed inspection and repair procedures.

DRUM BRAKES

♦ **Refer to illustrations 18.11 and 18.12**

9   Using a scribe or chalk, mark the drum and hub so the drum can be reinstalled in the same position on the hub.

10 Remove and discard the retaining clip (if equipped) and pull the brake drum off the hub and brake assembly. If this proves difficult, make sure the parking brake is released, then squirt some penetrating oil around the center hub area. Allow the oil to soak in and try to pull the drum off again.

11 If the drum still cannot be pulled off, the brake shoes will have to be adjusted in. This is done by first removing the cover in the brake drum. Pull the self-adjusting lever off the star wheel and use a small screwdriver to turn the wheel, which will move the shoes away from the drum (see illustration). With the drum removed, clean the brake assembly with brake system cleaner (see **Warning** above).

12 Note the thickness of the lining material on the brake shoes. If the material is worn to within 1/16-inch of the recessed rivets or metal backing, the shoes should be replaced (see illustration). The shoes should also be replaced if they are cracked, glazed, (shiny surface) or contaminated with brake fluid.

13 Check to make sure all the brake assembly springs are connected and in good condition.

14 Check the brake components for signs of fluid leakage. Carefully pry back the rubber cups on the wheel cylinder, located at the top of the backing plate. Any leakage is an indication that the wheel cylinders should be replaced or overhauled immediately (Chapter 9). Also check the hoses and connections for signs of leakage.

15 Wipe the inside of the drum with a clean rag and brake cleaner or denatured alcohol.

16 Check the inside of the drum for cracks, scoring, deep scratches and hard spots, which will appear as small discolored areas. If imperfections cannot be removed with sandpaper or emery cloth, the drum must be taken to a machine shop for resurfacing.

17 After the inspection process is complete, and if all the components are in good condition, reinstall the brake drums. Install the wheels and lower the vehicle to the ground.

PARKING BRAKE

18 The parking brake is operated by a hand lever and locks the rear brakes. The easiest, and perhaps most obvious, method of periodically checking the operation of the parking brake assembly is to park the vehicle on a steep hill with the parking brake set and the transaxle in Neutral (be sure to stay in the vehicle while performing this check.) If the parking brake cannot prevent the vehicle from rolling, it is in need of adjustment (see Chapter 9).

19 Fuel system check

✳✳ WARNING:

Certain precautions must be taken when inspecting or servicing fuel system components. Work in a well ventilated area and don't allow open flames (cigarettes, appliance pilot lights, etc.) near the work area. Mop up spills immediately and do not store fuel soaked rags where they could ignite. The fuel system is under pressure - nothing should be disconnected until the pressure is relieved (see Chapter 4).

1   The fuel system is most easily checked with the vehicle raised on a hoist so the components underneath the vehicle are readily visible and accessible.

2   If the smell of gasoline is noticed while driving or after the vehicle has been in the sun, the system should be thoroughly inspected immediately.

3   Remove the gas tank cap and check for damage, corrosion and an unbroken sealing imprint on the gasket. Replace the cap with a new one if necessary.

4   With the vehicle raised, inspect the gas tank and filler neck for punctures, cracks and other damage. The connection between the filler neck and tank is especially critical. Sometimes a rubber filler neck will leak due to loose clamps or deteriorated rubber, problems a home mechanic can usually rectify.

✳✳ WARNING:

Do not, under any circumstances, try to repair a fuel tank yourself (except rubber components). A welding torch or any open flame can easily cause the fuel vapors to explode if the proper precautions are not taken.

5   Carefully check all rubber hoses and metal lines leading away from the fuel tank. Check for loose connections, deteriorated hoses, crimped lines and other damage. Follow the lines to the front of the vehicle, carefully inspecting them all the way. Repair or replace damaged sections as necessary.

20 Air filter replacement

▶ **Refer to illustrations 20.2a and 20.2b**

1   At the specified intervals, the air filter should be replaced with a new one. The air filter housing is located on the left (driver's side) of the engine compartment.

2   On 1985 through 1991 models, remove the screws, lift the top cover off and withdraw the air filter (see illustration).

3   On 1992 through 1997 models, loosen the hose clamp and unhook the rear air intake duct from the throttle body. Remove the wing screws or release the clips, then remove the rear of the cover. Withdraw the air filter.

4   On 1998 through 2000 models, disconnect the electrical connector from the intake temperature sensor. Loosen the hose clamp and unhook the air intake duct from the throttle body. Remove the screws, then remove the front air intake duct from the housing. Withdraw the air filter. On 2001 and later models, just release the clips on the air cleaner cover, push the cover and the intake hose back toward the engine to compress the hose and remove the air filter.

5   While the filter housing cover is off, be careful not to drop anything down into the air duct or air filter assembly.

6   Wipe out the inside of the air filter housing with a clean rag.

7   Place a new air filter in the air filter housing. Make sure it seats properly in the bottom of the housing.

8   On 1985 through 1991 models, install the cover and tighten the screws.

9   On 1992 through 1997, install the cover and clips or tighten the screws. Install the rear air intake duct onto the throttle body and tighten the hose clamp securely.

10 On 1998 and later models, install the front air intake duct onto the housing and tighten the screws. Install the air intake duct onto the throttle body and tighten the hose clamp securely. Connect the electrical connector onto the intake temperature sensor.

20.2a On 1985 through 1991 models, remove the air cleaner cover screws (arrows) and detach the cover (two screws are located on the side, not visible in this photo) . . .

20.2b . . . then pull the air filter element out of the housing

21 Drivebelt check, adjustment and replacement

1985 MODELS

▶ **Refer to illustrations 21.3 and 21.4**

1   The drivebelts, or V-belts as they are often called, are located at the front of the engine and play an important role in the overall operation of the vehicle and its components. Due to their function and material make-up, the belts are prone to failure after a period of time and should be inspected and adjusted periodically to prevent major engine damage.

2   The number of belts used on a particular vehicle depends on the accessories installed. Drivebelts are used to turn the alternator, power steering pump, water pump and air-conditioning compressor. Depending on the pulley arrangement, more than one of these components may be driven by a single belt.

3   With the engine off, open the hood and locate the various belts at the front of the engine. Using your fingers (and a flashlight, if necessary), move along the belts checking for cracks and separation of the belt plies. Also check for fraying and glazing, which gives the belt a shiny appearance. Both sides of the belt should be inspected (see illustration).

4   The tension of each belt is checked by pushing on the belt at a distance halfway between the pulleys. Push firmly with your thumb and see how much the belt deflects. If the distance between pulley center to pulley center is between 7 and 11 inches, the belt should deflect 1/4 inch. If the belt travels between pulleys 12 to 16 inches apart, the belt should deflect 1/2 inch (see illustration).

5   If it is necessary to adjust the belt tension, either to make the belt tighter or looser, it is done by moving the belt driven accessory on its mounting brackets.

6   For each component there will be an adjusting bolt and a pivot bolt. Both bolts must be loosened slightly to enable you to move the component.

7   After the two bolts have been loosened, move the component away from engine to tighten the belt or toward the engine to loosen the belt. Hold the accessory in position and check the belt tension. If it is correct, tighten the two bolts until just snug, then recheck the tension. If the tension is correct, tighten the bolts securely.

8   It will often be necessary to use some sort of prybar to move the accessory while the belt is adjusted. If this must be done to gain the proper leverage, be very careful not to damage the component being moved or the part being pried against.

1986 AND LATER MODELS

▶ **Refer to illustrations 21.10, 21.12, 21.13, and 21.15**

9   A single serpentine drivebelt is located at the front of the engine and plays an important role in the overall operation of the engine and its components. Due to its function and material make up, the belt is prone to wear and should be periodically inspected. The serpentine belt drives the alternator, power steering pump, water pump and air conditioning compressor (if equipped).

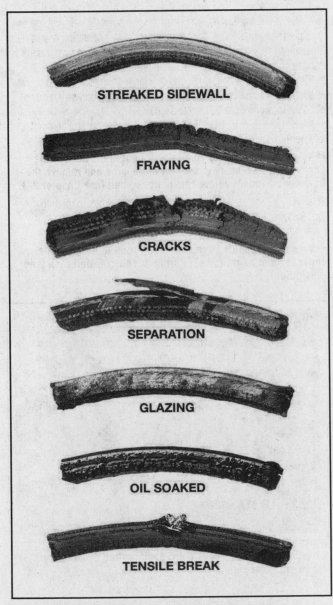

21.3 Here are some of the more common problems associated with drivebelts (check the belts carefully to prevent an untimely breakdown)

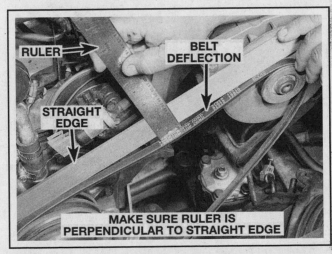

21.4 Drivebelt tension can be checked with a straightedge and ruler

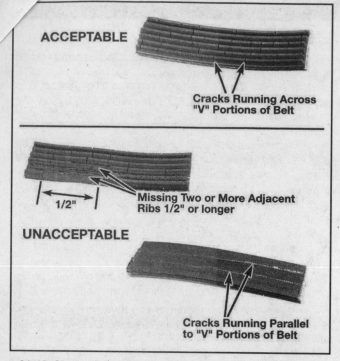

ACCEPTABLE

Cracks Running Across "V" Portions of Belt

1/2" Missing Two or More Adjacent Ribs 1/2" or longer

UNACCEPTABLE

Cracks Running Parallel to "V" Portions of Belt

21.10 Check the belt for signs of wear like these - if the belt looks worn, replace it

10 With the engine off, open the hood and use your fingers (and a flashlight, if necessary), to move along the belt checking for cracks and separation of the belt plies. Also check for fraying and glazing, which gives the belt a shiny appearance. Both sides of the belt should be inspected, which means you will have to twist the belt to check the underside (see illustration).

11 Check the ribs on the underside of the belt. They should all be the same depth, with none of the surface uneven.

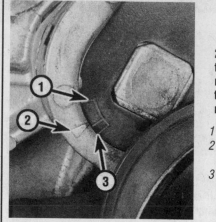

21.12 The belt tension indicator must be between the maximum and minimum marks

1 Minimum tension
2 Belt tension indicator pointer
3 Maximum tension

12 The tension of the belt is checked visually. Locate the belt tensioner at the front of the engine under the power steering pump on the right (passenger) side, then find the tensioner operating marks (see illustration) located on the side of the tensioner. If the indicator mark is outside of the operating range, the belt should be replaced.

13 To replace the belt, rotate the tensioner counterclockwise to release belt tension (see illustration). The tensioner will swing down once the tension of the belt is released.

14 Remove the belt from the auxiliary components and carefully release the tensioner.

→**Note: On models with an engine mount at the front of the engine, it is necessary to support the engine and remove the front engine mount before removing the belt (see Chapter 2A).**

15 Route the new belt over the various pulleys, again rotating the tensioner to allow the belt to be installed, then release the belt tensioner.

→**Note: Most models have a drivebelt routing decal on the upper radiator panel to help during drivebelt installation (see illustration).**

21.13 Rotate the tensioner counterclockwise to release tension on the drivebelt

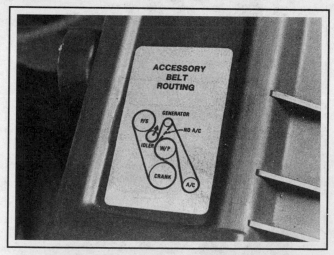

ACCESSORY BELT ROUTING

GENERATOR
P/S
NO A/C
IDLER W/P
CRANK A/C

21.15 Most models have a drivebelt routing decal on the upper radiator panel

22 Seat belt check

1 Check the seat belts, buckles, latch plates and guide loops for obvious damage and signs of wear.

2 See if the seat belt reminder light comes on when the key is turned to the Run or Start position. A chime should also sound.

3 The seat belts are designed to lock up during a sudden stop or impact, yet allow free movement during normal driving. Make sure the retractors return the belt against your chest while driving and rewind the belt fully when the buckle is unlatched.

4 If any of the above checks reveal problems with the seat belt system, replace parts as necessary.

23 Neutral start switch check

✳✳ WARNING:

During the following checks there's a chance the vehicle could lunge forward, possibly causing damage or injuries. Allow plenty of room around the vehicle, apply the parking brake and hold down the regular brake pedal during the checks.

1 Try to start the engine in each gear. The engine should crank only in Park or Neutral.

2 Make sure the steering column lock allows the key to go into the Lock position only when the shift lever is in Park.

3 The ignition key should come out only in the Lock position.

24 Seatback latch check (two-door models only)

1 It's important to periodically check the seatback latch mechanism to prevent the seatback from moving forward during a sudden stop or an accident.

2 Grasping the top of the seat, attempt to tilt it forward. It should tilt only when the latch on the rear of the seat is pulled up. Note that there is a certain amount of free play built into the latch mechanism.

3 When returned to the upright position, the seatback should latch securely.

25 Spare tire and jack check

1 Periodically checking the security and condition of the spare tire and jack will help to familiarize you with the procedures necessary for emergency tire replacement and also ensure that no components work loose during normal vehicle operation.

2 Following the instructions in your owner's manual or under *Jacking and towing* near the front of this manual, remove the spare tire and jack.

3 Using a reliable air pressure gauge, check the pressure in the spare tire. It should be kept at the pressure marked on the tire sidewall.

4 Make sure the jack operates freely and all components are undamaged.

5 When finished, make sure the wing nuts hold the jack and tire securely in place.

26 Fuel filter replacement

♦ Refer to illustrations 26.3 and 26.4

✳✳ WARNING:

Gasoline is extremely flammable, so extra precautions must be taken when working on any part of the fuel system. Don't smoke or allow open flames or unshielded light bulbs in or near the work area. Also, don't work in a garage where a gas appliance (clothes dryer or water heater) is present. Have a fire extinguisher rated for gasoline fires handy and know how to use it!

1985 THROUGH 1992 MODELS

1 Relieve the fuel system pressure (see Chapter 4).

2 Raise the vehicle and support it securely on jackstands.

3 Unscrew the fuel line-to-fuel filter fittings (see illustration). Use a back-up wrench to keep from twisting the line.

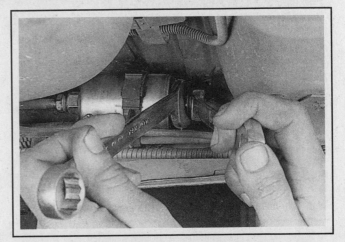

26.3 To remove the fuel filter, use two wrenches to unscrew the fuel line fittings from both ends of the filter

4   Remove the filter from the clip (see illustration).

5   Snap the new filter securely into the clip. Make sure the arrow on the filter points toward the engine.

6   Reattach the fuel lines to the filter, using new O-ring seals.

→Note: The fuel line and/or filter must be replaced with new ones if they are scratched or damaged during installation.

7   Start the engine and check for fuel leaks.

1993 AND LATER MODELS

8   Relieve the fuel system pressure (see Chapter 4).

9   Raise the vehicle and support it securely on jackstands.

10 Twist the connector 1/4 turn in each direction to loosen any debris within the connector, then blow out the debris from the connector. Squeeze the plastic tabs on the male end of the connector and pull the connector apart. Repeat for the other connector.

→Note: Some models are equipped with two quick connect fittings at each end of the fuel filter, while other models have one screw-in fitting (like earlier models) and one quick connect fitting. On models with screw-in type fittings refer to illustration 26.3 and unscrew the fuel line fitting.

11 Remove the filter mounting bolt, unhook the tab from the receptacle in the frame and remove the filter.

12 Wipe off the male end of both connectors, then apply a drop of clean engine oil onto the fuel filter male connectors where they are inserted into the female connector. Position the new filter so the arrow

26.4 Fuel filter mounting details

A   Fuel line fittings   *B   Mounting clip*

on the filter points toward the engine.

13 Insert the fuel line female connectors onto the fuel filter's male connectors and push them on until the plastic tabs snap and lock into place.

14 Pull on the female end of each connector to make sure it is securely locked into place.

15 Move the filter into place and install the bolt.

16 Start the engine and check for fuel leaks.

27   Automatic transaxle fluid and filter change

♦ Refer to illustration 27.11

1   At the specified time intervals, the transaxle fluid should be drained and replaced. Since the fluid will remain hot long after driving, perform this procedure only after everything has cooled down completely.

2   Before beginning work, purchase the specified transaxle fluid (see *Recommended lubricants and fluids* at the end of this Chapter) and a new filter.

3   Other tools necessary for this job include jackstands to support the vehicle in a raised position, a drain pan capable of holding several quarts, newspapers and clean rags.

4   Raise and support the vehicle on jackstands.

5   With a drain pan in place, remove the front and side oil pan mounting bolts.

6   Loosen the rear pan bolts approximately four turns.

7   Carefully pry the transaxle oil pan loose with a screwdriver, allowing the fluid to drain.

8   Remove the remaining bolts, pan and gasket. Carefully clean the gasket surface of the transaxle to remove all traces of the old gasket and sealant.

9   Drain the fluid from the transaxle oil pan, clean the pan with solvent and dry it with compressed air.

✖ WARNING:

Wear eye protection. Be careful not to lose the magnet.

10 Remove the filter from the mount inside the transaxle.

11 Install the new filter and O-ring or seal (see illustration). Apply a light coat of petroleum jelly (Vaseline) to the seal prior to installation.

Install the seal with the smaller diameter end going in first.

12 Make sure the gasket surface on the transaxle oil pan is clean, then install a new gasket. Put the pan in place against the transaxle and install the bolts. Working around the pan, tighten each bolt a little at a time until the final torque figure is reached.

13 Lower the vehicle and add about seven quarts of the specified type of automatic transaxle fluid through the filler tube (see Section 6).

14 With the shift lever in Park and the parking brake set, run the engine at a fast idle, but don't race it.

15 Move the shift lever through each gear and back to Park. Check the fluid level. You'll probably have to add about a pint of fluid.

16 Check under the vehicle for leaks during the first few trips.

27.11 Be sure to install a new transaxle filter pick-up tube seal

28 Cooling system servicing (draining, flushing and refilling)

▶ Refer to illustrations 28.6a, 28.6b and 28.6c

❋ WARNING:

Make sure the engine is completely cool before performing this procedure.

❋ CAUTION:

Never mix green-colored ethylene glycol anti-freeze and orange-colored "DEX-COOL" silicate-free coolant because doing so will destroy the efficiency of the "DEX-COOL" coolant which is designed to last for 100,000 miles or five years.

1   Periodically, the cooling system should be drained, flushed and refilled to replenish the antifreeze mixture and prevent formation of rust and corrosion, which can impair the performance of the cooling system and cause engine damage.

2   At the same time the cooling system is serviced, all hoses and the radiator cap should be inspected and replaced if defective (see Section 9).

3   Since antifreeze is a corrosive and poisonous solution, be careful not to spill any of the coolant mixture on the vehicle's paint or your skin. If this happens, rinse it off immediately with plenty of clean water. Consult local authorities about the dumping of antifreeze before draining the cooling system. In many areas, reclamation centers have been set up to collect automobile oil and drained antifreeze/water mixtures, rather than allowing them to be added to the sewage system.

4   With the engine cold, remove the radiator cap.

5   Move a large container under the radiator to catch the coolant as it's drained.

6   Drain the radiator by unscrewing the petcock (see illustration) at the bottom on the left side. If it's corroded and can't be turned easily, or if the radiator isn't equipped with a petcock, disconnect the lower radiator hose to allow the coolant to drain. Be careful not to get antifreeze on your skin or in your eyes.

➡ Note: To drain the engine block, remove the front and rear drain plugs (see illustrations).

7   Disconnect the hose from the coolant reservoir and remove the reservoir. Flush it out with clean water.

8   Place a garden hose in the radiator filler neck and flush the system until the water runs clear at all drain points.

9   In severe cases of contamination or clogging of the radiator, remove it (see Chapter 3) and reverse flush it. This involves inserting the hose in the bottom radiator outlet to allow the water to run against the normal flow, draining through the top. A radiator repair shop should be consulted if further cleaning or repair is necessary.

10 When the coolant is regularly drained and the system refilled with the correct antifreeze/water mixture, there should be no need to use chemical cleaners or descalers.

11 To refill the system, reconnect the radiator hoses and install the reservoir and the overflow hose.

12 Fill the radiator with the recommended mixture of antifreeze and water (see Section 4) to the base of the filler neck and then add more coolant to the reservoir until it reaches the lower mark.

13 With the radiator cap still removed, start the engine and run it until normal operating temperature is reached. With the engine idling, add additional coolant to the radiator and the reservoir. Install the radiator and reservoir caps.

14 Keep a close watch on the coolant level and the cooling system hoses during the first few miles of driving. Tighten the hose clamps and/or add more coolant as necessary. The coolant level should be at the FULL HOT mark with the engine at normal operating temperature.

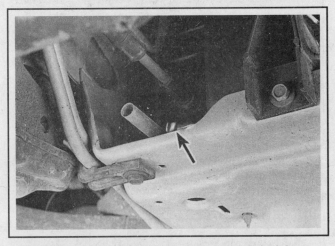

28.6a The radiator petcock (arrow) is located at the left lower corner of the radiator (viewed from below)

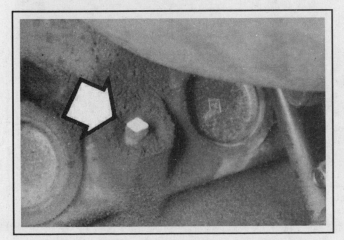

28.6b The front engine block drain plug (arrow) is located adjacent to the starter

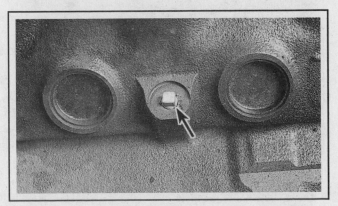

28.6c The rear engine block drain plug (arrow) is located near the middle of the block

29 Positive Crankcase Ventilation (PCV) valve check and replacement

1985 THROUGH 1991 MODELS

♦ **Refer to illustration 29.1**

1   With the engine idling at normal operating temperature, pull the valve (with hose attached) out of the rubber grommet in the rocker arm cover or intake manifold (see illustration).

2   Place your finger over the end of the valve. If there is no vacuum at the valve, check for a plugged hose, manifold port, or the valve itself. Replace any plugged or deteriorated hoses.

3   Turn off the engine and shake the PCV valve, listening for a rattle. If the valve doesn't rattle, replace it with a new one.

4   To replace the valve, pull it out of the end of the hose, noting its installed position and direction.

5   When purchasing a replacement PCV valve, make sure it's for your particular vehicle, model year and engine size. Compare the old

valve with the new one to make sure they are the same.

6   Push the valve into the end of the hose until it's seated.

7   Inspect the rubber grommet for damage and replace it with a new one if necessary.

8   Push the PCV valve and hose securely into position in the rocker arm cover or intake manifold.

1992 AND 1993 MODELS

♦ **Refer to illustration 29.10**

9   The PCV valve on these models is located in the intake manifold under a cover. Remove the three bolts and detach the cover.

10 Use needle-nose pliers to withdraw the PCV valve and spring from the manifold (see illustration). Be careful not to lose the O-ring at the end of the valve.

11 Shake the PCV valve, listening for a rattle. If the valve doesn't rattle, replace it with a new one. Make sure the new PCV valve is for your particular vehicle, model year and engine size. Compare the old valve with the new one to make sure they are the same.

12 Insert the new O-ring, PCV valve and spring into position and install the cover (be sure to use a new gasket) and bolts, tightening the bolts securely.

1994 AND LATER MODELS

♦ **Refer to illustrations 29.14, 29.15 and 29.16**

13 Remove the sight shield from the fuel rail and intake manifold. On 2001 and later models, also remove the heat shield from the PCV valve.

14 Disconnect the electrical connector from the MAP sensor (see illustration).

15 Press the PCV valve cover down and rotate it 1/4 of a turn counterclockwise (see illustration). Slowly lift the cover, the upper O-ring and the spring from the housing.

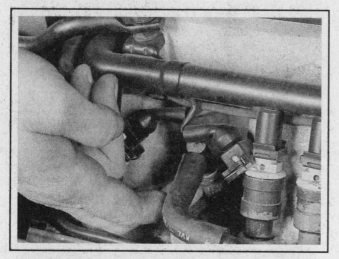

29.1 The PCV valve plugs into the rocker arm cover or intake manifold; pull it out, then feel for suction at the end of the valve and shake it, listening for a clicking sound

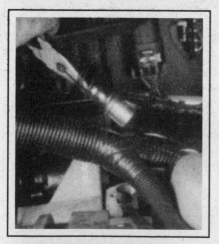

29.10 Use needle-nose pliers to remove the PCV and spring from the cavity in the intake manifold

29.14 The PCV valve on 1996 and later engines is located at the front of the intake manifold below the MAP sensor (arrow)

29.15 Press down on the PCV valve access cover, rotate it counterclockwise then pull up to remove it (be sure to check the O-ring and replace it if necessary)

16 Remove the PCV valve and the lower O-ring from the intake manifold (see illustration).

17 Shake the valve, listening for a rattle. If the valve doesn't rattle, replace it with a new one. Make sure the new valve is for your particular vehicle, model year and engine size. Compare the old valve with the new one to make sure they are the same.

18 Insert a new O-ring, valve and spring into position and install the cover. Rotate the cover clockwise and lock the cover in place.

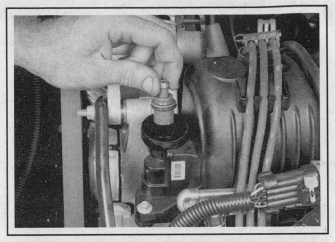

29.16 Remove the PCV valve and the lower O-ring from the intake manifold

30 Evaporative emissions control system check

▶ **Refer to illustration 30.2**

1   The function of the evaporative emissions control system is to draw fuel vapors from the gas tank and fuel system, store them in a charcoal canister and then burn them during normal engine operation.

2   The most common symptom of a fault in the evaporative emissions system is a strong fuel odor in the engine compartment. If a fuel odor is detected, inspect the charcoal canister, located at the front of the engine compartment (see illustration). Check the canister and all hoses for damage and deterioration.

3   The evaporative emissions control system is explained in more detail in Chapter 6.

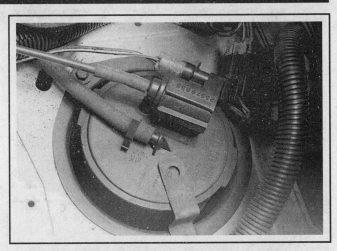

30.2 The charcoal canister is located at the left (driver's side) front corner of the engine compartment (shown with air duct removed for clarity

31 Exhaust Gas Recirculation (EGR) system check

CONVENTIONAL EGR VALVES

▶ **Refer to illustrations 31.2a and 31.2b**

➡ **Note: The following procedure applies only to models equipped with conventional-type EGR valves. Some models are equipped with a newer digital or linear EGR valve that does not have a diaphragm (see Chapter 6).**

1   The EGR valve is located on the intake manifold. Most of the time, when a problem develops in the emissions system, it's due to a stuck or corroded EGR valve.

2   With the engine cold to prevent burns, reach under the valve and push on the diaphragm (see illustrations). Using moderate pressure, you should be able to move the diaphragm.

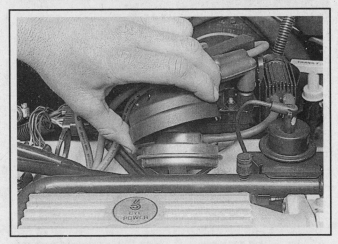

31.2a Some EGR valves have plastic covers which you can pull off with your fingers

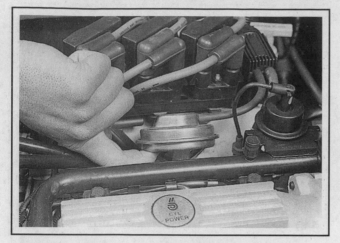

31.2b The diaphragm, reached from under the EGR valve, should move easily with finger pressure

3 If the diaphragm doesn't move or moves only with much effort, replace the EGR valve with a new one. If in doubt about the condition of the valve, compare the free movement with a new valve.

ELECTRONIC EGR VALVES

4 On these models, the EGR valve base is made of powdered metal, which should not be cleaned with abrasives, scrapers or solvents.

5 Test the pintle by depressing it with a piece of wood wrapped in soft cloth and replace the EGR valve if the pintle is either sticky or moves roughly.

6 Clean the end of the pintle only with a moist, soft cloth. Check for diagnostic trouble codes with a scan tool.

7 Refer to Chapter 6 for more information on the EGR system.

32 Spark plug replacement

▶ **Refer to illustrations 32.2, 32.5a, 32.5b, 32.6, 32.9 and 32.10**

1 Three spark plugs are located at the front and three at the rear (firewall) side of the engine.

2 In most cases, the tools necessary for spark plug replacement included a spark plug socket which fits onto a ratchet (spark plug sockets are padded inside to prevent damage to the porcelain insulators on the new plugs), various extensions and a gap gauge to check and adjust the gaps on the new plugs (see illustration). A special plug wire removal tool is available for separating the wire boots from the spark plugs, but it isn't absolutely necessary. A torque wrench should be used to tighten the new plugs. Because the aluminum cylinder heads used on these models are easily damaged, allow the engine to cool before removing or installing the spark plugs.

3 The best approach when replacing the spark plugs is to purchase the new ones in advance, adjust them to the proper gap and replace the plugs one at a time. When buying the new spark plugs, be sure to obtain the correct plug type for your particular engine. The plug type can be found on the Emission Control Information label located under the hood and on the reference chart at the store where you buy the plugs. If differences exist between the plug specified on the emissions label and other sources, assume the emissions label is correct.

4 Allow the engine to cool completely before attempting to remove any of the plugs. While you are waiting for the engine to cool, check the new plugs for defects and adjust the gaps.

5 The gap is checked by inserting the proper thickness gauge

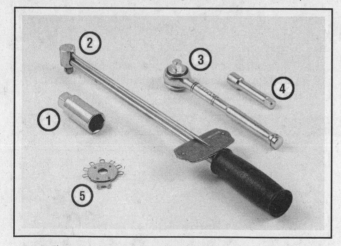

32.2 Tools required for changing spark plugs

1 *Spark plug socket* - *This will have special padding inside to protect the spark plug's porcelain insulator*

2 *Torque wrench* - *Although not mandatory, using this tool is the best way to ensure the plugs are tightened properly*

3 *Ratchet* - *Standard hand tool to fit the spark plug socket*

4 *Extension* - *Depending on model and accessories, you may need special extensions and universal joints to reach one or more of the plugs*

5 *Spark plug gap gauge* - *This gauge for checking the gap comes in a variety of styles. Make sure the gap for your engine is included*

32.5a Spark plug manufacturers recommend using a wire-type gauge when checking the gap - if the wire does not slide between the electrodes with a slight drag, adjustment is required

32.5b To change the gap, bend the side electrode only, as indicated by the arrows, and be very careful not to crack or chip the porcelain insulator surrounding the center electrode

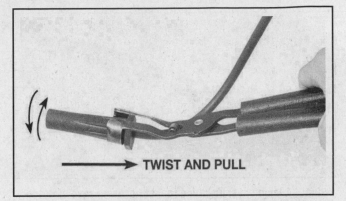

TWIST AND PULL

32.6 When removing the spark plug wires, pull only on the boot and twist it back-and-forth - don't pull on the wire

between the electrodes at the tip of the plug (see illustration). The gap between the electrodes should be the same as the one specified on the Emissions Control Information label. The wire should just slide between the electrodes with a slight amount of drag. If the gap is incorrect, use the adjuster on the gauge body to bend the curved side electrode slightly until the proper gap is obtained (see illustration). If the side electrode is not exactly over the center electrode, bend it with the adjuster until it is. Check for cracks in the porcelain insulator (if any are found, the plug should not be used).

6   With the engine cool, remove the spark plug wire from one spark plug. Pull only on the boot at the end of the wire - do not pull on the wire (see illustration). A plug wire removal tool should be used if available.

7   If compressed air is available, use it to blow any dirt or foreign material away from the spark plug hole. A common bicycle pump will also work. The idea here is to eliminate the possibility of debris falling into the cylinder as the spark plug is removed.

8   Place the spark plug socket over the plug and remove it from the engine by turning it in a counterclockwise direction.

9   Compare the spark plug to those shown in this illustration to get an indication of the general running condition of the engine.

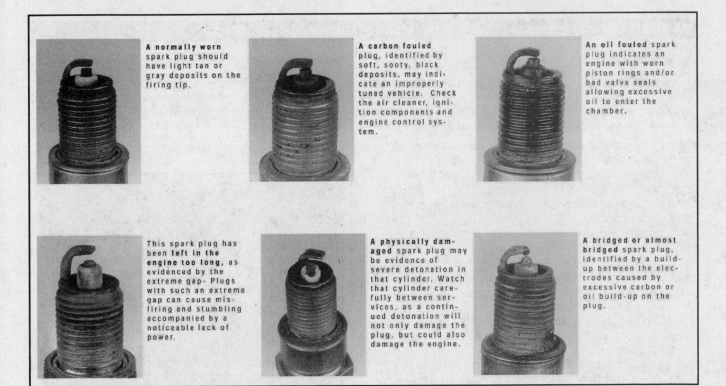

A **normally worn** spark plug should have light tan or gray deposits on the firing tip.

A **carbon fouled** plug, identified by soft, sooty, black deposits, may indicate an improperly tuned vehicle. Check the air cleaner, ignition components and engine control system.

An **oil fouled** spark plug indicates an engine with worn piston rings and/or bad valve seals allowing excessive oil to enter the chamber.

This spark plug has been **left in the engine too long,** as evidenced by the extreme gap- Plugs with such an extreme gap can cause misfiring and stumbling accompanied by a noticeable lack of power.

A **physically damaged** spark plug may be evidence of severe detonation in that cylinder. Watch that cylinder carefully between services, as a continued detonation will not only damage the plug, but could also damage the engine.

A **bridged or almost bridged** spark plug, identified by a build-up between the electrodes caused by excessive carbon or oil build-up on the plug.

32.9 Inspect the spark plug to determine engine running conditions

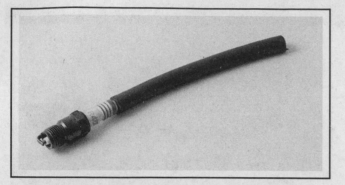

32.10 A length of snug-fitting rubber hose will save time and prevent damaged threads when installing the spark plugs

10 Thread one of the new plugs into the hole until you can no longer turn it with your fingers, then tighten it with a torque wrench (if available) or the ratchet. It might be a good idea to slip a short length of rubber hose over the end of the plug to use as a tool to thread it into place (see illustration). The hose will grip the plug well enough to turn it, but will start to slip if the plug begins to cross-thread in the hole - this will prevent damaged threads and the accompanying repair costs.

11 Before pushing the spark plug wire onto the end of the plug, inspect it following the procedures outlined in Section 33.

12 Attach the plug wire to the new spark plug, again using a twisting motion on the boot until it's seated on the spark plug.

13 Repeat the procedure for the remaining spark plugs, replacing them one at a time to prevent mixing up the spark plug wires.

33 Spark plug wire check and replacement

1 The spark plug wires should be checked at the recommended intervals and whenever new spark plugs are installed in the engine.

2 The wires should be inspected one at a time to prevent mixing up the order, which is essential for proper engine operation.

3 Disconnect the plug wire from the spark plug. To do this, grab the rubber boot, twist slightly and pull the wire off. Do not pull on the wire itself, only on the rubber boot.

4 Check inside the boot for corrosion, which will look like a white crusty powder. Push the wire and boot back onto the end of the spark plug. It should be a tight fit on the plug. If it isn't, remove the wire and use pliers to carefully crimp the metal connector inside the boot until it fits securely on the end of the spark plug.

5 Using a clean rag, wipe the entire length of the wire to remove any built-up dirt and grease. Once the wire is clean, check for burns, cracks and other damage. Do not bend the wire excessively or pull the wire lengthwise - the conductor inside might break.

6 Disconnect the wire from the coil or distributor. Again, pull only on the rubber boot. Check for corrosion and a tight fit in the same manner as the spark plug end. Replace the wire at the coil or distributor.

7 Check the remaining spark plug wires one at a time, making sure they are securely fastened at the ignition coil or distributor and the spark plug when the check is complete.

8 If new spark plug wires are required, purchase a set for your specific engine model. Wire sets are available pre-cut, with the rubber boots already installed. Remove and replace the wires one at a time to avoid mix-ups in the firing order.

34 Distributor cap and rotor check and replacement (1985 models only)

◆ **Refer to illustration 34.3**

1 At the distributor cap, disconnect the electrical connector to the coil.

2 Remove the distributor cap by placing a screwdriver on the slotted head of each latch, press down and turn each latch 90 degrees to release the hook. When all the latches are disengaged, separate the cap from the distributor with the spark plug wires still attached.

3 Inspect the cap for cracks and other damage. Closely examine the terminals on the inside of the cap for excessive corrosion (see illustration). Slight pitting is normal. Deposits on the terminals may be removed with a small file.

4 The rotor is visible, with the cap removed, at the top of the distributor shaft. Remove the two retaining screws and remove the rotor.

5 Inspect the rotor for cracks and other damage. Carefully check the condition of the rotor tip and the metal contact at the top of the rotor for excessive burning or pitting. Check the top of the rotor for carbon tracks, replace the rotor if carbon tracks are visible.

6 Install the rotor (either new or original). Note that the slot in the rotor engages the tab on the distributor shaft when the rotor is properly seated. Tighten the screws securely.

7 If a new cap is being installed, release the locking tabs to the spark plug wire retaining ring and remove the ring with the wires attached (this will prevent mixing the wires). If the ring is missing, or the wires have been previously replaced with a type that does not attach to the ring, make sure you label the wires and mark the cap before removing so they can be reinstalled in their original locations. Transfer the ignition coil to the new cap (see Chapter 5).

8 Install the cap to the distributor. Note that the tab on the cap engages the slot in the distributor housing when the cap is properly seated. Engage the latches, by pushing down and rotating 90 degrees. Connect the electrical connector to the coil.

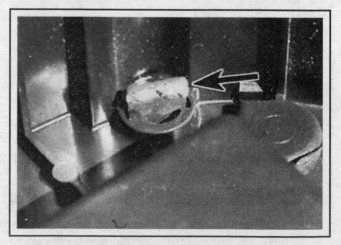

34.3 Inspect the distributor cap for cracks and carbon tracks and the terminals (arrow) for corrosion and damage

35 Ignition timing check and adjustment (1985 models only)

▶ **Refer to illustration 35.2**

➡**Note: It is imperative that the procedures included on the Vehicle Emission Control Information label be followed when adjusting the ignition timing. The label will include all information concerning preliminary steps to be performed before adjusting the timing as well as the timing specifications.**

1 Locate the VECI label under the hood and read through and perform the preliminary instructions.
2 Locate the timing scale, near the crankshaft pulley, and the notch on the crankshaft pulley (see illustration). Clean the timing scale and mark the notch with a dab of white paint.
3 Connect the pick-up lead of the timing light to the number one spark plug wire and the power leads to the battery, according to the manufacturer's instructions.
4 Start the engine and allow it to reach normal operating temperature. Aim the timing light at the timing scale and note at what point on the scale the notch is lined up with. The numbers on the scale represent degrees of advance (BTDC) or retard (ATDC), with the zero representing TDC.
5 If the notch is not lined up at the correct point, according to the specifications on the VECI label, Loosen the distributor hold-down bolt slightly, and while watching the scale rotate the distributor until the notch lines up with the correct mark.

35.2 The ignition timing scale (arrow) is located at the front of the engine

6 Tighten the distributor hold-down bolt and recheck the timing.
7 Turn off the engine, remove the timing light and return the engine to the normal operating mode.

36 Cabin air filter

1 Refer to Chapter 11 and remove the right-hand sound panel from below the instrument panel.
2 The cabin air filter is located above the blower motor, in the air inlet panel.

3 Lift the access cover over the filter and withdraw the old filter.
4 Installation is the reverse of the removal procedure. Install the new filter with the same side UP as on the old filter.

Recommended lubricants and fluids

➡Note: Listed here are manufacturer recommendations at the time this manual was written. Manufacturers occasionally upgrade their fluid and lubricant specifications, so check with your local auto parts store for current recommendations.

Engine oil	
Type	API grade SG, SH, SG/CD or SH/CD multigrade and fuel efficient oil
Viscosity	See accompanying chart
Automatic transaxle fluid	Dexron II, Dexron IIE or Dexron III ATF*
Engine coolant	50/50 mixture of water and the specified ethylene glycol-based (green color) antifreeze or "DEX-COOL," silicate-free (orange-color) coolant - DO NOT mix the two types (refer to Sections 4, 9 and 28)
Brake fluid	Delco Supreme II or DOT 3 fluid
Power steering fluid	GM power steering fluid or equivalent
Chassis lubrication	NLGI #2 multi-purpose chassis grease

On 2003 through 2005 models, look for "Approved for the H Specification" on the label.

Capacities*

Engine oil (with filter change)	4.0 to 4.5 qts
Fuel tank	18.0 to 19.5 gals
Cooling system	10 to 13.0 qts (depending on cooling system option)
Automatic transaxle (filter change and refill)	
1985 through 1996	6 qts
1997 and later	7.4 qts

All capacities approximate. Add as necessary to bring to the appropriate level.

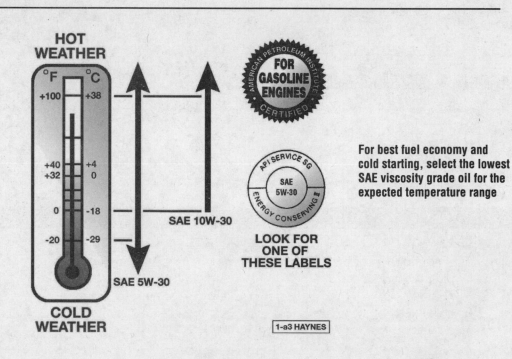

For best fuel economy and cold starting, select the lowest SAE viscosity grade oil for the expected temperature range

Ignition system

Firing order	1-6-5-4-3-2
Cylinder locations	
Left (front) bank	1-3-5
Right (rear) bank	2-4-6
Spark plug type and gap*	
1985	AC R44TS8 or equivalent @ 0.080 inch
1986 and 1987	AC R44LTS or equivalent @ 0.045 inch
1988	AC R44LTS6 or equivalent @ 0.060 inch
1989 through 1991	AC R45LTS6 or equivalent @ 0.060 inch
1992 and 1993	AC 41-600** or equivalent @ 0.060 inch
1994 and 1995	AC 41-601** or equivalent @ 0.060 inch
1996 through 2002	AC41-921 or equivalent @ 0.060 inch
2003 through 2005	AC41-101 or equivalent @ 0.060 inch
Ignition timing	
1985	Refer to Vehicle Emission Control Information label in the engine compartment
1986 and later	Non-adjustable

* *Refer to the Vehicle Emission Control Information label in the engine compartment and follow the information on the label if it differs from that shown here.*

** *Platinum type spark plugs*

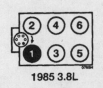 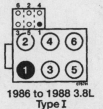 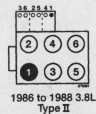 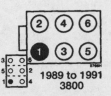 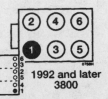

1985 3.8L   1986 to 1988 3.8L Type I   1986 to 1988 3.8L Type II   1989 to 1991 3800   1992 and later 3800

Cylinder location and coil terminal identification diagrams

General

Brake pad wear limit	1/8 in
Brake shoe wear limit	1/16 in
Radiator cap pressure rating	15 psi

Torque specifications   Ft-lbs (unless otherwise indicated)

Spark plugs	
Through 2001	20
2002 through 2005	132 in-lbs
Engine oil drain plug	20 to 25
Automatic transaxle oil pan bolts	96 to 120 in-lbs
Wheel lug nuts	100

Notes

2A

ENGINES

Section

Reference to other Chapters

1 General information

CAUTION:

If the vehicle is equipped with a Delco Loc II or Theftlock audio system, make sure you have the correct activation code before disconnecting the battery. See the information at the front of this manual for the radio re-activation procedure.

This Part of Chapter 2 is devoted to in-vehicle repair procedures for the 3.0 and 3.8 liter V6 engines. Some 3.8 liter engines (VIN code C, K and L) are equipped with a balance shaft; they will be referred to as "3800" engines to avoid confusion with the non-balance shaft equipped 3.8 liter engines.

Information concerning camshaft, balance shaft and engine removal and installation, as well as engine block and cylinder head overhaul, is in Part B of this Chapter.

The following repair procedures are based on the assumption the engine is installed in the vehicle. If the engine has been removed from the vehicle and mounted on a stand, many of the steps included in this Part of Chapter 2 will not apply.

The Specifications included in this Part of Chapter 2 apply only to the procedures in this Part. The Specifications necessary for rebuilding the block and cylinder heads are found in Part B.

2 Repair operations possible with the engine in the vehicle

Many major repair operations can be accomplished without removing the engine from the vehicle.

Clean the engine compartment and the exterior of the engine with some type of pressure washer before any work is done. A clean engine will make the job easier and will help keep dirt out of the internal areas of the engine.

Depending on the components involved, it may be a good idea to remove the hood to improve access to the engine as repairs are performed (refer to Chapter 11 if necessary).

If vacuum, exhaust, oil or coolant leaks develop, indicating a need for gasket or seal replacement, the repairs can generally be made with the engine in the vehicle. The intake and exhaust manifold gaskets, oil pan gasket and cylinder head gaskets are all accessible with the engine in place.

Exterior engine components such as the intake and exhaust manifolds, the oil pan, the oil pump, the water pump, the starter motor, the alternator and the fuel injection system can be removed for repair with the engine in place. The timing chain and sprockets can also be replaced with the engine in the vehicle, but the camshaft (and balance shaft; if equipped) cannot be removed with the engine in place.

Since the cylinder heads can be removed without pulling the engine, valve component servicing can also be accomplished with the engine in the vehicle.

In extreme cases caused by a lack of necessary equipment, repair or replacement of piston rings, pistons, connecting rods and rod bearings is possible with the engine in the vehicle. However, this practice is not recommended because of the cleaning and preparation work that must be done to the components involved.

3 Top Dead Center (TDC) for number one piston - locating

◆ **Refer to illustrations 3.6 and 3.8**

1   Top Dead Center (TDC) is the highest point in the cylinder that each piston reaches as it travels up-and-down when the crankshaft turns. Each piston reaches TDC on the compression stroke and again on the exhaust stroke, but TDC generally refers to piston position on the compression stroke.

2   Positioning the piston(s) at TDC is an essential part of certain procedures such as timing chain/sprocket removal and camshaft removal.

3   Before beginning this procedure, be sure to place the transaxle in Park, apply the parking brake and block the rear wheels. Raise the front of the vehicle and support it securely on jackstands.

4   Remove the spark plugs (see Chapter 1).

5   When looking at the drivebelt end of the engine, normal crankshaft rotation is clockwise. In order to bring any piston to TDC, the crankshaft must be turned with a socket and ratchet attached to the bolt threaded into the center of the vibration damper on the crankshaft.

MODELS WITH DISTRIBUTORLESS (C3I) IGNITION

6   Have an assistant turn the crankshaft with a socket and ratchet as described above while you hold a finger over the number one spark plug hole (see illustration).

➡**Note: See the cylinder numbering diagram in the specifications for this Chapter.**

7   When the piston approaches TDC, air pressure will be felt at the spark plug hole. Instruct your assistant to turn the crankshaft slowly.

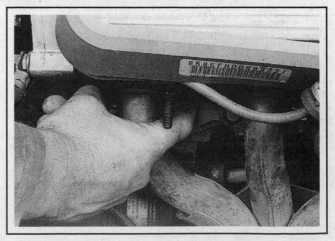

3.6 Hold a finger over the spark plug opening until you feel air escaping . . .

8   Insert a plastic pen into the spark plug hole (see illustration). As the piston rises the pen will be pushed out. Note the point where the pen stops moving out - this is TDC.

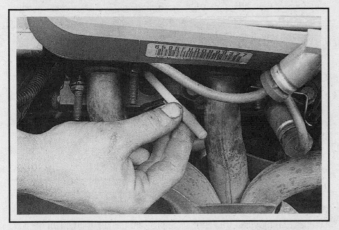

3.8 . . . then insert a soft plastic pen into the hole to detect piston movement

9   After the number one piston has been positioned at TDC on the compression stroke, TDC for any of the remaining pistons can be located by repeating the procedure described above and following the firing order.

MODELS WITH DISTRIBUTOR (HEI) IGNITION

10   Make a mark on the distributor housing directly below the number one spark plug wire terminal on the distributor cap.

➡**Note: The terminal numbers are usually marked on the spark plug wires, near the distributor. Remove the distributor cap.**

11   Turn the crankshaft with a socket and ratchet (see Step 5) until the line on the vibration damper is aligned with the zero mark on the timing scale.

12   The rotor should now be pointing directly at the mark on the distributor housing. If it isn't, the piston is at TDC on the exhaust stroke - to get the piston on the compression stroke, turn the crankshaft one complete turn (360-degrees) clockwise. When the rotor is pointing at the number one spark plug wire terminal in the distributor cap (which is indicated by the mark on the housing) and the ignition timing marks are aligned, the number one piston is at TDC on the compression stroke.

4   Rocker arm covers - removal and installation

▶ **Refer to illustrations 4.2, 4.3 and 4.4**

FRONT COVER REMOVAL

1   Disconnect the negative battery cable from the battery.

✳ CAUTION:

If the vehicle is equipped with a Delco Loc II or Theftlock audio system, make sure you have the correct activation code before disconnecting the battery. See the information at the front of this manual for the radio re-activation procedure.

2   Remove the spark plug wires from the spark plugs and remove the harness cover (see illustration). Number each wire before removal to ensure correct reinstallation.

3   Remove the PCV tube (see illustration) from the cover (except 3800 engine).

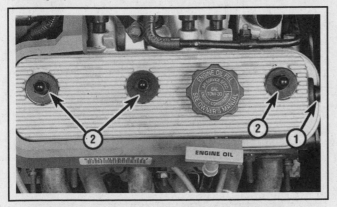

4.3   Rocker arm cover mounting details - early models

   1   PCV tube   *2   Mounting nuts*

4   On 3800 engines, remove the acoustic cover over the top of the intake plenum. On 1996 and later models it will be necessary to twist out the oil filler tube from the front valve cover, pull up the front of the

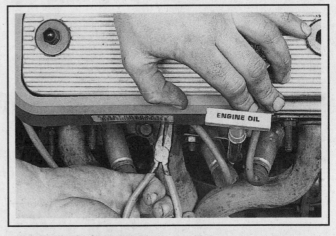

4.2   Unclip the harness cover

4.4   Remove the acoustic engine cover from the top of the plenum - on 1996 and later models, it will be necessary to twist out the oil filler neck first, then pull the cover up and forward to remove it

acoustic cover and pull it forward (towards the radiator) to release it from the bracket at the rear of the plenum (see illustration). On some model years it may also be necessary to remove the drivebelt (see Chapter 1) and the alternator brace.

5   Remove the rocker arm cover mounting bolts/nuts.
6   Detach the rocker arm cover.

➡**Note: If the cover sticks to the cylinder head, use a soft-face hammer to dislodge it.**

REAR COVER REMOVAL

7   Disconnect the negative battery cable from the battery.

✳✳ CAUTION:

If the vehicle is equipped with a Delco Loc II or Theftlock audio system, make sure you have the correct activation code before disconnecting the battery. See the information at the front of this manual for the radio re-activation procedure.

8   Remove the spark plug wires from the spark plugs and remove the wire holder. Be sure each wire is labeled before removal to ensure correct reinstallation.
9   Remove the serpentine drivebelt (see Chapter 1).
10   Remove the power steering pump, if necessary (see Chapter 10).

3.0 and 3.8 liter engines

11   Remove the ignition coil assembly (see Chapter 5).
12   Remove the drivebelt tensioner.
13   Remove the alternator, if necessary (see Chapter 5).
14   Remove the engine lift bracket and rear alternator brace.
15   Drain the coolant (see Chapter 1) and remove the throttle body coolant hoses.

3800 engines

16   Remove the acoustic engine cover (see Step 4) and the cover bracket from the rear exhaust manifold. Remove the EGR pipe, valve and adapter (see Chapter 6). On some models it will also be necessary to remove the air injection pipe and valve assembly from above the rear rocker arm cover. On 2001 and later models, disconnect and set aside the vacuum hose from the power brake booster.

All engines

17   Remove the rocker arm cover mounting bolts/nuts.
18   Detach the rocker arm cover.

➡**Note: If the cover sticks to the cylinder head, use a soft-face hammer to dislodge it.**

INSTALLATION

19   The mating surfaces of the cylinder head and rocker arm cover must be perfectly clean when the covers are installed. Use a gasket scraper to remove all traces of sealant or old gasket, then clean the mating surfaces with lacquer thinner or acetone (if there's sealant or oil on the mating surfaces when the cover is installed, oil leaks may develop). The rocker arm covers are made of aluminum, so be extra careful not to nick or gouge the mating surfaces with the scraper.

20   On 3800 engines, apply thread locking compound to the bolt threads before installing the cover. On all models, place the rocker arm cover and new gasket in position on the cylinder head. On 3800 engines, install the mounting bolts. On earlier models, install the new rubber grommets, cover washers and the mounting nuts.

21   Tighten the bolts/nuts in several steps to the torque listed in this Chapter's specifications.

22   Complete the installation by reversing the removal procedure. Be sure to add coolant if it was drained.

23   Start the engine and check for oil leaks at the rocker arm cover-to-head joints.

5   Rocker arms and pushrods - removal, inspection and installation

▸ **Refer to illustrations 5.2a, 5.2b, 5.3, 5.4 and 5.6**

➡**Note: Store each set of rocker components separately in a marked plastic bag to ensure that they're installed in their original locations.**

1   Refer to Section 4 and detach the rocker arm covers from the cylinder heads.

2   On models with rocker arm shafts, loosen the three bolts on each shaft a little at a time, working from the center out (see illustration). Keep track of the rocker arm positions, since they must be returned to the same locations (see illustrations).

5.2a Loosen the three rocker arm shaft bolts (arrows) and remove the rocker arms and shaft as an assembly

5.2b Remove the rocker arm retainers only if you have to inspect the shaft - the plastic retainers will probably break, so make sure you have new ones on hand

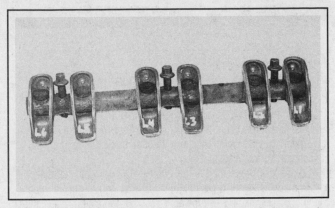

5.2c Reinstall the rockers (numbered with paint before removal) to the shaft in their original sequence

5.4 Later engines have individual rocker arm pivot bolts

➡Note: If the rocker assembly is being removed for access to other components (such as pushrod or cylinder head removal), do not remove the nylon retainers. Normal engine operation hardens the nylon to the point that the retainers will break when removed. Don't break them off unless you need to remove the rocker arms from the shaft.

3   On Vin K engines, loosen the rocker arm pivot bolts one at a time and detach the rocker arms and bolts.

4   On Vin C and VIN L engines, loosen the rocker arm pivot bolts one at a time and detach the rocker arms, bolts, pivots and pivot retainers (see illustration).

5   If necessary, remove the pushrod guide from the cylinder head.

6   Remove the pushrods and store them separately to make sure they don't get mixed up during installation (see illustration).

INSPECTION

7   Check each rocker arm for wear, cracks and other damage, especially where the pushrods and valve stems contact the rocker arm.

8   Check the pivot seat in each rocker arm and the pivot faces. Look for galling, stress cracks and unusual wear patterns. If the rocker arms are worn or damaged, replace them with new ones and install new pivots or shafts as well.

➡Note: On 2001 and later models, there is single, one-piece pedestal for all the rocker arms, rather than separate pedestals for each rocker arm.

9   Make sure the hole at the pushrod end of each rocker arm is open.

10 Inspect the pushrods for cracks and excessive wear at the ends. Roll each pushrod across a piece of plate glass to see if it's bent (if it wobbles, it's bent).

INSTALLATION

11 Lubricate the lower end of each pushrod with clean engine oil or moly-base grease and install them in their original locations. Make sure each pushrod seats completely in the lifter socket.

5.6 If more than one pushrod is being removed, store them in a perforated cardboard box to prevent mix-ups during installation - note the label indicating the front of the engine

12 Apply moly-base grease to the ends of the valve stems, the upper ends of the pushrods and to the pivot faces to prevent damage to the mating surfaces before engine oil pressure builds up.

13 If removed, install the pushrod guide onto the cylinder head.

14 Coat the rocker arm pivot bolt threads with a non-hardening thread locking compound.

15 On models with rocker arm shafts, tighten the three bolts on each shaft a little at a time, working from the center out until the torque listed in this Chapter's Specifications is reached.

16 On Vin K engines, install the rocker arms and bolts. On Vin C and VIN L engines, install the rocker arms, the pivot retainers, pivots, and bolts. Tighten the bolts to the torque listed in this Chapter's specifications.

17 Install the rocker arm covers (see Section 4).

6 Valve springs, retainers and seals - replacement

◆ **Refer to illustrations 6.4, 6.8, 6.14 and 6.17**

➡**Note:** Broken valve springs and defective valve stem seals can be replaced without removing the cylinder heads. Two special tools and a compressed air source are normally required to perform this operation, so read through this Section carefully and rent or buy the tools before beginning the job. If compressed air is not available, a length of nylon rope can be used to keep the valves from falling into the cylinder during this procedure.

1   Refer to Section 4 and remove the rocker arm cover from the affected cylinder head. If all of the valve stem seals are being replaced, remove both rocker arm covers.

2   Remove the spark plug from the cylinder that has the defective component. If all of the valve stem seals are being replaced, all of the spark plugs should be removed.

3   Turn the crankshaft until the piston in the affected cylinder is at Top Dead Center on the compression stroke (refer to Section 3 for instructions). If you're replacing all of the valve stem seals, begin with cylinder number one and work on the valves for one cylinder at a time. Move from cylinder-to-cylinder following the firing order sequence (1-6-5-4-3-2).

4   Thread an adapter into the spark plug hole (see illustration) and

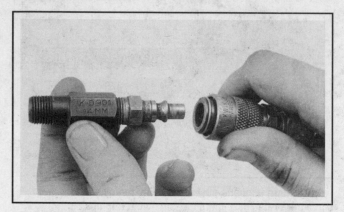

6.4 This is what the air hose adapter that threads into the spark plug hole looks like - they're commonly available from auto parts stores

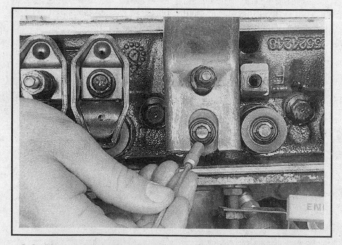

6.8 After compressing the valve spring, remove the keepers with a magnet (as shown here) or small needle-nose pliers

connect an air hose from a compressed air source to it. Most auto parts stores can supply the air hose adapter.

➡**Note:** Many cylinder compression gauges utilize a screw-in fitting that may work with your air hose quick-disconnect fitting.

5   Remove the bolt, pivot and rocker arm for the valve with the defective part and pull out the pushrod (see Section 5). If all of the valve stem seals are being replaced, all of the rocker arms and pushrods should be removed.

6   Apply compressed air to the cylinder. The valves should be held in place by the air pressure. If the valve faces or seats are in poor condition, leaks may prevent the air pressure from retaining the valves - refer to the alternative procedure below.

7   If you don't have access to compressed air, an alternative method can be used. Position the piston at a point approximately 45-degrees before TDC on the compression stroke, then feed a long piece of nylon rope through the spark plug hole until it fills the combustion chamber. Be sure to leave the end of the rope hanging out of the engine so it can be removed easily. Use a large breaker bar and socket to rotate the crankshaft in the normal direction of rotation until slight resistance is felt.

8   Stuff shop rags into the cylinder head holes above and below the valves to prevent parts and tools from falling into the engine, then use a valve spring compressor to compress the spring. Remove the keepers with small needle-nose pliers or a magnet (see illustration).

➡**Note:** A couple of different types of tools are available for compressing the valve springs with the head in place. One type grips the lower spring coils and presses on the retainer as the knob is turned, while the other type, shown here, utilizes the rocker arm bolt for leverage. Both types work very well, although the lever type is usually less expensive.

9   Remove the spring retainer and valve spring, then remove the guide seal.

➡**Note 1:** If air pressure fails to hold the valve in the closed position during this operation, the valve face or seat is probably damaged. If so, the cylinder head will have to be removed for additional repair operations.

➡**Note 2:** 3.0 and 3.8 liter engines have umbrella or positive type valve seals on the intake valves and O-ring type valve seals on the exhaust valves.

10 Wrap a rubber band or tape around the top of the valve stem so the valve won't fall into the combustion chamber, then release the air pressure.

➡**Note:** If a rope was used instead of air pressure, turn the crankshaft slightly in the direction opposite normal rotation.

11 Inspect the valve stem for damage. Rotate the valve in the guide and check the end for eccentric movement, which would indicate that the valve is bent.

12 Move the valve up-and-down in the guide and make sure it doesn't bind. If the valve stem binds, either the valve is bent or the guide is damaged. In either case, the head will have to be removed for repair.

13 Reapply air pressure to the cylinder to retain the valve in the closed position, then remove the tape or rubber band from the valve stem. If a rope was used instead of air pressure, rotate the crankshaft in the normal direction of rotation until slight resistance is felt.

14 Lubricate the valve stem with clean engine oil and install a new valve guide seal. Using a hammer and a deep socket or seal installation

6.14 Valve seals can be installed with an appropriate-size deep socket - tap them down until they seat on the valve guide

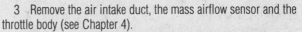

6.17 Apply a small dab of grease to each keeper as shown here before installation - it will hold them in place on the valve stem as the spring is released

tool, gently tap the seal into place until it's completely seated on the guide (see illustration).

15 Install the spring in position over the valve.

16 Install the valve spring retainer and compress the valve spring.

17 Position the keepers in the upper groove. Apply a small dab of grease to the inside of each keeper to hold it in place if necessary (see

illustration). Remove the pressure from the spring tool and make sure the keepers are seated.

18 Disconnect the air hose and remove the adapter from the spark plug hole. If a rope was used in place of air pressure, pull it out of the cylinder.

19 Install the rocker arm(s) and pushrod(s).

20 Install the spark plug(s) and connect the wire(s).

21 Refer to Section 4 and install the rocker arm cover(s).

22 Start and run the engine, then check for oil leaks and unusual sounds coming from the rocker arm cover area.

7 Intake manifold - removal and installation

▶ **Refer to illustrations 7.6, 7.10, 7.12, 7.13, 7.14, 7.31, 7.32, 7.36a and 7.36b**

REMOVAL (ONE-PIECE MANIFOLD)

1 Relieve the fuel system pressure (see Chapter 4).

2 Disconnect the negative battery cable from the battery.

✳✳ CAUTION:

If the vehicle is equipped with a Delco Loc II or Theftlock audio system, make sure you have the correct activation code before disconnecting the battery. See the information at the front of this manual for the radio re-activation procedure.

3 Remove the air intake duct, the mass airflow sensor and the throttle body (see Chapter 4).

4 Remove the power steering pump-to-manifold brace.

5 Remove the fuel injectors (see Chapter 4).

6 Remove the control cables from their brackets (see illustration).

7 Remove the drivebelt (see Chapter 1).

8 Disconnect and remove the alternator and brackets (see Chapter 5).

9 Drain the cooling system (see Chapter 1).

10 Disconnect the coolant tubes and hoses at the manifold (see illustration).

7.6 Pinch the tabs (arrows) together with pliers to detach the cables from the bracket

7.10 Remove the hoses (arrows) and tubes from the intake manifold (3.8 liter shown, others similar)

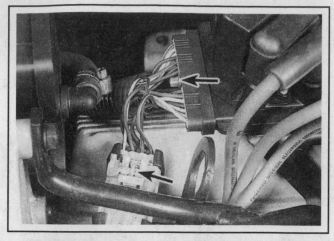

7.12 Unplug the electrical connectors (arrows) and move the harness aside

7.13 Intake manifold mounting bolt locations (arrows) (one-piece manifold shown)

11 Remove the ignition module assembly or distributor cap and spark plug wires (see Chapters 1 and 5).

12 Label and disconnect the fuel and vacuum lines and electrical wires at the manifold (see illustration). When disconnecting fuel line fittings, be prepared to catch some fuel, then cap the fittings to prevent contamination.

13 Remove the manifold mounting bolts (see illustration).

➡**Note: On 3.0 and 3.8 liter models, one bolt requires a Torx T-50 drive bit.**

14 Separate the manifold from the engine (see illustration). Do not pry between the manifold and heads, as damage to the gasket sealing surfaces may result. If you're installing a new manifold, transfer all fittings and sensors to the new manifold.

REMOVAL (TWO-PIECE MANIFOLD)

➡**Note: On VIN L engines, it is not necessary to disassemble the upper intake plenum from the lower intake manifold unless the gasket is damaged and requires replacement. On VIN K engines, it is necessary to first remove the upper manifold prior to removing the lower manifold.**

15 Relieve the fuel system pressure (see Chapter 4).

16 Disconnect the negative battery cable from the battery.

❊❊ CAUTION:

If the vehicle is equipped with a Delco Loc II or Theftlock audio system, make sure you have the correct activation code before disconnecting the battery. See the information at the front of this manual for the radio re-activation procedure.

17 Remove the engine cover/fuel injector sight shield and the air intake duct.

18 Disconnect the spark plug wires from the engine and set them aside.

19 Remove the fuel rail and injector assembly (see Chapter 4).

20 Remove the exhaust heat shield. On later models, remove the heat shield over the wiring harness.

21 Remove the ignition coil/module assembly (see Chapter 5).

22 Disconnect all hoses from the manifold assembly. Remove the MAP sensor and the PCV valve (see Chapter 1).

23 Disconnect the throttle cable and remove the cable bracket from

7.14 Carefully pry up on a casting boss - don't pry between gasket surfaces

the cylinder head.

24 Remove the drivebelt (see Chapter 1).

25 Remove the drivebelt tensioner.

26 Disconnect and remove the alternator and brackets (see Chapter 5).

27 Drain the cooling system (see Chapter 1).

28 Disconnect the coolant hoses and tubes at the manifold. When disconnecting fuel line fittings, be prepared to catch some fuel, then cap the fittings to prevent contamination.

29 Remove the EGR valve outlet pipe.

30 On VIN L engines, perform the following:

   a) *Remove the manifold assembly mounting bolts.*

   b) *Separate the manifold assembly from the engine. Do not pry between the manifold and heads, as damage to the gasket sealing surfaces may result. If you're installing a new manifold assembly, transfer all fittings and sensors to the new manifold assembly.*

❊❊ CAUTION:

On VIN K engines, do not try to remove the lower manifold without first removing the upper manifold as there are two additional lower manifold mounting bolts located under the upper manifold.

7.31 Remove the mounting bolts (not seen in this view) for the intake plenum (arrow) - there are 3 bolts at the front of the plenum, 3 on the rear, one stud and one bolt at the throttle body end and two bolts at the timing chain end

31 On VIN K engines, perform the following:

a) *Remove the upper intake plenum mounting bolts (see illustration).*

b) *Separate the upper manifold from the lower manifold. Do not pry between the upper and lower manifolds, as damage to the gasket sealing surfaces may result.*

c) *Remove lower intake manifold mounting bolts.*

d) *Separate the lower manifold from the engine. Do not pry between the manifold and heads, as damage to the gasket sealing surfaces may result. If you're installing new manifolds, transfer all fittings and sensors to the new manifold assembly.*

INSTALLATION

➡**Note: The mating surfaces of the cylinder heads, block and manifold must be perfectly clean when the manifold is installed. Gasket removal solvents in aerosol cans are available at most auto parts stores and may be helpful when removing old gasket material that's stuck to the heads and manifold (since the manifold is made of aluminum, aggressive scraping can cause damage). Be sure to follow the directions printed on the container.**

32 Use a gasket scraper to remove all traces of sealant and old gasket material (see illustration), then clean the mating surfaces with lacquer thinner or acetone. If there's old sealant or oil on the mating surfaces when the manifold is installed, oil or vacuum leaks may develop. Use a vacuum cleaner to remove any gasket material that falls into the intake ports or the lifter valley.

33 Use a tap of the correct size to chase the threads in the bolt holes, then use compressed air (if available) to remove the debris from the holes.

✳✳ **WARNING:**

Wear safety glasses or a face shield to protect your eyes when using compressed air.

34 If steel manifold gaskets are used, apply gasket sealant to both sides of the gaskets before positioning them on the heads. Apply RTV sealant to the ends of the new manifold-to-block seals, then install them. Make sure the pointed end of the seal fits snugly against both the head and block.

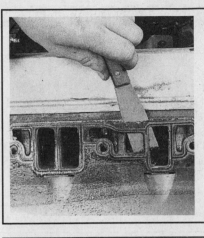

7.32 Remove all traces if gasket material, but don't gouge the soft aluminum

7.36a Intake manifold tightening sequence - 1993 and earlier

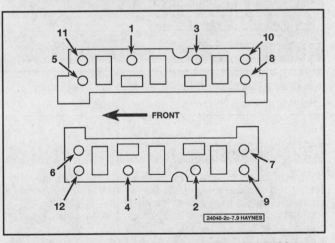

7.36b Intake manifold tightening sequence - 1994 and later

35 Carefully lower the manifold into place. Apply a non-hardening thread locking compound to the mounting bolt threads and install the bolts finger tight.

36 Tighten the mounting bolts in two stages, following the recommended sequence (see illustrations), until they're all at the torque listed in this Chapter's Specifications.

37 Install the remaining components in the reverse order of removal.

38 Change the oil and filter and fill the cooling system (see Chapter 1). Start the engine and check for oil and vacuum leaks.

8   Valve lifters - removal, inspection and installation

♦ **Refer to illustrations 8.6a, 8.6b, 8.6c, 8.7, 8.8, 8.11a, 8.11b, 8.11c and 8.12**

1   A noisy valve lifter can be isolated when the engine is idling. Hold a mechanic's stethoscope or a length of hose near the position of each valve while listening at the other end.

2   The most likely causes of noisy valve lifters are dirt trapped between the plunger and the lifter body or lack of oil flow, viscosity or pressure. Before condemning the lifters, we recommend checking the oil for fuel contamination, proper level, cleanliness and correct viscosity.

REMOVAL

3   Remove the rocker arm cover(s) as described in Section 4.
4   Remove the rocker arms and pushrods (see Section 5).
5   Remove the intake manifold(s) as described in Section 7.
6   Earlier models have conventional lifters. Later production engines have roller lifters. Roller lifters are kept from turning by a guide plate and lifter guide (see illustrations), which must be removed to access the lifters.

7   There are several ways to extract the lifters from the bores. A special tool designed to grip and remove lifters is manufactured by many tool companies and is widely available, but it may not be required in every case (see illustration). On newer engines without a lot of varnish buildup, the lifters can often be removed with a small magnet or even with your fingers. A machinist's scribe with a bent end can be used to pull the lifters out by positioning the point under the retainer ring in the top of each lifter.

✳ CAUTION:

Don't use pliers to remove the lifters unless you intend to replace them with new ones. The pliers will damage the precision machined and hardened lifters, rendering them useless.

8   Before removing the lifters, arrange to store them in a clearly labeled box to ensure that they're reinstalled in their original locations. Remove the lifters and store them where they won't get dirty (see illustration).

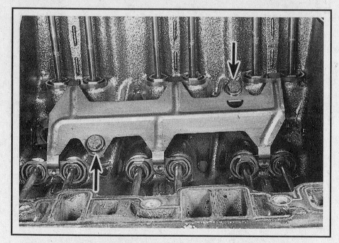

8.6a The guide plate is held in place by two bolts (arrows)

8.6b The lifter guides slip over the lifters

8.6c On 1991 and later engines the roller lifters are retained by guides bolted to the sides of the lifter valley - remove the four bolts (arrows) to take off the guides and remove the lifters

8.7 Many times you can remove the lifters by hand

8.8 The lifters on an engine that has accumulated many miles may have to be removed with a special tool - be sure to store the lifters in an organized manner to make sure they're reinstalled in their original locations

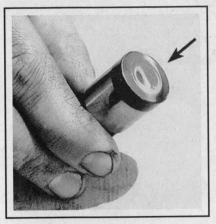

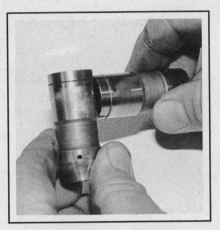

8.11a If the bottom (foot) of any lifter is worn concave, scratched or pitted, replace the entire set with new lifters

8.11b The bottom (foot) of each lifter should be slightly convex - the side of another lifter can be used as a straightedge to check it (if it appears flat, it's worn and should be discarded)

8.11c Check the pushrod seat (arrow) in the top of each lifter for wear

INSPECTION

9   Parts for valve lifters are not available separately. The work required to remove them from the engine again if repair is unsuccessful outweighs any potential savings from repairing them.

10   Clean the lifters with solvent and dry them thoroughly without mixing them up.

Conventional lifters

11   Check each lifter wall, pushrod seat and foot for scuffing, score marks and uneven wear. Each lifter foot (the surface that rides on the cam lobe) must be slightly convex, although this can be difficult to determine by eye. If the base of the lifter is concave (see illustrations), the lifters and camshaft must be replaced. If the lifter walls are damaged or worn (which isn't very likely), inspect the lifter bores in the engine block as well. If the pushrod seats (see illustration) are worn, check the pushrod ends.

Roller lifters

12   Check the rollers carefully for wear and damage and make sure they turn freely without excessive play (see illustration). The inspection procedure for conventional lifters also applies to roller lifters.

INSTALLATION

13   Used roller lifters can be reinstalled with a new camshaft and the

8.12 The roller on roller lifters must turn freely - check for wear and excessive play as well

original camshaft can be used if new lifters are installed.

14   If the conventional lifters are worn, they must be replaced with new ones and the camshaft must be replaced as well - never install used conventional lifters with a new camshaft or new lifters with a used camshaft.

15   When reinstalling used lifters, make sure they're replaced in their original bores. Soak new lifters in oil to remove trapped air. Coat all lifters with moly-base grease or engine assembly lube prior to installation.

16   The remaining installation steps are the reverse of removal.

17   Run the engine and check for oil leaks.

9   Exhaust manifolds - removal and installation

✳✳ WARNING:

Allow the engine to cool completely before beginning this procedure.

➡Note: Exhaust system fasteners are frequently difficult to remove - they get frozen in place because of the heating/cooling cycle to which they're constantly exposed. To ease removal,

apply penetrating oil to the threads of all exhaust manifold and exhaust pipe fasteners and allow it to soak in.

REMOVAL - FRONT MANIFOLD

◆ Refer to illustrations 9.4 and 9.8

1   Disconnect the negative battery cable.

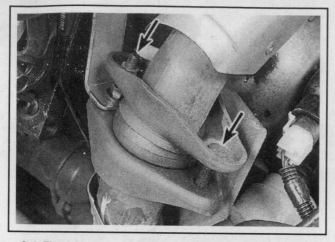

9.4 The exhaust manifolds are bolted to the crossover pipe (arrows)

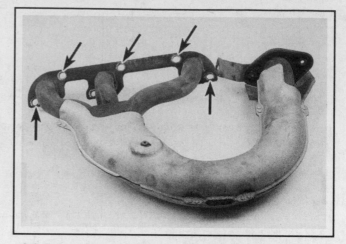

9.8 Front exhaust manifold mounting bolt locations

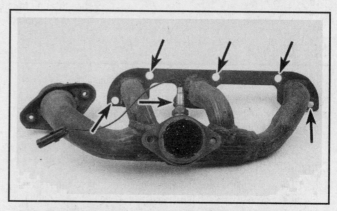

9.23 Rear exhaust manifold mounting bolt locations

❊❊ CAUTION:

If the vehicle is equipped with a Delco Loc II or Theftlock audio system, make sure you have the correct activation code before disconnecting the battery. See the information at the front of this manual for the radio re-activation procedure.

2   Remove the cooling fan for access (see Chapter 3).

3   Remove the mass airflow sensor, air duct and crankcase ventilation pipe, if necessary (see Chapter 4).

4   Unbolt the manifold from the crossover pipe (see illustration).

5   Remove the dipstick tube hold-down nut and wiggle the dipstick tube out of the block.

6   Detach the spark plug wires from the front spark plugs (see Chapter 1).

7   Remove the heat shield, if equipped. On 2003 and later models, remove the engine lift bracket.

8   Unbolt and remove the exhaust manifold (see illustration).

REMOVAL - REAR MANIFOLD

▶ **Refer to illustration 9.23**

9   Disconnect the negative battery cable.

❊❊ CAUTION:

If the vehicle is equipped with a Delco Loc II or Theftlock audio system, make sure you have the correct activation code before disconnecting the battery. See the information at the front of this manual for the radio re-activation procedure.

➡**Note: On 2003 and later models, the only Steps applicable are 12 and 13, then remove the engine lift bracket and the sight shield bracket.**

10 Drain the coolant from the radiator (see Chapter 1).

11 Remove the mass airflow sensor, air duct and crankcase ventilation pipe, if necessary (see Chapter 4).

12 Remove the two nuts attaching the crossover pipe to the rear exhaust manifold.

13 Disconnect the spark plug wires from the rear spark plugs (see Chapter 1).

14 Remove the EGR pipe and transaxle dipstick tube if it's in the way (see Chapters 6 and 7).

15 Unbolt the power steering pump (some models) without removing the hoses and hold it to one side with wire (see Chapter 10).

16 Remove the ignition module bracket nuts and studs (except 3800). On 3800 models, detach the alternator support bracket. Refer to Chapter 5 as necessary.

17 Remove the heater hoses and coolant tube brackets and tubing above the exhaust manifold.

18 Remove the manifold heat shield, if equipped. On 2002 and later models, move the AIR pipe and valve, then remove the power brake booster heatshield.

19 Detach the IAC wiring from the throttle body.

20 Set the parking brake, block the rear wheels and raise the front of the vehicle, supporting it securely on jackstands.

21 Working under the vehicle, remove the two exhaust pipe-to-manifold bolts.

22 Disconnect the oxygen sensor wire (the sensor is threaded into the manifold), then lower the vehicle.

23 Remove the six bolts and detach the manifold from the head (see illustration).

INSTALLATION

24 Clean the mating surfaces of the manifold and cylinder head to remove all traces of old gasket material, then check the manifold for

warpage and cracks. If the manifold gasket was blown, take the manifold to an automotive machine shop for resurfacing.

25 Place the manifold in position with a new gasket and install the bolts finger tight.

26 Starting in the middle and working out toward the ends, tighten the mounting bolts a little at a time until all of them are at the specified torque.

27 Install the remaining components in the reverse order of removal.

28 Start the engine and check for exhaust leaks between the manifold and cylinder head and between the manifold and exhaust pipe.

10 Cylinder heads - removal and installation

▸ Refer to illustrations 10.3a, 10.3b, 10.3c, 10.3d, 10.9, 10.10, 10.13, 10.16 and 10.19

➡ Note: Most of these engines require updated torque-to-yield cylinder head bolts as a replacement. Consult with a dealer parts department for information.

REMOVAL

1 Disconnect the negative battery cable at the battery.

10.3a The alternator bracket is attached to the cylinder head with two nuts or bolts (arrows)

10.3c On some models the air conditioner compressor bracket is attached to the cylinder head with one Torx T-50 bolt (arrow)

✳✳ CAUTION:

If the vehicle is equipped with a Delco Loc II or Theftlock audio system, make sure you have the correct activation code before disconnecting the battery. See the information at the front of this manual for the radio re-activation procedure.

2 Disconnect the spark plug wires and remove the spark plugs (see Chapter 1). Be sure to label the plug wires to simplify reinstallation.

3 Remove the nuts/bolts holding the brackets to the head (see illustrations).

4 Disconnect all wires and hoses from the cylinder head(s). Be sure to label them to simplify reinstallation.

5 Remove the intake manifold as described in Section 7.

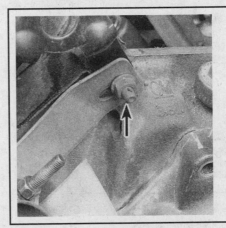

10.3b Remove the nut from the exhaust support bracket (3.8 liter shown)

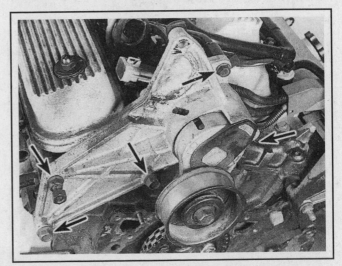

10.3d Drivebelt tensioner bolt locations (arrows) (typical)

10.9 Cylinder head bolt LOOSENING sequence

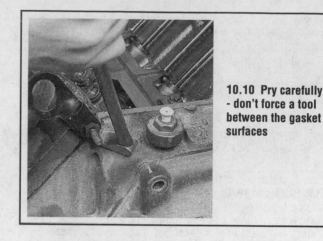

10.10 Pry carefully - don't force a tool between the gasket surfaces

6 Detach the exhaust manifold(s) from the cylinder head(s) being removed (see Section 9).

7 Remove the rocker arm cover(s) (see Section 4).

8 Remove the rocker arms and pushrods (see Section 5).

9 Loosen the head bolts in 1/4-turn increments until they can be removed by hand. Work from bolt-to-bolt as shown (see illustration).

10 Lift the head off the engine. If resistance is felt, don't pry between the head and block as damage to the mating surfaces will result. Recheck for head bolts that may have been overlooked, then use a hammer and block of wood to tap the head and break the gasket seal. Be careful because there are locating dowels in the block which position each head. As a last resort, pry each head up at the rear corner only and be careful not to damage anything (see illustration). After removal, place the head on blocks of wood to prevent damage to the gasket surfaces.

11 Refer to Chapter 2, Part B, for cylinder head disassembly, inspection and valve service procedures.

INSTALLATION

12 The mating surfaces of the cylinder heads and block must be perfectly clean when the heads are installed.

13 Use a gasket scraper to remove all traces of carbon and old gasket material (see illustration), then clean the mating surfaces with lacquer thinner or acetone. If there's oil on the mating surfaces when the heads are installed, the gaskets may not seal correctly and leaks may develop. When working on the block, it's a good idea to cover the lifter valley with shop rags to keep debris out of the engine. Use a shop rag or vacuum cleaner to remove any debris that falls into the cylinders.

14 Check the block and head mating surfaces for nicks, deep scratches and other damage. If damage is slight, it can be removed with a file; if it's excessive, machining may be the only alternative.

15 Use a tap of the correct size to chase the threads in the head bolt holes. Dirt, corrosion, sealant and damaged threads will affect torque readings.

16 Position the new gaskets over the dowel pins in the block. If steel gaskets are used, apply the proper type of gasket sealant. Install "non-retorquing" type gaskets dry (no sealant), unless the manufacturer states otherwise. Most gaskets are marked UP or TOP (see illustration) because they must be installed a certain way.

➡**Note: On later models, both gaskets have an arrow that should point to the timing chain end of the engine, and the right-side gasket also has an "R" marking.**

17 Carefully position the heads on the block without disturbing the gaskets.

18 Use NEW head bolts - don't reinstall the old ones - and apply Teflon pipe sealant to the threads and the undersides of the bolt heads.

10.13 Carefully remove all traces of old gasket material

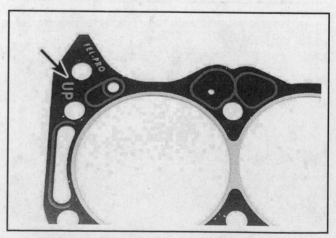

10.16 Look for gasket marks (arrow) to ensure correct installation

19 Tighten the bolts as directed in this Chapter's specifications in the sequence shown (see illustration). This must be done in three steps, following the sequence each time.

※※ **CAUTION:**

On 1986 through 1988 models, if you reach 60 ft-lbs. at any time during the final two steps, stop immediately - DO NOT complete the balance of the 90-degree turn. Go on to the next bolt in the sequence.

20 The remaining installation steps are the reverse of removal.
21 Change the oil and filter (see Chapter 1).

10.19 Cylinder head bolt TIGHTENING sequence

11 Vibration damper - removal and installation

▶ Refer to illustrations 11.7a, 11.7b and 11.9

※※ **CAUTION:**

The vibration damper is serviced as an assembly. Do not attempt to separate the pulley from the balancer hub.

1 Disconnect the negative cable from the battery.

※※ **CAUTION:**

If the vehicle is equipped with a Delco Loc II or Theftlock audio system, make sure you have the correct activation code before disconnecting the battery. See the information at the front of this manual for the radio re-activation procedure.

2 Loosen the lug nuts on the right front wheel.

3 Raise the vehicle and support it securely on jackstands.
4 Remove the right front wheel.
5 Remove the right front fender inner splash shield.
6 Remove the drivebelt (see Chapter 1).
7 Remove the lower bellhousing cover plate and position a large screwdriver in the ring gear teeth (see illustration) to keep the crankshaft from turning while an assistant removes the crankshaft balancer bolt (see illustration). The bolt is normally quite tight, so use a large breaker bar and a six-point socket.
8 The damper should pull off the crankshaft by hand. Leave the Woodruff key in place in the end of the crankshaft.

※※ **CAUTION:**

Be careful not to damage the crankshaft position sensor or its shield during damper removal/installation.

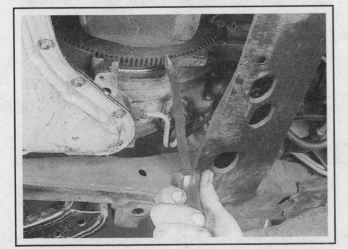

11.7a Wedge a large screwdriver in the ring gear teeth

11.7b Remove the bolt in the center of the hub (arrow)

11.9 Be sure to align the keyway (arrow) with the key in the crankshaft

9 Installation is the reverse of removal. Align the keyway with the key (see illustration) and avoid bending the metal tabs. Be sure to apply moly-base grease to the seal contact surface on the back side of the damper (if it isn't lubricated, the seal lip could be damaged and oil leakage would result).

10 Apply sealant to the threads and tighten the crankshaft bolt to the torque listed in this Chapter's Specifications.

11 Reinstall the remaining components in the reverse order of removal.

12 Crankshaft front oil seal - replacement

♦ **Refer to illustrations 12.2 and 12.5**

1 Remove the vibration damper (see Section 11). If there's a groove worn into the seal contact surface on the vibration damper, sleeves are available that fit over the groove, restoring the contact surface to like-new condition. These sleeves are sometimes included with the seal kit. Check with your parts supplier for details.

2 Pry the old oil seal out with a seal removal tool or a screwdriver (see illustration). Be very careful not to nick or otherwise damage the crankshaft in the process.

3 Apply a thin coat of RTV sealant to the outer edge of the new seal. Lubricate the seal lip with clean engine oil.

4 Place the seal squarely in position in the bore and press it into place with a seal driver. Make sure the seal enters the bore squarely and seats completely.

5 If the special tool is unavailable, carefully guide the seal into place with a large socket or piece of pipe and a hammer (see illustration). The outer diameter of the socket or pipe should be the same size as the seal outer diameter.

6 Install the vibration damper (see Section 11).

7 Reinstall the remaining parts in the reverse order of removal.

8 Start the engine and check for oil leaks at the seal.

12.2 Pry out the old seal with a seal removal tool (shown here) or a screwdriver

12.5 Gently drive the new seal into place with a hammer and large socket

13 Timing chain cover - removal and installation

♦ **Refer to illustrations 13.9, 13.10, 13.11 and 13.12**

➡**Note:** A special tool for aligning the crankshaft position sensor is required to complete this procedure on models equipped with distributorless ignition systems - read through all steps before beginning.

1 Disconnect the negative battery cable from the battery.

✷✷ CAUTION:

If the vehicle is equipped with a Delco Loc II or Theftlock audio system, make sure you have the correct activation code before disconnecting the battery. See the information at the front of this manual for the radio re-activation procedure.

13.9 Timing chain cover bolt locations (arrows)

13.10 On 1990 and earlier models, remove the camshaft button and spring from the end of the camshaft and inspect it for wear and damage

2 Set the parking brake and put the transmission in Park. Raise the front of the vehicle and support it securely on jackstands. Remove the splash shield from the right inner fender. On 2001 and later models, support the engine from below and remove the right engine mount (see Section 22).

3 Drain the oil and coolant (see Chapter 1). On later models remove the passenger side engine mount.

4 Remove the coolant hoses from the timing chain cover.

5 Remove the large (8 mm) water pump-to-block bolts, leaving the smaller (6 mm) diameter bolts in place.

➡**Note: The timing cover may be removed with the water pump still attached to it.**

6 Remove the vibration damper (see Section 11).

7 Unplug the connectors from the oil pressure, camshaft and crankshaft sensors on distributorless ignition models or remove the distributor, if equipped (see Chapter 5). Also remove the crankshaft sensor shield if equipped.

8 Remove the oil pan (see Section 17).

➡**Note: This Step is only necessary if the oil pan gasket is swollen or broken where it meets the bottom of the timing cover. If the gasket appears intact, upon cover removal, it can be cleaned and coated with a thin layer of RTV sealant when the cover is reinstalled.**

9 Remove the timing chain cover-to-engine block bolts (see illustration). Note that several of the bolts also secure the crankshaft sensor. Lift the sensor off when removing these bolts.

10 Separate the cover from the front of the engine. On 1990 and earlier models, remove the camshaft thrust spring and button that controls camshaft endplay (see illustration). If they are missing, look for them in the oil pan.

11 Use a gasket scraper to remove all traces of old gasket material and sealant from the cover and engine block (see illustration). The cover is made of aluminum, so be careful not to nick or gouge it. Clean the gasket sealing surfaces with lacquer thinner or acetone.

12 On 1990 and earlier models, check the camshaft thrust surface in the cover for excessive wear (see illustration). If it's worn, a new cover will be required (see the **Note** at the beginning of this Section).

13 On 1986 and later models, the oil pump cover must be removed and the cavity packed with petroleum jelly as described in Section 15 before the cover is installed.

13.11 Remove all traces of old gasket material

13.12 If the camshaft thrust surface (arrow) on the cover is worn a new cover must be installed (1990 and earlier models)

14 Apply a thin layer of RTV sealant to both sides of the new gasket, then position the gasket on the engine block (the dowel pins should keep it in place). Make sure the spring and button are in place in the end of the camshaft (hold them in place with grease), then attach the

cover to the engine. The oil pump drive must engage with the crankshaft or distributor gear.

15 Apply Teflon pipe sealant to the bolt threads, then install them finger tight. Install the crankshaft sensor, but leave the bolts finger tight until Step 16. Follow a criss-cross pattern when tightening the other bolts and work up to the torque listed in this Chapter's specifications in three steps to avoid warping the cover.

16 Position the crankshaft sensor on the timing chain cover. A special tool is available for this purpose - check with your local auto parts store. If the tool is not available, temporarily install the vibration damper and adjust the position of the sensor so it straddles the vanes on the back of the damper, with equal spacing on either side of the vane, then tighten the bolts securely.

✳✳ CAUTION:

If this is not done, the sensor may contact the vanes on the damper, which will break the sensor and damage the vibration damper. Tighten the bolts to the torque listed in this Chapter's specifications for the timing chain cover.

17 The remainder of installation is the reverse of removal.
18 Add oil and coolant, start the engine and check for leaks.

14 Oil filter adapter and pressure regulator valve - removal and installation (1986 and later models only)

♦ Refer to illustrations 14.3, 14.4 and 14.6

1 Remove the oil filter (see Chapter 1).
2 Remove the timing chain cover (see Section 13).
3 Remove the four bolts holding the oil filter adapter to the timing chain cover (see illustration). The cover is spring loaded, so remove the bolts while keeping pressure on the cover, then release the spring pressure carefully.
4 Remove the pressure regulator valve and spring (see illustration). Use a gasket scraper to remove all traces of the old gasket.
5 Clean all parts with solvent and dry them with compressed air (if available).

✳✳ WARNING:

Wear eye protection. Check for wear, score marks and valve binding.

6 Installation is the reverse of removal. Be sure to use a new gasket.

✳✳ CAUTION:

If a new timing chain cover is being installed on the engine, make sure the oil pressure relief valve supplied with the new cover is used. If the old style relief valve is installed in a new cover, oil pressure problems will result (see illustration).

14.3 Removing the oil filter adapter

7 Tighten the bolts to the torque listed in this Chapter's Specifications.
8 Run the engine and check for oil leaks.

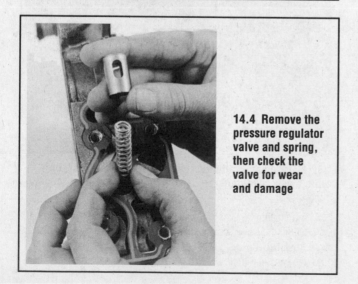

14.4 Remove the pressure regulator valve and spring, then check the valve for wear and damage

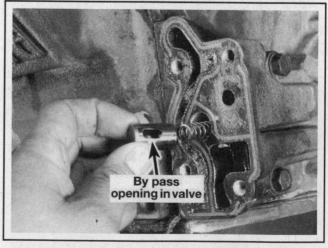

By pass opening in valve

14.6 If the timing chain cover is replaced with a new one, be sure to use the correct pressure relief valve - an early valve will not work in a later timing chain cover! Early model shown; later models use a solid valve

15 Oil pump - removal, inspection and installation

1985 MODELS

Removal

1 Remove the oil filter.
2 Remove the oil pump cover-to-timing chain cover attaching bolts.
3 Lift off the cover, oil pump drive gear and driven gears as an assembly.
4 Do not attempt to remove the oil filter bypass valve and spring - they are staked in place.

Inspection

5 Clean all the components thoroughly in solvent.
6 Inspect all components for wear or scoring.
7 Temporarily install the oil pump gears back into the body and measure the gear lash and side clearance. Place a straight edge across the gears and pump body and measure the end clearance.
8 Compare the measurements to the specifications, replacing any worn or damaged components.
➥**Note: Considering that a malfunctioning oil pump will cause sever engine damage, we recommend replacing the oil pump gears, cover and body if there's any doubt to their condition.**

Installation

9 Pack the pump body with petroleum jelly and install the gears forcing the petroleum jelly into every cavity. Failure to do so could cause the pump to lose its prime when the engine is started, resulting in engine damage from lack of oil pressure.
10 Install the pump cover, using a new gasket. Install the oil filter and add clean oil to the engine as necessary.

1986 AND LATER MODELS

▶ **Refer to illustrations 15.14, 15.20 and 15.21**

Removal

11 Remove the oil filter (see Chapter 1).
12 Remove the oil filter adapter, pressure regulator valve and spring (see Section 14).
13 Remove the timing chain cover (see Section 13).

15.14 The oil pump cover is attached to the inside of the timing chain cover - a T-30 Torx driver is required for removal of the screws

14 Remove the oil pump cover-to-timing chain cover bolts (see illustration).
15 Lift out the cover and oil pump gears as an assembly.

Inspection

16 Clean the parts with solvent and dry them with compressed air (if available).

❋❋ WARNING:

Wear eye protection!

17 Inspect all components for wear and score marks. Replace any worn out or damaged parts.
18 Refer to Section 14 for bypass valve information.
19 Reinstall the gears in the timing chain cover.
20 Measure the outer gear-to-housing clearance with a feeler gauge (see illustration).
21 Measure the inner gear-to-outer gear clearance at several points (see illustration).
22 Use a dial indicator or straightedge and feeler gauges to measure

15.20 Measuring the outer gear-to-housing clearance with a feeler gauge

15.21 Measuring the inner gear-to-outer gear clearance with a feeler gauge

the gear end clearance (distance from the gear to the gasket surface of the cover).

23 Check for pump cover warpage by laying a precision straightedge across the cover and trying to slip a feeler gauge between the cover and straightedge.

24 Compare the measurements to this Chapter's Specifications. Replace all worn or damaged components with new ones.

Installation

25 Remove the gears and pack the pump cavity with petroleum jelly.

26 Install the gears - make sure petroleum jelly is forced into every cavity. Failure to do so could cause the pump to lose its prime when the engine is started, causing damage from lack of oil pressure.

27 Install the pump cover, using a new gasket.

28 Install the pressure regulator spring and valve.

29 Install the timing chain cover.

30 Install the oil filter and check the oil level. Start and run the engine and check for correct oil pressure, then look carefully for oil leaks at the timing chain cover.

16 Timing chain and sprockets - removal and installation

▶ Refer to illustrations 16.1, 16.3, 16.4, 16.5a, 16.5b and 16.6

REMOVAL

1 Remove the timing chain cover (see Section 13), then slide the shim off the nose of the crankshaft (see illustration).

2 The timing chain should be replaced with a new one if the total free play midway between the sprockets exceeds one inch. Failure to replace the timing chain may result in erratic engine performance, loss of power and lowered fuel mileage.

3 Temporarily install the vibration damper bolt and turn the crankshaft clockwise to align the timing marks on the crankshaft and cam-

shaft sprockets directly opposite each other (see illustration).

4 Remove the timing chain damper.

➡ Note: On some earlier models it will be necessary to detach the spring and rotate the damper out of the way (see illustration). On all other models simply remove the damper bolt to detach it from the block.

16.3 The marks on the crankshaft and camshaft sprockets (arrows) must be aligned adjacent to each other as shown here (the mark on the camshaft sprocket is a dimple - the one on the crankshaft sprocket is a raised dot)

16.1 The shim has a notch in it to fit over the key in the crankshaft

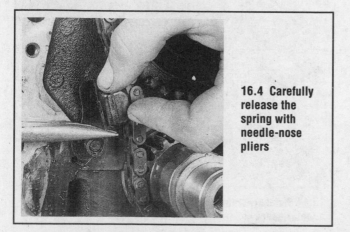

16.4 Carefully release the spring with needle-nose pliers

16.5a Earlier models use two bolts (arrow) to retain the camshaft sprocket while later models use a single bolt to retain the sprocket.

5   Remove the camshaft sprocket bolts (see illustrations). Try not to turn the camshaft in the process (if you do, realign the timing marks after the bolts are loosened).

6   Alternately pull the camshaft sprocket and then the crankshaft sprocket forward and remove the sprockets and timing chain as an assembly (see illustration).

7   On 3800 engines, inspect the balance shaft drive and driven gears for wear and damage.

8   Clean the timing chain components with solvent and dry them with compressed air (if available).

✳✳ WARNING:

Wear eye protection.

9   Inspect the components for wear and damage. Look for teeth that are deformed, chipped, pitted, polished or discolored.

INSTALLATION

➡**Note: If the crankshaft has been disturbed, install the sprocket temporarily and turn the crankshaft until the mark on the crankshaft sprocket is exactly at the top. If the camshaft was disturbed, install the sprocket temporarily and turn the camshaft until the timing mark is at the bottom, opposite the mark on the crankshaft sprocket (see illustration 16.3).**

10 Assemble the timing chain on the sprockets, then slide the sprocket and chain assembly onto the shafts with the timing marks aligned as shown in illustration 16.3.

➡**Note: On 3800 engines, make sure the timing marks on the balance shaft gears are aligned before installing the camshaft timing chain sprocket. Alignment of the balance shaft gears is covered in Chapter 2B.**

11 Install the camshaft sprocket bolt(s) and tighten them to the torque listed in this Chapter's specifications.

12 Attach the timing chain damper assembly to the block and install the spring.

16.5b On 3.0 and 3.8 liter engines, the camshaft position sensor is activated by a small magnet held in place by a special copper-colored bolt - note its location for correct installation later

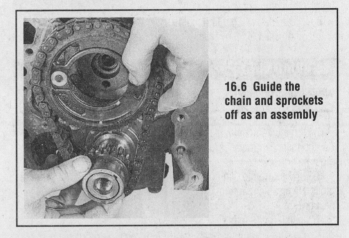

16.6 Guide the chain and sprockets off as an assembly

13 Install the camshaft thrust button and spring. Hold it in place with grease.

14 Lubricate the chain and sprocket with clean engine oil and install the timing chain cover (see Section 13).

17   Oil pan - removal and installation

▶ **Refer to illustration 17.4**

1   Disconnect the cable from the negative battery terminal.

✳✳ CAUTION:

If the vehicle is equipped with a Delco Loc II or Theftlock audio system, make sure you have the correct activation code before disconnecting the battery. See the information at the front of this manual for the radio re-activation procedure.

2   Raise the vehicle and place it securely on jackstands. Drain the engine oil and replace the oil filter (refer to Chapter 1 if necessary).

3   Remove the driveplate inspection cover and starter, if necessary for access (refer to Chapter 5). On 1992 and later models, disconnect the electrical connector from the oil lever sensor. On 2001 and later models, support the engine from above with an engine support fixture or crane, then remove the right engine mount and bracket (see Section 22).

4   Remove the oil pan mounting bolts (see illustration) and carefully separate the oil pan from the block. Don't pry between the block and the pan or damage to the sealing surfaces may result and oil leaks may develop. Instead, tap the pan with a soft-face hammer to break the gasket seal.

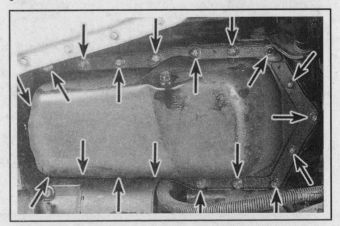

17.4 The oil pan bolts (arrows) are located around the perimeter of the oil pan (viewed from below)

5   Clean the pan with solvent and remove all old sealant and gasket material from the block and pan mating surfaces. Clean the mating surfaces with lacquer thinner or acetone and make sure the bolt holes in the block are clear. Check the oil pan flange for distortion, particularly around the bolt holes. If necessary, place the pan on a block of wood and use a hammer to flatten and restore the gasket surface.

6   Always use a new gasket whenever the oil pan is installed.

7   Place the oil pan in position on the block and install the bolts.

8   After the bolts are installed, tighten them to the torque listed in this Chapter's specifications. Starting at the center, follow a criss-cross pattern and work up to the final torque in three steps.

9   The remaining steps are the reverse of the removal procedure.

10 Refill the engine with oil. Run the engine until normal operating temperature is reached and check for leaks.

18 Oil pump pickup tube and screen assembly - removal and installation

▶ **Refer to illustration 18.2**

1   Remove the oil pan (see Section 17).

2   Unbolt the oil pump pickup tube and screen assembly and detach it from the engine (see illustration).

3   Clean the screen and housing assembly with solvent and dry it with compressed air, if available.

✱✱ WARNING:

Wear eye protection.

4   If the oil screen is damaged or has metal chips in it, replace it. An abundance of metal chips indicates a major engine problem which must be corrected.

5   Make sure the mating surfaces of the pipe flange and the engine block are clean and free of nicks and install the screen assembly with a new gasket.

6   Install the oil pan (see Section 17).

7   Be sure to refill the engine with oil before starting it.

18.2 Remove the bolts and lower the pickup tube and screen assembly

19 Driveplate - removal and installation

19.2 Place a prybar or large screwdriver through one of the holes in the driveplate to keep the crankshaft from turning as the bolts are loosened/tightened

▶ **Refer to illustration 19.2**

1   Refer to Chapter 7 and remove the transaxle.

2   Place a prybar or large screwdriver through a hole in the driveplate to keep the crankshaft from turning, then remove the mounting bolts (see illustration).

3   Pull straight back on the driveplate to detach it from the crankshaft. Inspect the driveplate for cracks and chipped teeth. Replace it if it's damaged.

4   Reinstall the driveplate. The holes in the driveplate are staggered to ensure correct positioning on the crankshaft. Use non-hardening thread locking compound on the bolt threads and tighten them to the specified torque in a criss-cross pattern.

5   Reinstall the transaxle (see Chapter 7).

20 Crankshaft rear oil seal - replacement

1985 THROUGH 1990 MODELS

♦ **Refer to illustrations 20.3a, 20.3b, 20.10 and 20.13**

➥Note: Braided fabric seals inserted into grooves in the engine block and main bearing cap are used to seal against oil leakage around the crankshaft. The upper rear main bearing oil seal can be replaced only with the crankshaft removed (Chapter 2, Part B) but it can be repaired with the crankshaft in place. Two piece rubber seals are available from aftermarket suppliers.

1 Remove the oil pan (see Section 17).

2 Remove the rear main bearing cap.

3 Using a special tool (available at most auto parts stores, or you can fabricate one yourself), drive the upper seal gently back into the groove in the engine block, packing it tightly. It will pack in to a depth of between 1/4-inch and 3/4-inch (see illustrations).

4 Repeat the procedure on the other end of the seal.

5 Measure how far the seal was driven up in the groove on each side and add 1/16-inch. Remove the old seal from the main bearing cap.

6 Use the main bearing cap as a fixture and cut two pieces of the old seal to the predetermined lengths.

7 Using the packing tool, work the short pieces of the previously cut seal into the engine block groove. Lubricate the seal with oil to ease installation.

8 Remove the guide tool.

9 Place a new seal in the main bearing cap groove with both ends projecting an equal amount above the parting surface of the cap.

10 Use the handle of a hammer or similar tool to force the seal into the groove until it projects no more than 1/16-inch. Cut the ends of the seal flush with the surface of the cap with a single-edge razor blade (see illustration).

11 Soak the neoprene seals which fit into the side grooves in the bearing cap in light oil or kerosene for one or two minutes.

12 Install the neoprene seals in the groove between the bearing cap and the block. The seals are slightly undersize and swell in the presence of heat and oil. They are slightly longer than the groove in the bearing cap and must be cut to fit.

13 Apply a small amount of RTV sealant to the joint where the bearing cap meets the block to help eliminate oil leakage. A very thin coat is all that's necessary (see illustration).

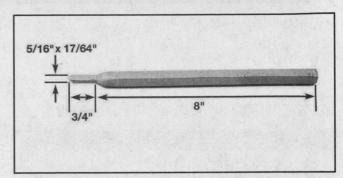

20.3a Grind a piece of 1/2-inch diameter brass or aluminum rod to these dimensions as a rear seal driver

20.3b Packing the seal into the engine block groove with the special tool

14 Install the main bearing cap on the block. Force the seals up into the bearing cap with a blunt instrument to be sure of a good seal at the upper parting line. Install the bolts and tighten them to the torque listed in the Chapter 2B specifications.

15 Install the oil pan.

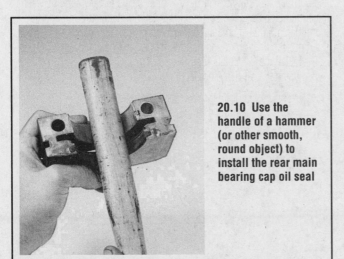

20.10 Use the handle of a hammer (or other smooth, round object) to install the rear main bearing cap oil seal

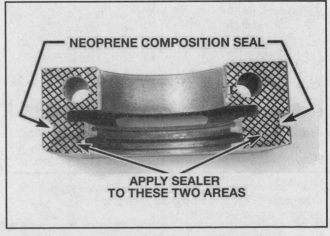

20.13 Sealant must be applied to the main bearing cap parting line

1991 AND LATER VIN C AND VIN L MODELS

▶ **Refer to illustration 20.18**

16 Remove the transaxle (see Chapter 7).
17 Remove the driveplate (see Section 19).
18 Using a thin screwdriver or seal removal tool, carefully remove

20.18 Carefully pry the old oil seal out

the oil seal from the engine block (see illustration). Be very careful not to damage the crankshaft surface while prying the seal out.
19 Clean the bore in the block and the seal contact surface on the crankshaft. Check the seal contact surface on the crankshaft for scratches and nicks that could damage the new seal lip and cause oil leaks - if the crankshaft is damaged, the only alternative is a new or different crankshaft. Inspect the seal bore for nicks and scratches. Carefully smooth it with a fine file if necessary, but don't nick the crankshaft in the process.
20 A special tool, available at some auto parts stores, is recommended to install the new oil seal. Lubricate the lips of the seal with clean engine oil. Slide the seal onto the mandril until the dust lip bottoms squarely against the collar of the tool.

➡**Note: If the special tool isn't available, carefully work the seal lip over the crankshaft and tap it into place with a hammer and punch.**

21 Align the dowel pin on the tool with the dowel pin hole in the crankshaft and attach the tool to the crankshaft by hand-tightening the bolts.
22 Turn the tool handle until the collar bottoms against the case, seating the seal.
23 Loosen the tool handle and remove the bolts. Remove the tool.
24 Check the seal and make sure it's seated squarely in the bore.
25 Install the driveplate (see Section 19).
26 Install the transaxle (see Chapter 7).

21 Rear main oil seal carrier (VIN K engine) - removal and installation

✳ **CAUTION:**

On 2001 and later models, use no lubricant on the oil seal during installation, and do not touch the seal lip or it may be damaged. New seals are marked "this side out" on one side.

➡**Note: The rear main oil seal carrier on these engines does not have to be removed for oil seal replacement. Rear main oil seal replacement is the same as on all other models.**

REMOVAL

1 Remove the oil pan (see Section 17).
2 Remove the driveplate (see Section 19).
3 Remove the rear main oil seal carrier mounting bolts.
4 Separate the rear main oil seal carrier from the engine.

INSTALLATION

✳ **CAUTION:**

Correct alignment of the rear main oil seal carrier to the back surface of the engine is critical to make sure the rear main oil seal and oil pan seal correctly.

➡**Note: The rear main seal carrier on these engines does not have to be removed for oil seal replacement. Rear main seal replacement is the same procedure as for earlier engines, except that the seal is pressed into the carrier, not the block.**

5 If still in place, remove the rear main oil seal from the carrier and install a new one.
6 Install the rear main oil seal carrier and mounting bolts. The lips of the oil seal should center the carrier, but the alignment of the carrier must be checked. Tighten the bolts to the torque listed in this Chapter's specifications.
7 Place a straightedge on the cylinder block oil pan mating surface and the rear main oil seal carrier flange. Place a feeler gauge between the straightedge and the sealing surfaces and make sure there is no more than a 0.004-inch step, either high or low, between the two parts. If out of specification, reposition the carrier and measure again.
8 Install the driveplate (see Section 19).
9 Install the oil pan (see Section 17).

22 Engine mount - check and replacement

▶ **Refer to illustration 22.4**

❊ WARNING:

A special engine support fixture is available to support the engine during repair operations. These fixtures are available from rental yards. Improper lifting methods or devices are hazardous and could result in severe injury or death. DO NOT place any part of your body under the engine/transaxle when it's supported only by a jack. Failure of the lifting device could result in serious injury or death.

➡**Note: See Chapter 7 for transaxle mount information.**

1   Engine mounts seldom require attention, but broken or deteriorated mounts should be replaced immediately or the added strain placed on the driveline components may cause damage or wear.

CHECK

2   During the check, the engine must be raised slightly to remove the weight from the mounts.

22.4 Broken engine mount (removed from vehicle for clarity)

3   Raise the vehicle and support it securely on jackstands, then position a jack under the engine oil pan. Place a large block of wood between the jack head and the oil pan, then carefully raise the engine just enough to take the weight off the mounts.

❊ WARNING:

DO NOT place any part of your body under the engine when it's supported only by a jack!

4   Check the mounts to see if the rubber is cracked, hardened or separated from the metal plates (see illustration). Sometimes the rubber will split right down the center.
5   Check for relative movement between the mount plates and the engine or frame (use a large screwdriver or prybar to attempt to move the mounts). If movement is noted, lower the engine and tighten the mount fasteners.
6   Rubber preservative may be applied to the mounts to slow deterioration.

REPLACEMENT

7   Disconnect the negative battery cable from the battery, then raise the vehicle and support it securely on jackstands (if not already done).

❊ CAUTION:

If the vehicle is equipped with a Delco Loc II or Theftlock audio system, make sure you have the correct activation code before disconnecting the battery. See the information at the front of this manual for the radio re-activation procedure.

8   Raise the engine slightly with a jack or hoist. Install an engine support as described in the **Warning** above. Remove the fasteners and detach the mount from the frame bracket.
9   Remove the mount-to-block bracket bolts/nuts and detach the mount.
10 Installation is the reverse of removal. Use thread locking compound on the threads and be sure to tighten everything securely.

Specifications

General

Cylinder numbers (drivebelt end-to-transaxle end)	
Front bank (radiator side)	1-3-5
Rear bank	2-4-6
Firing order	1-6-5-4-3-2

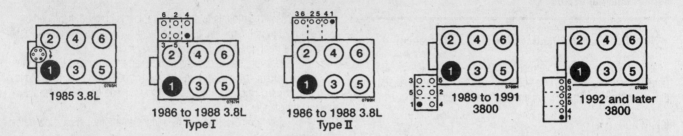

Cylinder location and coil terminal identification diagrams

Valve lifters

Diameter	0.8420 to 0.8427 in
Lifter-to-bore clearance	0.0008 to 0.0025 in

Oil pump

1985 models	
Gear lash	0.0015 to 0.003
Side clearance	0.003 to 0.005
End clearance	0.002 to 0.006
1986 and later models	
Outer gear-to-housing clearance	0.008 to 0.015 in
Inner gear-to-outer gear clearance	0.006 in
Gear end clearance	0.001 to 0.0035 in
Pump cover warpage limit	0.002 in

Torque specifications · Ft-lbs (unless otherwise indicated)

Camshaft sprocket bolts	
1985 through 1987	20
1988 through 1991	26
1992	
Step one	74
Step two	Turn an additional 105-degrees
1993 and later	
Step one	74
Step two	Turn an additional 90-degrees

Torque specifications   Ft-lbs (unless otherwise indicated)

Cylinder head bolts
  1985   80*
  1986 through 1988
    Step one   25
    Step two   Turn an additional 90-degrees
    Step three   Turn an additional 90-degrees
  1989 and 1990
    Step one   35
    Step two   Turn an additional 130-degrees
    Step three   Turn an additional 30-degrees
  1991 through 1995
    Step one   35
    Step two   Turn an additional 130-degrees
    Step three   Turn the center four bolts an additional 30-degrees
  1996 and 1997
    Step one   37
    Step two   Turn an additional 130-degrees
    Step three   Turn the center four bolts an additional 30-degrees
  1998 through 2000
    Step one   37
    Step two   Turn an additional 120-degrees
    Step three   Turn the center four bolts an additional 30-degrees
  2001 and later
    Step 1   37
    Step 2   Turn an additional 120 degrees

Driveplate-to-crankshaft bolts
  1985 through 1991   60
  1992 and later
    Step one   132 in-lbs
    Step two   Turn an additional 50-degrees

Exhaust manifold-to-cylinder head bolts
  1985 and 1986   25
  1987   37
  1988 through 1990   41
  1991 through 1995   38
  1996 and later   22

Intake manifold-to-cylinder head bolts/nuts
  1985 and 1986   47
  1987   32
  1988 through 1990   120 in-lbs
  1991 through 1995   96 in-lbs
  1996 and later   132 in-lbs

Intake manifold, upper to lower bolts
  1994 and earlier   132 in-lbs
  1995 and later   89 in-lbs

Oil pan bolts
  1985 through 1987   96 in-lbs
  1988 and later   120 to 125 in-lbs

Torque specifications Ft-lbs (unless otherwise indicated)

Oil filter adapter-to-timing chain cover bolts
    1986   30
    1987 through 1997   24
    1988 and later
        Step one   132 in-lbs
        Step two   Turn an additional 50-degrees
Oil pump cover-to-timing chain cover bolts   96 to 98 in-lbs
Oil pump pickup tube and screen assembly bolts
    1995 and earlier   96 in-lbs
    1996 and later   132 in-lbs
Rear main oil seal corner bolts   22
Rocker arm cover nuts/bolts   88 in-lbs
Rocker arm pivot bolts
    1985   30
    1986 and 1987   43
    1988 through 1992   28
    1993 and 1994
        Step one   18
        Step two   Turn an additional 70-degrees
    1995 and later
        Step one   132 in-lbs
        Step two   Turn an additional 90-degrees
Timing chain cover bolts
    1995 and earlier   22
    1996 through 1997
        Step one   11
        Step two   Turn an additional 40-degrees
    1998 and later
        Step one   15
        Step two   Turn an additional 40-degrees
Vibration damper-to-crankshaft bolt
    1985 and 1986   200
    1987 through 1990   219
    1991
        Step one   105
        Step two   Turn an additional 56-degrees
    1992 and later
        Step one   110
        Step two   Turn an additional 76-degrees

  * *GM recommends using replacement torque-to-yield cylinder head bolts. Follow the torque angle specifications for 1986 through 1988 models when using these updated head bolts.*

Section

2B

GENERAL ENGINE OVERHAUL PROCEDURES

1 Engine overhaul - general information

→**Note: On models equipped with the Delco Loc II audio system, be sure the lockout feature is turned off before performing any procedure which requires disconnecting the battery.**

Included in this portion of Chapter 2 are the general overhaul procedures for the cylinder heads and internal engine components.

The information ranges from advice concerning preparation for an overhaul and the purchase of replacement parts to detailed, step-by-step procedures covering removal and installation of internal engine components and the inspection of parts.

The following Sections have been written based on the assumption the engine has been removed from the vehicle. For information concerning in-vehicle engine repair, as well as removal and installation of the external components necessary for the overhaul, see Part A of this Chapter and Section 7 of this Part.

The Specifications included in this Part are only those necessary for the inspection and overhaul procedures which follow. Refer to Part A for additional Specifications.

It's not always easy to determine when, or if, an engine should be completely overhauled, as a number of factors must be considered.

High mileage isn't necessarily an indication an overhaul is needed, while low mileage doesn't preclude the need for an overhaul. Frequency of servicing is probably the most important consideration. An engine that's had regular and frequent oil and filter changes, as well as other required maintenance, will most likely give many thousands of miles of reliable service. Conversely, a neglected engine may require an overhaul very early in its life.

Excessive oil consumption is an indication that the piston rings, valve seals and/or valve guides are in need of attention. Make sure oil leaks aren't responsible before deciding the rings and/or guides are bad. Perform a cylinder compression check to determine the extent of the work required (see Section 3).

Remove the oil pressure sending unit and check the oil pressure with a gauge installed in its place (see Section 2). Compare the results to this Chapter's Specifications. As a general rule, engines should have ten psi oil pressure for every 1,000 rpm. If the pressure is extremely low, the bearings and/or oil pump are probably worn out.

Loss of power, rough running, knocking or metallic engine noises, excessive valve train noise and high fuel consumption rates may also point to the need for an overhaul, especially if they're all present at the same time. If a complete tune-up doesn't remedy the situation, major mechanical work is the only solution.

An engine overhaul involves restoring the internal parts to the specifications of a new engine. During an overhaul, the piston rings are replaced and the cylinder walls are reconditioned (rebored and/or honed). If a rebore is done by an automotive machine shop, new oversize pistons will also be installed. The main bearings, connecting rod bearings and camshaft bearings are generally replaced with new ones and, if necessary, the crankshaft may be reground to restore the journals. Generally, the valves are serviced as well, since they're usually in less-than-perfect condition at this point. While the engine is being overhauled, other components, such as the starter and alternator, can be rebuilt as well. The end result should be a like new engine that will give many trouble free miles.

→**Note: Critical cooling system components such as the hoses, drivebelts, thermostat and water pump MUST be replaced with new parts when an engine is overhauled. The radiator should be checked carefully to ensure it isn't clogged or leaking (see Chapter 3). Also, we don't recommend overhauling the oil pump - always install a new one when an engine is rebuilt.**

Before beginning the engine overhaul, read through the entire procedure to familiarize yourself with the scope and requirements of the job. Overhauling an engine isn't particularly difficult, if you follow all of the instructions carefully, have the necessary tools and equipment and pay close attention to all specifications; however, it can be time consuming. Plan on the vehicle being tied up for a minimum of two weeks, especially if parts must be taken to an automotive machine shop for repair or reconditioning. Check on availability of parts and make sure any necessary special tools and equipment are obtained in advance. Most work can be done with typical hand tools, although a number of precision measuring tools are required for inspecting parts to determine if they must be replaced. Often an automotive machine shop will handle the inspection of parts and offer advice concerning reconditioning and replacement.

→**Note: Always wait until the engine has been completely disassembled and all components, especially the engine block, have been inspected before deciding what service and repair operations must be performed by an automotive machine shop. Since the block's condition will be the major factor to consider when determining whether to overhaul the original engine or buy a rebuilt one, never purchase parts or have machine work done on other components until the block has been thoroughly inspected. As a general rule, time is the primary cost of an overhaul, so it doesn't pay to install worn or substandard parts.**

As a final note, to ensure maximum life and minimum trouble from a rebuilt engine, everything must be assembled with care in a spotlessly clean environment.

2 Oil pressure check

♦ **Refer to illustrations 2.2 and 2.3**

1   Low engine oil pressure can be a sign of an engine in need of rebuilding. A "low oil pressure" indicator (often called an "idiot light") is not a test of the oiling system. Such indicators only come on when the oil pressure is dangerously low. Even a factory oil pressure gauge in the instrument panel is only a relative indication, although much better for driver information than a warning light. A better test is with a mechanical (not electrical) oil pressure gauge. When used in conjunction with an accurate tachometer, an engine's oil pressure performance can be compared to factory Specifications for that year and model.

2   Locate the oil pressure sending unit (see illustration).

3   Remove the oil pressure sending unit and install a fitting which will allow you to directly connect your hand-held, mechanical oil pressure gauge (see illustration). Use Teflon tape or sealant on the threads of the adapter and the fitting on the end of your gauge's hose.

2.2 The oil pressure sending unit is located on top of the oil filter housing (arrow) - remove the sending unit . . .

2.3 . . . and install a gauge to check the engine oil pressure

4   Connect an accurate tachometer to the engine, according to the tachometer manufacturer's instructions.

5   Check the oil pressure with the engine running (normal operating temperature) at the specified engine speed, and compare it to this Chapter's Specifications. If it's extremely low, the bearings and/or oil pump are probably worn out.

3   Cylinder compression check

▶ Refer to illustration 3.6

1   A compression check will tell you what mechanical condition the upper end (pistons, rings, valves, head gaskets) of the engine is in. Specifically, it can tell you if the compression is down due to leakage caused by worn piston rings, defective valves and seats or a blown head gasket.

➡ **Note: The engine must be at normal operating temperature and the battery must be fully charged for this check.**

2   Begin by cleaning the area around the spark plugs before you remove them. Compressed air should be used, if available, otherwise a small brush or even a bicycle tire pump will work. The idea is to prevent dirt from getting into the cylinders as the compression check is being done.

3   Remove all of the spark plugs from the engine (see Chapter 1).

4   Block the throttle wide open.

5   Disable the fuel and ignition systems by removing the ECM fuse and the IGN or C3I fuse (see Chapter 6). As an added precaution, detach the electrical connector from the coil pack (see Chapter 5).

6   Install the compression gauge in the number one spark plug hole (see illustration).

7   Crank the engine over at least seven compression strokes and watch the gauge. The compression should build up quickly in a healthy engine. Low compression on the first stroke, followed by gradually increasing pressure on successive strokes, indicates worn piston rings. A low compression reading on the first stroke, which doesn't build up during successive strokes, indicates leaking valves or a blown head gasket (a cracked head could also be the cause). Deposits on the undersides of the valve heads can also cause low compression. Record the highest gauge reading obtained.

8   Repeat the procedure for the remaining cylinders and compare the results to this Chapter's Specifications.

9   If the readings are below normal, add some engine oil (about

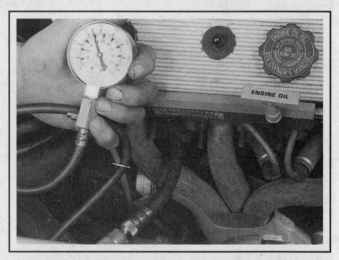

3.6 A compression gauge with a threaded fitting for the spark plug hole is preferred over the type that requires hand pressure to maintain the seal

three squirts from a plunger-type oil can) to each cylinder, through the spark plug hole, and repeat the test.

10   If the compression increases after the oil is added, the piston rings are definitely worn. If the compression doesn't increase significantly, the leakage is occurring at the valves or head gasket. Leakage past the valves may be caused by burned valve seats and/or faces or warped, cracked or bent valves.

11   If two adjacent cylinders have equally low compression, there's a strong possibility the head gasket between them is blown. The appearance of coolant in the combustion chambers or the crankcase would verify this condition.

12   If one cylinder is about 20 percent lower than the others, and the

engine has a slightly rough idle, a worn exhaust lobe on the camshaft could be the cause.

13 If the compression is unusually high, the combustion chambers are probably coated with carbon deposits. If that's the case, the cylinder heads should be removed and decarbonized.

14 If compression is way down or varies greatly between cylinders, it would be a good idea to have a leak-down test performed by an automotive repair shop. This test will pinpoint exactly where the leakage is occurring and how severe it is.

15 Install the ECM fuse and drive the vehicle to restore the block learn memory.

4 Vacuum gauge diagnostic checks

▶ **Refer to illustration 4.5**

1 A vacuum gauge provides valuable information about what is going on in the engine at a low cost. You can check for worn rings or cylinder walls, leaking head or intake manifold gaskets, vacuum leaks in the intake manifold, restricted exhaust, stuck or burned valves, weak valve springs, improper valve timing, and ignition problems. Vacuum gauge readings are easy to misinterpret, however, so they should be used in conjunction with other tests to confirm the diagnosis.

2 Both the absolute readings and the rate of needle movement are important for accurate interpretation. Most gauges measure vacuum in inches of mercury (in-Hg). The following references to vacuum assume the diagnosis is being performed at sea level. As elevation increases (or atmospheric pressure decreases), the reading will decrease. For every 1,000 foot increase in elevation above approximately 2000 feet, the gauge readings will decrease about one inch of mercury.

3 Connect the vacuum gauge directly to intake manifold vacuum, not to ported (throttle body) vacuum. Be sure no hoses are left disconnected during the test or false readings will result.

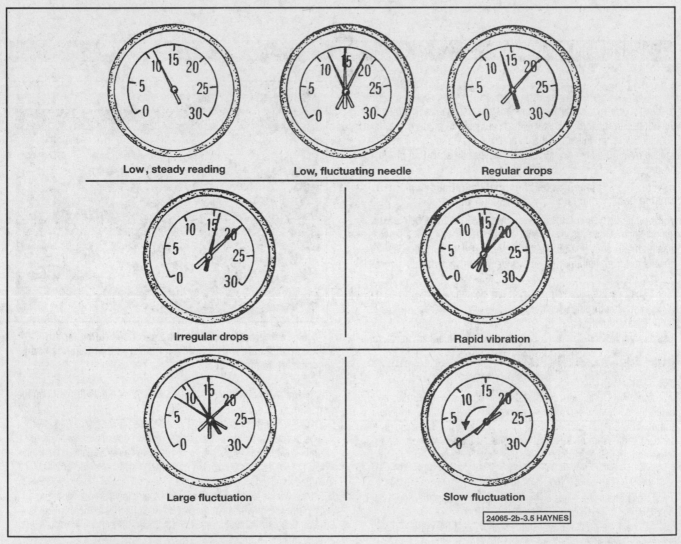

Low, steady reading   Low, fluctuating needle   Regular drops

Irregular drops   Rapid vibration

Large fluctuation   Slow fluctuation

24065-2b-3.5 HAYNES

4.5 Typical vacuum gauge diagnostic readings

➡**Note: Do not disconnect engine sensors or vacuum solenoids to connect the vacuum gauge. Disconnected engine control components can affect engine operation and produce abnormal vacuum gauge readings.**

4   Before you begin the test, warm the engine up completely. Block the wheels and set the parking brake. With the transmission in Park, start the engine and allow it to run at normal idle speed.

❊❊ WARNING:

Carefully inspect the fan blades for cracks or damage before starting the engine. Keep your hands and the vacuum gauge clear of the fan and do not stand in front of the vehicle or in line with the fan when the engine is running.

5   Read the vacuum gauge; an average, healthy engine should normally produce about 17 to 22 inches of vacuum with a fairly steady gauge needle at idle. Refer to the following vacuum gauge readings and what they indicate about the engine's condition (see illustration).

6   A low steady reading usually indicates a leaking intake manifold gasket. this could be at one of the cylinder heads, between the upper and lower manifolds, or at the throttle body. Other possible causes are a leaky vacuum hose or incorrect camshaft timing.

7   If the reading is 3 to 8 inches below normal and it fluctuates at that low reading, suspect an intake manifold gasket leak at an intake port or a faulty fuel injector.

8   If the needle regularly drops about two to four inches at a steady rate, the valves are probably leaking. Perform a compression check or leakdown test to confirm this.

9   An irregular drop or downward flicker of the needle can be caused by a sticking valve or an ignition misfire. Perform a compression check or leakdown test and inspect the spark plugs to identify the faulty cylinder.

10 A rapid needle vibration of about four inches at idle combined with exhaust smoke indicates worn valve guides. Perform a leakdown test to confirm this. If the rapid vibration occurs with an increase in engine speed, check for a leaking intake manifold gasket or head gasket, weak valve springs, burned valves, or ignition misfire.

11 A slight fluctuation - one inch up and down - may mean ignition problems. Check all the usual tune-up items and, if necessary, run the engine on an ignition analyzer.

12 If there is a large fluctuation, perform a compression or leakdown test to look for a weak or dead cylinder or a blown head gasket.

13 If the needle moves slowly through a wide range, check for a clogged PCV system or intake manifold gasket leaks.

14 Check for a slow return of the gauge to a normal idle reading after revving the engine by quickly snapping the throttle open until the engine reaches about 2,500 rpm and let it shut. Normally the reading should drop to near zero, rise about 5 inches above normal idle reading, and then return to the previous idle reading. If the vacuum returns slowly and doesn't peak when the throttle is snapped shut, the rings may be worn. If there is a long delay, look for a restricted exhaust system (often the muffler or catalytic converter). One way to check this is to temporarily disconnect the exhaust ahead of the suspected part and repeat the test.

5   Engine - removal and installation

❊❊ WARNING:

Gasoline is extremely flammable, so take extra precautions when disconnecting any part of the fuel system. Don't smoke or allow open flames or bare light bulbs in or near the work area and don't work in a garage where a gas appliance (such as a clothes dryer or water heater) is installed. Since gasoline is carcinogenic, wear fuel-resistant gloves whenever working on the fuel system. If you spill gasoline on your skin, rinse it off immediately. Have a fire extinguisher rated for gasoline fires handy and know how to use it! Also, the air conditioning system is under high pressure - have a dealer service department or service station discharge the system before disconnecting any of the hoses or fittings.

➡**Note: Read through the following steps carefully and familiarize yourself with the procedure before beginning work.**

ENGINE REMOVAL - METHODS AND PRECAUTIONS

If you've decided the engine must be removed for overhaul or major repair work, several preliminary steps should be taken.

Locating a suitable place to work is extremely important. Adequate work space, along with storage space for the vehicle, will be needed. If a shop or garage isn't available, at the very least a flat, level, clean work surface made of concrete or asphalt is required.

Cleaning the engine compartment and engine before beginning the removal procedure will help keep tools clean and organized.

An engine hoist or A-frame will also be necessary. Make sure the equipment is rated in excess of the combined weight of the engine and transaxle. Safety is of primary importance, considering the potential hazards involved in lifting the engine out of the vehicle.

If the engine is being removed by a novice, a helper should be available. Advice and aid from someone more experienced would also be helpful. There are many instances when one person cannot simultaneously perform all of the operations required when lifting the engine out of the vehicle.

Plan the operation ahead of time. Arrange for or obtain all of the tools and equipment you'll need prior to beginning the job. Some of the equipment necessary to perform engine removal and installation safely and with relative ease are (in addition to an engine hoist) a heavy duty floor jack, complete sets of wrenches and sockets as described in the front of this manual, wooden blocks and plenty of rags and cleaning solvent for mopping up spilled oil, coolant and gasoline. If the hoist must be rented, be sure to arrange for it in advance and perform all of the operations possible without it beforehand. This will save you money and time.

Plan for the vehicle to be out of use for quite a while. A machine shop will be required to perform some of the work which the do-it-yourselfer can't accomplish without special equipment. These shops often have a busy schedule, so it would be a good idea to consult them before removing the engine in order to accurately estimate the amount of time required to rebuild or repair components that may need work.

Always be extremely careful when removing and installing the engine. Serious injury can result from careless actions. Plan ahead, take your time and a job of this nature, although major, can be accomplished successfully.

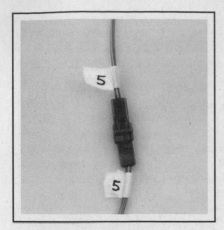

5.4 Label each wire before unplugging the connector

5.11 Unbolt the power steering pump and move it aside, then use wire or rope to hold it in place

5.12 Unbolt the compressor and set it aside without disconnecting the hoses

REMOVAL

▶ **Refer to illustrations 5.4, 5.11, 5.12, 5.15, 5.19, 5.24a and 5.24b**

1   Refer to Chapter 4 and relieve the fuel system pressure, then disconnect the negative cable from the battery.

✳✳ CAUTION:

If the vehicle is equipped with a Delco Loc II or Theftlock audio system, make sure you have the correct activation code before disconnecting the battery. See the information at the front of this manual for the radio re-activation procedure.

2   Cover the fenders and cowl and remove the hood (see Chapter 11). Special pads are available to protect the fenders, but an old bedspread or blanket will also work.

3   Remove the air cleaner assembly and mass airflow sensor (see Chapter 4).

4   Remove the plastic firewall cover. Label the vacuum lines, emissions system hoses, wiring connectors, ground straps and fuel lines to ensure correct reinstallation, then detach them. Pieces of masking tape with numbers or letters written on them work well (see illustration). If there's any possibility of confusion, make a sketch of the engine com-

partment and clearly label the lines, hoses and wires.

5   Raise the vehicle and support it securely on jackstands. Drain the cooling system (see Chapter 1).

6   Label and detach all coolant hoses from the engine.

7   Remove the windshield washer reservoir, cooling fan(s) and radiator (see Chapter 3).

8   Remove the drivebelt (see Chapter 1).

9   Disconnect the fuel lines running from the engine to the chassis (see Chapter 4). Plug or cap all open fittings/lines.

10   Disconnect the throttle linkage (and TV linkage/speed control cable, if equipped) from the engine (see Chapters 4 and 7).

11   Unbolt the power steering pump and set it aside (see Chapter 10). Leave the lines/hoses attached and make sure the pump is kept in an upright position in the engine compartment (see illustration).

➡**Note: On some models, you may have to detach the pump.**

12   Unbolt the air conditioning compressor (see Chapter 3) and set it aside; do not disconnect the hoses (see illustration).

13   Drain the engine oil and remove the filter (see Chapter 1).

14   Remove the starter and the alternator (see Chapter 5). On 2001 and later engines, unbolt the cruise control module from the firewall and set it aside.

15   On earlier models, disconnect the engine shock absorber (see illustration).

5.15 Remove the engine shock absorber through bolt (arrow)

5.19 Attach the chain or hoist cable to the engine brackets (arrows)

5.24a Raise the engine enough to clear the mounts, then separate it from the transaxle

16 Disconnect the exhaust system from the engine (see Chapter 4).

17 Support the transaxle with a jack. Position a block of wood on the jack head to prevent damage to the transaxle.

18 Attach an engine sling or a length of chain to the lifting brackets on the engine.

19 Roll the hoist into position and connect the sling to it (see illustration). Take up the slack in the sling or chain, but don't lift the engine.

❋❋ WARNING:

DO NOT place any part of your body under the engine when it's supported only by a hoist or other lifting device.

20 Refer to Chapter 7 and remove the torque converter-to-driveplate fasteners.

21 Remove the transaxle-to-engine block bolts. Refer to Chapter 7.

22 Remove the engine mount-to-chassis bolts/nuts.

23 Recheck to be sure nothing is still connecting the engine to the vehicle. Disconnect anything still remaining.

24 Raise the engine slightly to disengage the mounts. Carefully separate the engine from the transaxle. Be sure the torque converter stays in place (clamp a pair of vise-grips to the housing to keep the converter from sliding out). Slowly raise the engine out of the vehicle (see illustrations). Check carefully to make sure nothing is hanging up as the hoist is raised.

25 Remove the driveplate and mount the engine on an engine stand.

INSTALLATION

26 Check the engine and transaxle mounts. If they're worn or damaged, replace them.

5.24b Lift the engine off the mounts and guide it carefully around any obstacles, as an assistant raises the hoist, until it clears the front of the vehicle

❋❋ CAUTION:

DO NOT use the transaxle-to-engine bolts to force the transaxle and engine together. Take great care when installing the torque converter, following the procedure outlined in Chapter 7.

28 Carefully lower the engine into the engine compartment - make sure the mounts line up. Reinstall the remaining components in the reverse order of removal. Double-check to make sure everything is hooked up right.

29 Add coolant, oil, power steering and transmission fluid as needed.

30 Run the engine and check for leaks and proper operation of all accessories, then install the hood and test drive the vehicle.

6   Engine rebuilding alternatives

The home mechanic is faced with a number of options when performing an engine overhaul. The decision to replace the engine block, piston/connecting rod assemblies and crankshaft depends on a number of factors, with the number one consideration being the condition of the block. Other considerations are cost, access to machine shop facilities, parts availability, time required to complete the project and the extent of prior mechanical experience.

Some of the rebuilding alternatives include:

Individual parts - If the inspection procedures reveal the engine block and most engine components are in reusable condition, purchasing individual parts may be the most economical alternative. The block, crankshaft and piston/connecting rod assemblies should all be inspected carefully. Even if the block shows little wear, the cylinder bores should be surface honed.

Short block - A short block consists of an engine block with a crankshaft and piston/connecting rod assemblies already installed. All new bearings are incorporated and all clearances will be correct. The existing camshaft, valve train components, cylinder heads and external parts can be bolted to the short block with little or no machine shop work necessary.

Long block - A long block consists of a short block plus an oil pump, oil pan, cylinder heads, rocker arm covers, camshaft and valve train components, timing sprockets and chain and timing chain cover. All components are installed with new bearings, seals and gaskets incorporated throughout. The installation of manifolds and external parts is all that's necessary. Give careful thought to which alternative is best for you and discuss the situation with local automotive machine shops, auto parts dealers and experienced rebuilders before ordering or purchasing replacement parts.

Low mileage used engines - Some auto recycling companies offer low mileage used engines, which can be a very cost-effective way to get your vehicle up and running again. These engines often come from vehicles that have been totaled in accidents. A low mileage used engine may also have a warranty like the newly remanufactured engines.

Give careful thought to which alternative is best for you and discuss the situation with local automotive machine shops, auto parts dealers and experienced rebuilders before ordering or purchasing replacement parts.

7.3a Engine - drivebelt end (typical)

7   Engine overhaul - disassembly sequence

▶ **Refer to illustrations 7.3a, 7.3b and 7.3c**

1   It's much easier to disassemble and work on the engine if it's mounted on a portable engine stand. A stand can often be rented quite cheaply from an equipment rental yard. Before it's mounted on a stand, the flywheel/driveplate should be removed from the engine.

2   If a stand isn't available, it's possible to disassemble the engine with it blocked up on the floor. Be extra careful not to tip or drop the engine when working without a stand.

3   If you're going to obtain a rebuilt engine, all external components (see illustrations) must come off first, to be transferred to the replacement engine, just as they will if you're doing a complete engine overhaul yourself. These include:

Alternator and brackets
Emissions control components
Ignition coil/module assembly, spark plugwires and spark plugs
Thermostat and housing cover

Water pump
EFI components
Intake/exhaust manifolds
Oil filter
Engine mounts
Driveplate
Engine rear plate (if equipped)

➡**Note: When removing the external components from the engine, pay close attention to details that may be helpful or important during installation. Note the installed position of gaskets, seals, spacers, pins, brackets, washers, bolts and other small items.**

4   If you're obtaining a short block, which consists of the engine block, crankshaft, pistons and connecting rods all assembled, then the cylinder heads, oil pan and oil pump will have to be removed as well. See *Engine rebuilding alternatives* for additional information regarding the different possibilities to be considered.

7.3b Engine - radiator side (typical)

7.3c Engine - firewall side (typical)

5   If you're planning a complete overhaul, the engine must be disassembled and the internal components removed in the following general order:

Rocker arm covers
Intake and exhaust manifolds
Rocker arms and pushrods
Valve lifters
Cylinder heads
Timing chain cover and oil pump
Timing chain and sprockets
Camshaft
Balance shaft (3800 engine only)
Oil pan
Piston/connecting rod assemblies
Crankshaft and main bearings

6   Before beginning the disassembly and overhaul procedures, make sure the following items are available. Also, refer to *Engine overhaul - reassembly sequence* for a list of tools and materials needed for engine reassembly.

Common hand tools
Small cardboard boxes or plastic bags for
   storing parts
Gasket scraper
Ridge reamer
Vibration damper puller
Micrometers
Telescoping gauges
Dial indicator set
Valve spring compressor
Cylinder surfacing hone
Piston ring groove cleaning tool
Electric drill motor
Tap and die set
Wire brushes
Oil gallery brushes
Cleaning solvent

8 Cylinder head - disassembly

▶ **Refer to illustrations 8.2, 8.3 and 8.4**

➡**Note: New and rebuilt cylinder heads are commonly available for most engines at dealerships and auto parts stores. Due to the fact that some specialized tools are necessary for the disassembly and inspection procedures, and replacement parts aren't always readily available, it may be more practical and economical for the home mechanic to purchase replacement heads rather than taking the time to disassemble, inspect and recondition the originals.**

1 Cylinder head disassembly involves removal of the intake and exhaust valves and related components. Remove the rocker arm bolts, pivots and rocker arms from the cylinder heads. Label the parts or store them separately so they can be reinstalled in their original locations.

2 Before the valves are removed, arrange to label and store them, along with their related components, so they can be kept separate and reinstalled in their original locations (see illustration).

3 Compress the springs on the first valve with a spring compressor and remove the keepers (see illustration). Carefully release the valve spring compressor and remove the retainer, the spring and the spring seat (if used).

4 Pull the valve out of the head, then remove the oil seal from the guide. If the valve binds in the guide (won't pull through), push it back into the head and deburr the area around the keeper groove with a fine file or whetstone (see illustration).

5 Repeat the procedure for the remaining valves. Remember to keep all the parts for each valve together so they can be reinstalled in the same locations.

6 Once the valves and related components have been removed and stored in an organized manner, the head should be thoroughly cleaned and inspected. If a complete engine overhaul is being done, finish the engine disassembly procedures before beginning the cylinder head cleaning and inspection process.

8.2 A small plastic bag, with an appropriate label, can be used to store the valve train components so they can be kept together and reinstalled in the original position

8.3 Use a valve spring compressor to compress the spring, then remove the keepers from the valve stem with a magnet or small needle-nose pliers

8.4 If the valve won't pull through the guide, deburr the edge of the stem end and the area around the top of the keeper groove with a file or whetstone

9 Cylinder head - cleaning and inspection

1 Thorough cleaning of the cylinder heads and related valve train components, followed by a detailed inspection, will enable you to decide how much valve service work must be done during the engine overhaul.

➡**Note: If the engine was severely overheated, the cylinder head is probably warped (see Step 12).**

CLEANING

2 Scrape all traces of old gasket material and sealant off the head gasket, intake manifold and exhaust manifold mating surfaces. Be very careful not to gouge the cylinder head. Special gasket removal solvents that soften gaskets and make removal much easier are available at auto parts stores.

3 Remove all built up scale from the coolant passages.

4 Run a stiff wire brush through the various holes to remove deposits that may have formed in them.

5 Run an appropriate size tap into each of the threaded holes to remove corrosion and thread sealant that may be present. If compressed air is available, use it to clear the holes of debris produced by this operation.

❊❊ WARNING:

Wear eye protection when using compressed air!

6 Clean the rocker arm pivot stud threads with a wire brush.

7 Clean the cylinder head with solvent and dry it thoroughly. Compressed air will speed the drying process and ensure that all holes and recessed areas are clean.

➡**Note: Decarbonizing chemicals are available and may prove very useful when cleaning cylinder heads and valve train components. They're very caustic and should be used with caution. Be sure to follow the instructions on the container.**

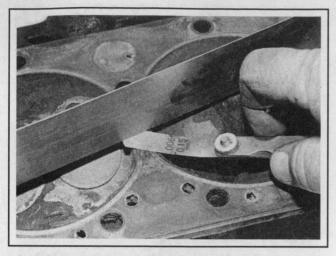

9.12 Check the cylinder head gasket surface for warpage by trying to slip a feeler gauge under the straightedge (see this Chapter's Specifications for the maximum warpage allowed and use a feeler gauge of that thickness)

8   Clean the rocker arms, pivots, bolts and pushrods with solvent and dry them thoroughly (don't mix them up during the cleaning process). Compressed air will speed the drying process and can be used to clean out the oil passages.

9   Clean all the valve springs, spring seats, rotators, keepers and retainers with solvent and dry them thoroughly. Do the components from one valve at a time to avoid mixing up the parts.

10   Scrape off any heavy deposits that may have formed on the valves, then use a motorized wire brush to remove deposits from the valve heads and stems. Again, make sure the valves don't get mixed up.

INSPECTION

▶ **Refer to illustrations 9.12, 9.14, 9.15, 9.16, 9.17 and 9.18**

➡**Note: Be sure to perform all of the following inspection procedures before concluding machine shop work is required. Make a list of the items that need attention.**

CYLINDER HEAD

11   Inspect the head very carefully for cracks, evidence of coolant leakage and other damage. If cracks are found, check with an automotive machine shop concerning repair. If repair isn't possible, a new cylinder head should be obtained.

12   Using a straightedge and feeler gauge, check the head gasket mating surface for warpage (see illustration). If the warpage exceeds the limit in this Chapter's Specifications, it can be resurfaced at an automotive machine shop.

➡**Note: If the heads are resurfaced, the intake manifold flanges will also require machining.**

13   Examine the valve seats in each of the combustion chambers. If they're pitted, cracked or burned, the head will require valve service that's beyond the scope of the home mechanic.

14   Check the valve stem-to-guide clearance by measuring the lateral movement of the valve stem with a dial indicator attached securely to the head (see illustration). The valve must be in the guide and approximately 1/16-inch off the seat. The total valve stem movement indicated by the gauge needle must be divided by two to obtain the actual

9.14 A dial indicator can be used to determine the valve stem-to-guide clearance (move the valve stem as indicated by the arrows)

clearance. After this is done, if there's still some doubt regarding the condition of the valve guides, they should be checked by an automotive machine shop (the cost should be minimal).

VALVES

15   Carefully inspect each valve face for uneven wear, deformation, cracks, pits and burned areas. Check the valve stem for scuffing and galling and the neck for cracks. Rotate the valve and check for any obvious indication that it's bent. Look for pits and excessive wear on the end of the stem. The presence of any of these conditions (see illustration) indicates the need for valve service by an automotive machine shop.

16   Measure the margin width on each valve (see illustration). Any valve with a margin narrower than specified in this Chapter will have to be replaced with a new one.

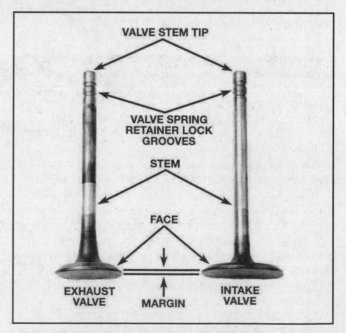

9.15 Check for valve wear at the points shown here

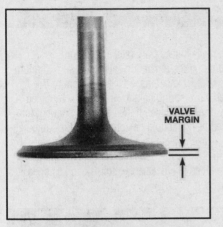

9.16 The margin width on each valve must be as specified (if no margin exists, the valve cannot be reused)

9.17 Measure the free length of each valve spring with a dial or vernier caliper

9.18 Check each valve spring for squareness

VALVE COMPONENTS

17 Check each valve spring for wear (on the ends) and pits. Measure the free length and compare it to this Chapter's Specifications (see illustration). Any springs that are shorter than specified have sagged and shouldn't be reused. The tension of all springs should be checked with a special fixture before deciding they're suitable for use in a rebuilt engine (take the springs to an automotive machine shop for this check).

18 Stand each spring on a flat surface and check it for squareness (see illustration). If any of the springs are distorted or sagged, replace all of them with new parts.

19 Check the spring retainers and keepers for obvious wear and cracks. Any questionable parts should be replaced with new ones, as extensive damage will occur if they fail during engine operation.

ROCKER ARM COMPONENTS

20 Check the rocker arm faces (the areas that contact the pushrod ends and valve stems) for pits, wear, galling, score marks and rough spots. Check the rocker arm pivot contact areas and pivots as well. Look for cracks in each rocker arm and bolt.

21 Inspect the pushrod ends for scuffing and excessive wear. Roll each pushrod on a flat surface, like a piece of plate glass, to determine if it's bent.

22 Check the rocker arm bolt holes in the cylinder heads for damaged threads.

23 Any damaged or excessively worn parts must be replaced with new ones.

ALL COMPONENTS

24 If the inspection process indicates the valve components are in generally poor condition and worn beyond the limits specified, which is usually the case in an engine that's being overhauled, reassemble the valves in the cylinder head and refer to Section 10 for valve servicing recommendations.

10 Valves - servicing

1 Because of the complex nature of the job and the special tools and equipment needed, servicing of the valves, the valve seats and the valve guides, commonly known as a valve job, should be done by a professional.

2 The home mechanic can remove and disassemble the head, do the initial cleaning and inspection, then reassemble and deliver it to a dealer service department or an automotive machine shop for the actual service work. Doing the inspection will enable you to see what condition the head and valvetrain components are in and will ensure that you know what work and new parts are required when dealing with an automotive machine shop.

3 The dealer service department, or automotive machine shop, will remove the valves and springs, recondition or replace the valves and valve seats, recondition the valve guides, check and replace the valve springs, rotators, spring retainers and keepers (as necessary), replace the valve seals with new ones, reassemble the valve components and make sure the installed spring height is correct. The cylinder head gasket surface will also be resurfaced if it's warped.

4 After the valve job has been performed by a professional, the head will be in like new condition. When the head is returned, be sure to clean it again before installation on the engine to remove any metal particles and abrasive grit that may still be present from the valve service or head resurfacing operations. Use compressed air, if available, to blow out all the oil holes and passages.

11 Cylinder head - reassembly

▸ **Refer to illustrations 11.4 and 11.6**

1   Regardless of whether or not the head was sent to an automotive repair shop for valve servicing, make sure it's clean before beginning reassembly.

2   If the head was sent out for valve servicing, the valves and related components will already be in place. Begin the reassembly procedure with Step 8.

3   Install the spring seats before the valve seals.

4   Install new seals on each of the valve guides.

➥**Note: 3.0 and 3.8 liter engines have seals only on the intake valves. Using a hammer and a deep socket or seal installation tool, gently tap each seal into place until it's completely seated on the guide (see illustration). Don't twist or cock the seals during installation or they won't seal properly on the valve stems.**

5   Beginning at one end of the head, lubricate and install the first valve. Apply moly-base grease or clean engine oil to the valve stem.

6   Position the valve springs (and shims, if used) over the valves. Compress the springs with a valve spring compressor and carefully install the keepers in the groove, then slowly release the compressor and make sure the keepers seat properly. Apply a small dab of grease to

each keeper to hold it in place if necessary (see illustration).

7   Repeat the procedure for the remaining valves. Be sure to return the components to their original locations - don't mix them up!

8   Check the installed valve spring height with a ruler graduated in 1/32-inch increments or a dial caliper. If the head was sent out for service work, the installed height should be correct (but don't automatically assume it is). The measurement is taken from the top of each spring seat, rotator or top shim to the bottom of the retainer. If the height is greater than specified in this Chapter, shims can be added under the springs to correct it.

❋❋ CAUTION:

Do not, under any circumstances, shim the springs to the point where the installed height is less than specified.

9   Apply moly-base grease to the rocker arm faces and the pivots, then install the rocker arms and pivots on the cylinder heads. Tighten the bolts finger-tight.

11.4 Make sure the new valve stem seals are seated against the tops of the valve guides

11.6 Apply a small dab of grease to each keeper as shown here before installation - it'll hold them in place on the valve stem as the spring is released

12 Camshaft, balance shaft and bearings - removal and inspection

➥**Note: Since there isn't enough room to remove the camshaft with the engine in the vehicle, the engine must be out of the vehicle and mounted on a stand for this procedure.**

CAMSHAFT LOBE LIFT CHECK

With cylinder head installed

▸ **Refer to illustration 12.3**

1   In order to determine the extent of cam lobe wear, the lobe lift should be checked prior to camshaft removal. Refer to Part A and

remove the rocker arm covers.

2   Position the number one piston at TDC on the compression stroke (see Part A).

3   Beginning with the number one cylinder valves, mount a dial indicator on the engine and position the plunger against the top surface of the first rocker arm. The plunger should be directly above and in line with the pushrod (see illustration).

4   Zero the dial indicator, then very slowly turn the crankshaft in the normal direction of rotation (clockwise) until the indicator needle stops and begins to move in the opposite direction. The point at which it stops indicates maximum cam lobe lift.

12.3 When checking the camshaft lobe lift, the dial indicator plunger must be positioned directly above and in-line with the pushrod

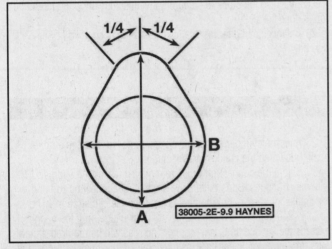

12.9 To verify camshaft lobe lift, measure the major (A) and minor (B) diameters of each lobe with a micrometer or vernier caliper - subtract each minor diameter from the major diameter to arrive at the lobe lift

5   Record this figure for future reference, then reposition the piston at TDC on the compression stroke.

6   Move the dial indicator to the remaining number one cylinder rocker arm and repeat the check. Be sure to record the results for each valve.

7   Repeat the check for the remaining valves. Since each piston must be at TDC on the compression stroke for this procedure, work from cylinder-to-cylinder following the firing order sequence.

8   After the check is complete, compare the results to this Chapter's Specifications. If camshaft lobe lift is less than specified, cam lobe wear has occurred and a new camshaft should be installed.

With cylinder head removed

▶ **Refer to illustration 12.9**

9   If the cylinder heads have already been removed, an alternate method of lobe measurement can be used. Remove the camshaft as described below. Using a micrometer, measure the lobe at its highest point. Then measure the base circle perpendicular (90-degrees) to the lobe (see illustration). Do this for each lobe and record the results.

10 Subtract the base circle measurement from the lobe height. The difference is the lobe lift. See Step 8 above.

REMOVAL

▶ **Refer to illustrations 12.11, 12.12 and 12.13**

11 Refer to the appropriate Sections in Part A and remove the timing chain and sprockets, lifters and pushrods. On 3800 engines, detach the balance shaft drive gear from the end of the camshaft and remove the camshaft thrust plate-to-block bolts (see illustration).

12 Thread long bolts into the camshaft sprocket bolt holes to use as a handle when removing the camshaft from the block (see illustration).

13 Carefully pull the camshaft out. Support the cam near the block so the lobes don't nick or gouge the bearings as it's withdrawn (see illustration).

➡ **Note: The balance shaft (3800 engine only) requires special tools for removal and installation. If the bearings are bad, have the shaft assembly replaced by an automotive machine shop (see step 19).**

12.11 Remove the retaining bolts and the camshaft thrust plate

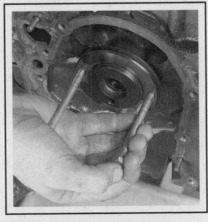

12.12 Thread long bolts into the sprocket bolt holes to use as a handle when removing and installing the camshaft

12.13 Support the camshaft near the block to avoid damaging the bearings

INSPECTION

♦ **Refer to illustration 12.15**

14 After the camshaft has been removed from the engine, cleaned with solvent and dried, inspect the bearing journals for uneven wear, pitting and evidence of seizure. If the journals are damaged, the bearing inserts in the block are probably damaged as well. Both the camshaft and bearings will have to be replaced.

15 Measure the bearing journals with a micrometer (see illustration) to determine if they're excessively worn or out-of-round.

16 Check the camshaft lobes for heat discoloration, score marks, chipped areas, pitting and uneven wear. If the lobes are in good condition and if the lobe lift measurements are as specified, the camshaft can be reused.

17 Check the bearings in the block for wear and damage. Look for galling, pitting and discolored areas.

18 The inside diameter of each bearing can be determined with a small hole gauge and outside micrometer or an inside micrometer. Subtract the camshaft bearing journal diameter(s) from the corresponding bearing inside diameter(s) to obtain the bearing oil clearance. If it's excessive, new bearings will be required regardless of the condition of the originals.

19 Balance shaft and camshaft bearing replacement requires spe-

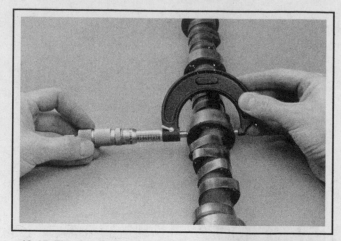

12.15 The camshaft bearing journal diameters are checked to pinpoint excessive wear and out-of-round conditions

cial tools and expertise that place it outside the scope of the home mechanic. Take the block to an automotive machine shop to ensure the job is done correctly.

20 Balance shaft gear installation and timing is covered in Section 24.

13 Pistons and connecting rods - removal

♦ **Refer to illustrations 13.1, 13.3, 13.4 and 13.6**

➡ **Note: Prior to removing the piston/connecting rod assemblies, remove the cylinder heads, the oil pan and the oil pump by referring to the appropriate Sections in Part A of Chapter 2.**

1 Use your fingernail to feel if a ridge has formed at the upper limit of ring travel (about 1/4-inch down from the top of each cylinder). If carbon deposits or cylinder wear have produced ridges, they must be completely removed with a special tool (see illustration). Follow the manufacturer's instructions provided with the tool. Failure to remove the ridges before attempting to remove the piston/connecting rod assemblies may result in piston breakage.

2 After the cylinder ridges have been removed, turn the engine upside-down so the crankshaft is facing up.

3 Before the connecting rods are removed, check the endplay with feeler gauges. Slide them between the first connecting rod and the crankshaft throw until the play is removed (see illustration). The endplay is equal to the thickness of the feeler gauge(s). If the endplay exceeds the service limit, new connecting rods will be required. If new rods (or a new crankshaft) are installed, the endplay may fall under the minimum specified in this Chapter (if it does, the rods will have to be machined to restore it - consult an automotive machine shop for advice if necessary). Repeat the procedure for the remaining connecting rods.

4 Check the connecting rods and caps for identification marks. If

13.1 A ridge reamer is required to remove the ridge from the top of each cylinder - do this before removing the pistons

13.3 Check the connecting rod endplay with a feeler gauge as shown

13.4 Mark the caps with a center punch

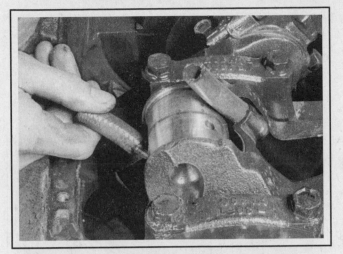

13.6 To prevent damage to the crankshaft journals and cylinder walls, slip sections of rubber or plastic hose over the rod bolts before removing the pistons

they aren't plainly marked, use a small center punch (see illustration) to make the appropriate number of indentations on each rod and cap (1, 2, 3, etc., depending on the cylinder they're associated with).

5   Loosen each of the connecting rod cap nuts 1/2-turn at a time until they can be removed by hand. Remove the number one connecting rod cap and bearing insert. Don't drop the bearing insert out of the cap.

6   Slip a short length of plastic or rubber hose over each connecting rod cap bolt to protect the crankshaft journal and cylinder wall as the piston is removed (see illustration).

7   Remove the bearing insert and push the connecting rod/piston assembly out through the top of the engine. Use a wooden or plastic hammer handle to push on the upper bearing surface in the connecting rod. If resistance is felt, double-check to make sure all of the ridge was removed from the cylinder.

8   Repeat the procedure for the remaining cylinders.

9   After removal, reassemble the connecting rod caps and bearing inserts in their respective connecting rods and install the cap nuts finger tight. Leaving the old bearing inserts in place until reassembly will help prevent the connecting rod bearing surfaces from being accidentally nicked or gouged.

10   Don't separate the pistons from the connecting rods (see Section 18 for additional information).

14   Crankshaft - removal

▶ **Refer to illustrations 14.1, 14.3, 14.4a, 14.4b and 14.4c**

➡**Note: The crankshaft can be removed only after the engine has been removed from the vehicle. It's assumed the driveplate, vibration damper, timing chain, oil pan, oil pump and piston/ connecting rod assemblies have already been removed.**

1   Before the crankshaft is removed, check the endplay. Mount a dial indicator with the stem in line with the crankshaft and just touching one of the crank throws (see illustration).

2   Push the crankshaft all the way to the rear and zero the dial indicator. Next, pry the crankshaft to the front as far as possible and check the reading on the dial indicator. The distance it moves is the endplay.

If it's greater than specified in this Chapter, check the crankshaft thrust surfaces for wear. If no wear is evident, new main bearings should correct the endplay.

3   If a dial indicator isn't available, feeler gauges can be used. Gently pry or push the crankshaft all the way to the front of the engine. Slip feeler gauges between the crankshaft and the front face of the thrust main bearing to determine the clearance (see illustration).

4   On VIN K engines, loosen, then remove, the side bolts securing the main bearing caps to the sides of the cylinder block.

5   Check the main bearing caps to see if they're marked to indicate their locations. They should be numbered consecutively from the front of the engine to the rear. If they aren't, mark them with number stamping dies or a center punch (see illustrations). Main bearing caps gener-

14.1 Check crankshaft endplay with a dial indicator . . .

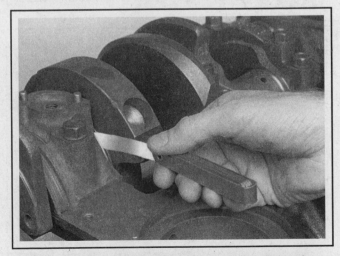

14.3 . . . or slip feeler gauge between the crankshaft and main bearing (number two) thrust surfaces - the endplay is equal to the feeler gauge thickness

ally have a cast-in arrow, which points to the front of the engine (see illustration). Loosen the main bearing cap bolts 1/4-turn at a time each, until they can be removed by hand. Note if any stud bolts are used and make sure they're returned to their original locations when the crankshaft is reinstalled.

6 Gently tap the caps with a soft-face hammer, then separate them from the engine block. If necessary, use the bolts as levers to remove

the caps. Try not to drop the bearing inserts if they come out with the caps.

7 Carefully lift the crankshaft out of the engine. It may be a good idea to have an assistant available, since the crankshaft is quite heavy. With the bearing inserts in place in the engine block and main bearing caps, return the caps to their respective locations on the engine block and tighten the bolts finger tight.

14.5a Use a center punch or number stamping dies to mark the main bearing caps to ensure installation in their original locations on the block - make the punch marks near one of the bolt heads

14.5b Mark the caps in order from the front of the engine to the rear (one mark for the front cap, two for the second one and so on)

14.5c The arrow on the main bearing cap indicates the front of the engine

15 Engine block - cleaning

▶ **Refer to illustrations 15.4a, 15.4b, 15.8 and 15.10**

1 Remove the main bearing caps and separate the bearing inserts from the caps and the engine block. Tag the bearings, indicating which cylinder they were removed from and whether they were in the cap or the block, then set them aside.

2 Using a gasket scraper, remove all traces of gasket material from the engine block. Be very careful not to nick or gouge the gasket seal-

ing surfaces.

3 Remove all of the covers and threaded oil gallery plugs from the block. The plugs are usually very tight - they may have to be drilled out and the holes retapped. Use new plugs when the engine is reassembled.

4 Remove the core plugs from the engine block. To do this, knock one side of the plugs into the block with a hammer and punch, then grasp them with large pliers and pull them out (see illustrations).

15.4a A hammer and large punch can be used to knock the core plugs sideways in their bores

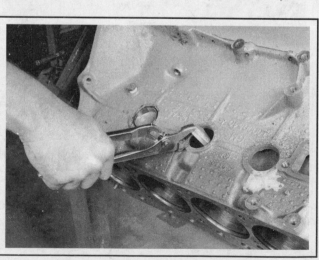

15.4b Pull the core plugs from the block with pliers

15.8 Clean and restore all threaded holes in the block - especially the main bearing cap and head bolt holes - with a tap (be sure to remove debris from the holes when you're done)

15.10 A large socket on an extension can be used to drive the new core plugs into the bores

5   If the engine is extremely dirty, it should be taken to an automotive machine shop to be steam cleaned or hot tanked.

6   After the block is returned, clean all oil holes and oil galleries one more time. Brushes specifically designed for this purpose are available at most auto parts stores. Flush the passages with warm water until the water runs clear, dry the block thoroughly and wipe all machined surfaces with a light, rust preventive oil. If you have access to compressed air, use it to speed the drying process and blow out all the oil holes and galleries.

❊❊ WARNING:

Wear eye protection when using compressed air!

---

7   If the block isn't extremely dirty or sludged up, you can do an adequate cleaning job with hot soapy water and a stiff brush. Take plenty of time and do a thorough job. Regardless of the cleaning method used, be sure to clean all oil holes and galleries very thoroughly, dry the block completely and coat all machined surfaces with

light oil.

8   The threaded holes in the block must be clean to ensure accurate torque readings during reassembly. Run the proper size tap into each of the holes to remove rust, corrosion, thread sealant or sludge and restore damaged threads (see illustration). If possible, use compressed air to clear the holes of debris produced by this operation. Now is a good time to clean the threads on the head bolts and the main bearing cap bolts as well.

9   Reinstall the main bearing caps and tighten the bolts finger tight.

10   After coating the sealing surfaces of the new core plugs with a hard-setting sealant (such as Permatex no. 1), install them in the engine block (see illustration). Make sure they're driven in straight and seated properly or leakage could result. Special tools are available for this purpose, but a large socket, with an outside diameter that will just slip into the core plug, a 1/2-inch drive extension and a hammer will work just as well.

11   Apply non-hardening sealant (such as Permatex no. 2 or Teflon pipe sealant) to the new oil gallery plugs and thread them into the holes in the block. Make sure they're tightened securely.

12   If the engine isn't going to be reassembled right away, cover it with a large plastic trash bag to keep it clean.

---

16   Engine block - inspection

▶ **Refer to illustrations 16.4a, 16.4b and 16.4c**

➡**Note: The manufacturer recommends checking the block deck and transaxle mounting bolt hole bosses for warpage and the main bearing bore concentricity and alignment. Since special measuring tools are needed, the checks should be done by an automotive machine shop.**

1   Before the block is inspected, it should be cleaned as described in Section 15.

2   Visually check the block for cracks, rust and corrosion. Look for stripped threads in the threaded holes. It's also a good idea to have the

block checked for hidden cracks by an automotive machine shop that has the special equipment to do this type of work. If defects are found, have the block repaired, if possible, or replaced.

3   Check the cylinder bores for scuffing and scoring.

4   Measure the diameter of each cylinder at the top (just under the ridge area), center and bottom of the cylinder bore, parallel to the crankshaft axis (see illustrations).

➡**Note: These measurements should not be made with the bare block mounted on an engine stand - the cylinders will be distorted and the measurements will be inaccurate.**

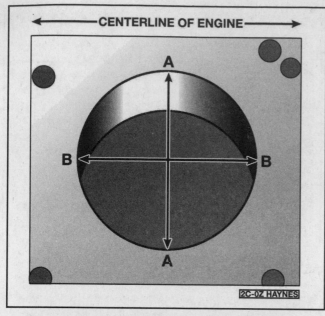

CENTERLINE OF ENGINE

2C-0Z HAYNES

16.4a Measure the diameter of each cylinder at a right angle to the engine centerline (A), and parallel to the engine centerline (B) - out-of-round is the difference between A and B; taper is the difference between the diameter at the top of the cylinder and the diameter at the bottom of the cylinder

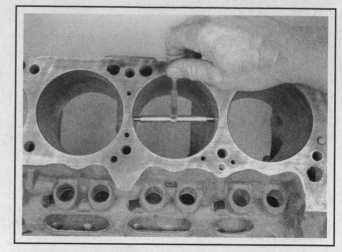

16.4b The ability to "feel" when the telescoping gauge is at the correct point will be developed over time, so work slowly and repeat the check until you're satisfied the bore measurement is accurate

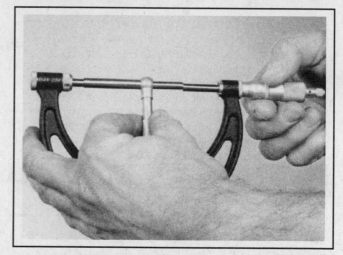

16.4c The gauge is then measured with a micrometer to determine the bore size

5   Next, measure each cylinder's diameter at the same three locations across the crankshaft axis. Compare the results to this Chapter's Specifications.

6   If the required precision measuring tools aren't available, the piston-to-cylinder clearances can be obtained, though not quite as accurately, using feeler gauge stock. Feeler gauge stock comes in 12-inch lengths and various thicknesses and is generally available at auto parts stores.

7   To check the clearance, select a feeler gauge and slip it into the cylinder along with the matching piston. The piston must be positioned exactly as it normally would be. The feeler gauge must be between the piston and cylinder on one of the thrust faces (90-degrees to the piston pin bore).

8   The piston should slip through the cylinder (with the feeler gauge in place) with moderate pressure.

9   If it falls through or slides through easily, the clearance is excessive and a new piston will be required. If the piston binds at the lower end of the cylinder and is loose toward the top, the cylinder is tapered. If tight spots are encountered as the piston/feeler gauge is rotated in the cylinder, the cylinder is out-of-round.

10 Repeat the procedure for the remaining pistons and cylinders.

11 If the cylinder walls are badly scuffed or scored, or if they're

out-of-round or tapered beyond the limits given in this Chapter's Specifications, have the engine block rebored and honed at an automotive machine shop. If a rebore is done, oversize pistons and rings will be required.

12 If the cylinders are in reasonably good condition and not worn to the outside of the limits, and if the piston-to-cylinder clearances can be maintained properly, they don't have to be rebored. Honing is all that's necessary (see Section 17).

17 Cylinder honing

▶ **Refer to illustrations 17.3a and 17.3b**

1   Prior to engine reassembly, the cylinder bores must be honed so the new piston rings will seat correctly and provide the best possible combustion chamber seal.

➡**Note: If you don't have the tools or don't want to tackle the honing operation, most automotive machine shops will do it for a reasonable fee.**

2   Before honing the cylinders, install the main bearing caps and tighten the bolts to the specified torque.

3   Two types of cylinder hones are commonly available - the flex hone or "bottle brush" type and the more traditional surfacing hone with spring-loaded stones. Both will do the job, but for the less experienced mechanic the "bottle brush" hone will probably be easier to use. You'll also need some honing oil (kerosene will work if honing oil isn't available), rags and an electric drill motor. Proceed as follows:

a) *Mount the hone in the drill motor, compress the stones and slip it into the first cylinder (see illustration). Be sure to wear safety goggles or a face shield!*

b) *Lubricate the cylinder with plenty of honing oil, turn on the drill and move the hone up-and-down in the cylinder at a pace that will produce a fine crosshatch pattern on the cylinder walls. Ideally, the crosshatch lines should intersect at approximately a 45-degree angle (see illustration). Be sure to use plenty of lubricant and don't take off any more material than is absolutely necessary to produce the desired finish.*

➡**Note: Piston ring manufacturers may specify a different crosshatch angle - read and follow any instructions included with the new rings.**

c) *Don't withdraw the hone from the cylinder while it's running. Instead, shut off the drill and continue moving the hone up-and-down in the cylinder until it comes to a complete stop, then compress the stones and withdraw the hone. If you're using a "bottle brush" type hone, stop the drill motor, then turn the chuck in the normal direction of rotation while withdrawing the hone from the cylinder.*

d) *Wipe the oil out of the cylinder and repeat the procedure for the remaining cylinders.*

4   After the honing job is complete, chamfer the top edges of the cylinder bores with a small file so the rings won't catch when the pistons are installed. Be very careful not to nick the cylinder walls with the end of the file.

5   The entire engine block must be washed again very thoroughly with warm, soapy water to remove all traces of the abrasive grit produced during the honing operation.

➡**Note: The bores can be considered clean when a lint-free white cloth - dampened with clean engine oil - used to wipe them out doesn't pick up any more honing residue, which will show up as gray areas on the cloth. Be sure to run a brush through all oil holes and galleries and flush them with running water.**

6   After rinsing, dry the block and apply a coat of light rust preventive oil to all machined surfaces. Wrap the block in a plastic trash bag to keep it clean and set it aside until reassembly.

17.3a A "bottle brush" hone is the easiest type of cylinder hone to use

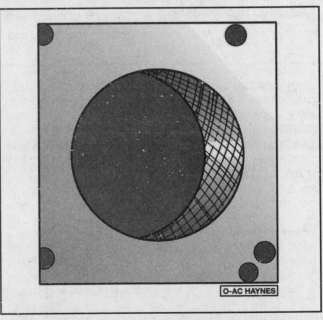

17.3b The cylinder hone should leave a smooth, crosshatch pattern with the lines intersecting at approximately a 60-degree angle

18 Pistons and connecting rods - inspection

▶ **Refer to illustrations 18.4a, 18.4b, 18.10 and 18.11**

1   Before the inspection process can be carried out, the piston/connecting rod assemblies must be cleaned and the original piston rings removed from the pistons.

➡**Note: Always use new piston rings when the engine is reassembled.**

2   Using a piston ring installation tool, carefully remove the rings from the pistons. Be careful not to nick or gouge the pistons in the process.

3   Scrape all traces of carbon from the top of the piston. A hand held wire brush or a piece of fine emery cloth can be used once the majority of the deposits have been scraped away. Do not, under any circumstances, use a wire brush mounted in a drill motor to remove deposits from the pistons. The piston material is soft and may be eroded away by the wire brush.

4   Use a piston ring groove cleaning tool to remove carbon deposits from the ring grooves. If a tool isn't available, a piece broken off the old ring will do the job. Be very careful to remove only the carbon deposits - don't remove any metal and do not nick or scratch the sides of the ring grooves (see illustrations).

5   Once the deposits have been removed, clean the piston/rod assemblies with solvent and dry them with compressed air (if available).

✳✳ WARNING:

Wear eye protection. Make sure the oil return holes in the back sides of the ring grooves are clear.

6   If the pistons and cylinder walls aren't damaged or worn excessively, and if the engine block isn't rebored, new pistons won't be necessary. Normal piston wear appears as even vertical wear on the piston thrust surfaces and slight looseness of the top ring in its groove. New piston rings, however, should always be used when an engine is rebuilt.

7   Carefully inspect each piston for cracks around the skirt, at the pin bosses and at the ring lands.

8   Look for scoring and scuffing on the thrust faces of the skirt, holes in the piston crown and burned areas at the edge of the crown. If

18.4a The piston ring grooves can be cleaned with a special tool, as shown here . . .

the skirt is scored or scuffed, the engine may have been suffering from overheating and/or abnormal combustion, which caused excessively high operating temperatures. The cooling and lubrication systems should be checked thoroughly. A hole in the piston crown is an indication that abnormal combustion (preignition) was occurring. Burned areas at the edge of the piston crown are usually evidence of spark knock (detonation). If any of the above problems exist, the causes must be corrected or the damage will occur again. The causes may include intake air leaks, incorrect fuel/air mixture, low octane fuel, ignition timing and EGR system malfunctions.

9   Corrosion of the piston, in the form of small pits, indicates coolant is leaking into the combustion chamber and/or the crankcase. Again, the cause must be corrected or the problem may persist in the rebuilt engine.

10 Measure the piston ring side clearance by laying a new piston ring in each ring groove and slipping a feeler gauge in beside it (see illustration). Check the clearance at three or four locations around each groove. Be sure to use the correct ring for each groove - they are different. If the side clearance is greater than specified in this Chapter, new pistons will have to be used.

18.4b . . . or a section of a broken ring

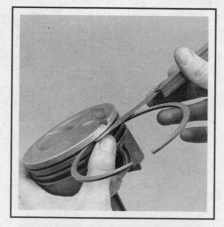

18.10 Check the ring side clearance with a feeler gauge at several points around the groove

18.11 Measure the piston diameter at a 90-degree angle to the piston pin and in line with it

11 Check the piston-to-bore clearance by measuring the bore (see Section 16) and the piston diameter. Make sure the pistons and bores are correctly matched. Measure the piston across the skirt, at a 90-degree angle to the piston pin (see illustration). The measurement must be taken at a specific point, depending on the engine type, to be accurate.

a) *The piston diameter on 3800 engines (with balance shaft) is measured directly in line with the piston pin centerline.*

b) *3.0 and 3.8 liter engine pistons are measured 3/4-inch (19 mm) below the center of the piston pin.*

12 Subtract the piston diameter from the bore diameter to obtain the clearance. If it's greater than specified, the block will have to be rebored and new pistons and rings installed.

13 Check the piston-to-rod clearance by twisting the piston and rod in opposite directions. Any noticeable play indicates excessive wear, which must be corrected. The piston/connecting rod assemblies should be taken to an automotive machine shop to have the pistons and rods resized and new pins installed.

14 If the pistons must be removed from the connecting rods for any reason, they should be taken to an automotive machine shop. While they are there have the connecting rods checked for bend and twist, since automotive machine shops have special equipment for this purpose.

➡ **Note: Unless new pistons and/or connecting rods must be installed, do not disassemble the pistons and connecting rods.**

15 Check the connecting rods for cracks and other damage. Temporarily remove the rod caps, lift out the old bearing inserts, wipe the rod and cap bearing surfaces clean and inspect them for nicks, gouges and scratches. After checking the rods, replace the old bearings, slip the caps into place and tighten the nuts finger tight.

➡ **Note: If the engine is being rebuilt because of a connecting rod knock, be sure to install new rods.**

19 Crankshaft - inspection

▶ **Refer to illustrations 19.1, 19.2, 19.4, 19.6 and 19.8**

1 Remove all burrs from the crankshaft oil holes with a stone, file or scraper (see illustration).

2 Clean the crankshaft with solvent and dry it with compressed air (if available).

✳✳ WARNING:

Wear eye protection when using compressed air. Be sure to clean the oil holes with a stiff brush (see illustration) and flush them with solvent.

3 Check the main and connecting rod bearing journals for uneven wear, scoring, pits and cracks.

4 Rub a penny across each journal several times (see illustration). If a journal picks up copper from the penny, it's too rough and must be reground.

19.1 The oil holes should be chamfered so sharp edges don't gouge or scratch the new bearings

19.2 Use a wire or stiff plastic bristle brush to clean the oil passages in the crankshaft

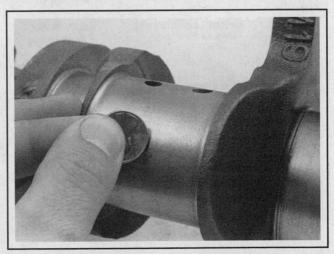

19.4 Rubbing a penny lengthwise on each journal will reveal its condition - if copper rubs off and is embedded in the crankshaft, the journals should be reground

ENGINE BEARING ANALYSIS

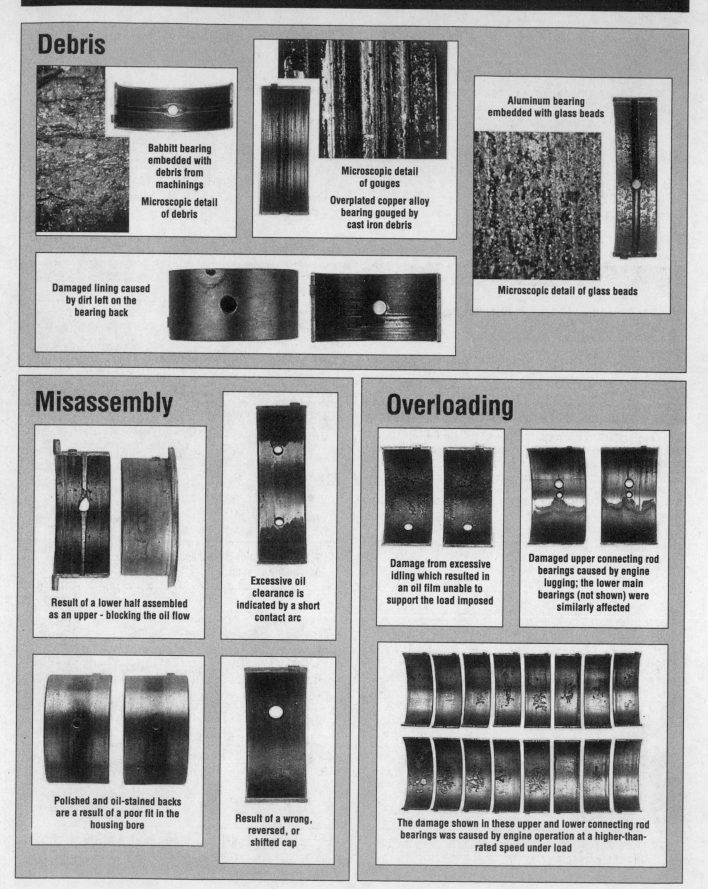

Debris

Babbitt bearing embedded with debris from machinings

Microscopic detail of debris

Microscopic detail of gouges

Overplated copper alloy bearing gouged by cast iron debris

Aluminum bearing embedded with glass beads

Microscopic detail of glass beads

Damaged lining caused by dirt left on the bearing back

Misassembly

Result of a lower half assembled as an upper - blocking the oil flow

Excessive oil clearance is indicated by a short contact arc

Polished and oil-stained backs are a result of a poor fit in the housing bore

Result of a wrong, reversed, or shifted cap

Overloading

Damage from excessive idling which resulted in an oil film unable to support the load imposed

Damaged upper connecting rod bearings caused by engine lugging; the lower main bearings (not shown) were similarly affected

The damage shown in these upper and lower connecting rod bearings was caused by engine operation at a higher-than-rated speed under load

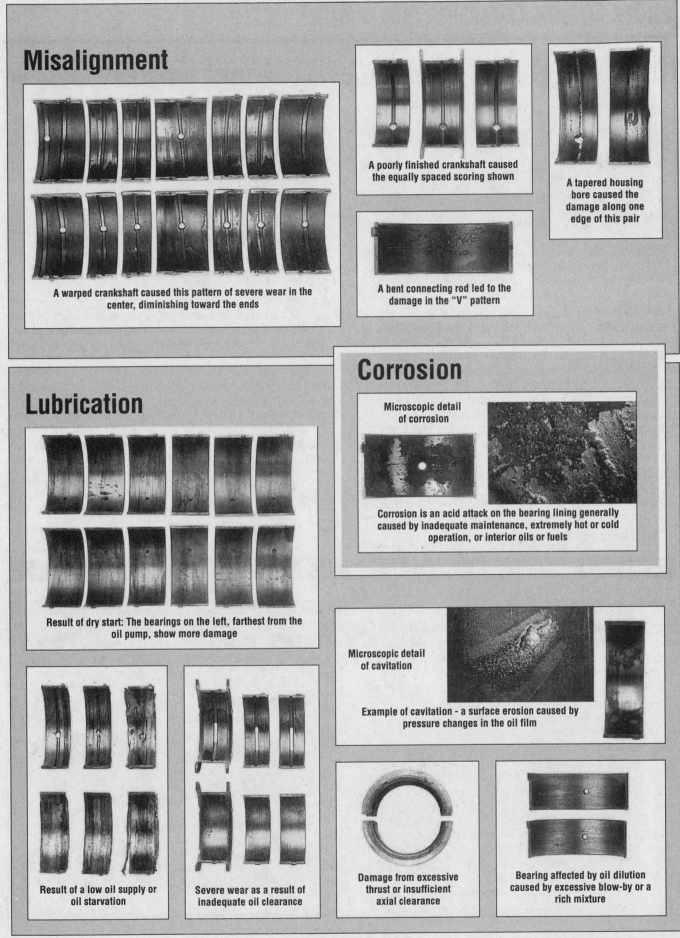

Misalignment

A warped crankshaft caused this pattern of severe wear in the center, diminishing toward the ends

A poorly finished crankshaft caused the equally spaced scoring shown

A tapered housing bore caused the damage along one edge of this pair

A bent connecting rod led to the damage in the "V" pattern

Lubrication

Result of dry start: The bearings on the left, farthest from the oil pump, show more damage

Result of a low oil supply or oil starvation

Severe wear as a result of inadequate oil clearance

Corrosion

Microscopic detail of corrosion

Corrosion is an acid attack on the bearing lining generally caused by inadequate maintenance, extremely hot or cold operation, or interior oils or fuels

Microscopic detail of cavitation

Example of cavitation - a surface erosion caused by pressure changes in the oil film

Damage from excessive thrust or insufficient axial clearance

Bearing affected by oil dilution caused by excessive blow-by or a rich mixture

19.6 Measure the diameter of each crankshaft journal at several points to detect taper and out-of-round conditions

5   Check the rest of the crankshaft for cracks and other damage. It should be magnafluxed to reveal hidden cracks - an automotive machine shop will handle the procedure.

6   Using a micrometer, measure the diameter of the main and connecting rod journals and compare the results to this Chapter's Specifications (see illustration). By measuring the diameter at a number of points around each journal's circumference, you'll be able to determine whether or not the journal is out-of-round. Take the measurement at each end of the journal, near the crank throws, to determine if the journal is tapered.

7   If the crankshaft journals are damaged, tapered, out-of-round or worn beyond the limits given in the Specifications, have the crankshaft

19.8 If the seals have worn grooves in the crankshaft journals, or if the seal contact surfaces are nicked or scratched, the new seals will leak

reground by an automotive machine shop. Be sure to use the correct size bearing inserts if the crankshaft is reconditioned.

8   Check the oil seal journals at each end of the crankshaft for wear and damage. If the seal has worn a groove in the journal, or if it's nicked or scratched (see illustration), the new seal may leak when the engine is reassembled. In some cases, an automotive machine shop may be able to repair the journal by pressing on a thin sleeve. If repair isn't feasible, a new or different crankshaft should be installed.

9   Refer to Section 20 and examine the main and rod bearing inserts.

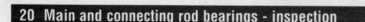

20 Main and connecting rod bearings - inspection

▶ **Refer to illustration 20.1**

1   Even though the main and connecting rod bearings should be replaced with new ones during the engine overhaul, the old bearings should be retained for close examination, as they may reveal valuable information about the condition of the engine (see illustration).

2   Bearing failure occurs because of lack of lubrication, the presence of dirt or other foreign particles, overloading the engine and corrosion. Regardless of the cause of bearing failure, it must be corrected before the engine is reassembled to prevent it from happening again.

3   When examining the bearings, remove them from the engine block, the main bearing caps, the connecting rods and the rod caps and lay them out on a clean surface in the same general position as their location in the engine. This will enable you to match any bearing problems with the corresponding crankshaft journal.

4   Dirt and other foreign particles get into the engine in a variety of ways. It may be left in the engine during assembly, or it may pass through filters or the PCV system. It may get into the oil, and from there into the bearings. Metal chips from machining operations and normal engine wear are often present. Abrasives are sometimes left in engine components after reconditioning, especially when parts aren't thoroughly cleaned using the proper cleaning methods. Whatever the source, these foreign objects often end up embedded in the soft bearing material and are easily recognized. Large particles won't embed in the bearing and will score or gouge the bearing and journal. The best pre-

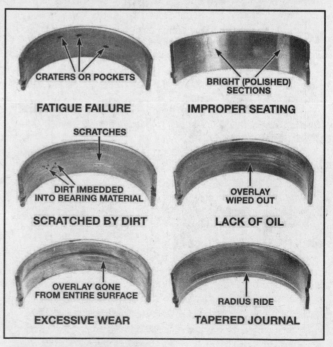

20.1 Typical bearing failures

vention for this cause of bearing failure is to clean all parts thoroughly and keep everything spotlessly clean during engine assembly. Frequent and regular engine oil and filter changes are also recommended.

5   Lack of lubrication (or lubrication breakdown) has a number of interrelated causes. Excessive heat (which thins the oil), overloading (which squeezes the oil from the bearing face) and oil leakage or throw off (from excessive bearing clearances, worn oil pump or high engine speeds) all contribute to lubrication breakdown. Blocked oil passages, which usually are the result of misaligned oil holes in a bearing shell, will also oil starve a bearing and destroy it. When lack of lubrication is the cause of bearing failure, the bearing material is wiped or extruded from the steel backing of the bearing. Temperatures may increase to the point where the steel backing turns blue from overheating.

6   Driving habits can have a definite effect on bearing life. Full throttle, low speed operation (lugging the engine) puts very high loads on bearings, which tends to squeeze out the oil film. These loads cause the bearings to flex, which produces fine cracks in the bearing face (fatigue failure). Eventually the bearing material will loosen in pieces and tear away from the steel backing. Short trip driving leads to corrosion of bearings because insufficient engine heat is produced to drive off the condensed water and corrosive gases. These products collect in the engine oil, forming acid and sludge. As the oil is carried to the engine bearings, the acid attacks and corrodes the bearing material.

7   Incorrect bearing installation during engine assembly will lead to bearing failure as well. Tight fitting bearings leave insufficient oil clearance and will result in oil starvation. Dirt or foreign particles trapped behind a bearing insert result in high spots on the bearing which lead to failure.

21 Engine overhaul - reassembly sequence

1   Before beginning engine reassembly, make sure you have all the necessary new parts, gaskets and seals as well as the following items on hand:

> Common hand tools
> Torque wrench (1/2-inch drive)
> Piston ring installation tool
> Piston ring compressor
> Vibration damper installation tool
> Short lengths of rubber or plastic hose to fit over
>   connecting rod bolts
> Plastigage
> Feeler gauges
> Fine-tooth file
> New engine oil
> Engine assembly lube or moly-base grease
> Gasket sealant
> Thread locking compound

2   In order to save time and avoid problems, engine reassembly must be done in the following general order:

> Crankshaft and main bearings
> Piston/connecting rod assemblies
> Balance shaft (3800 engine only)
> Camshaft
> Timing chain and sprockets
> Timing chain cover and oil pump
> Oil pan
> Cylinder heads
> Valve lifters
> Rocker arms and pushrods
> Intake and exhaust manifolds
> Rocker arm covers

22 Piston rings - installation

▶ **Refer to illustrations 22.3, 22.4, 22.5, 22.9a, 22.9b and 22.12**

1   Before installing the new piston rings, the ring end gaps must be checked. It's assumed the piston ring side clearance has been checked and verified correct (see Section 18).

2   Lay out the piston/connecting rod assemblies and the new ring sets so the ring sets will be matched with the same piston and cylinder during the end gap measurement and engine assembly.

3   Insert the top (number one) ring into the first cylinder and square it up with the cylinder walls by pushing it in with the top of the piston (see illustration). The ring should be near the bottom of the cylinder, at the lower limit of ring travel.

4   To measure the end gap, slip feeler gauges between the ends of the ring until a gauge equal to the gap width is found (see illustration). The feeler gauge should slide between the ring ends with a slight amount of drag. Compare the measurement to this Chapter's Specifications. If the gap is larger or smaller than specified, double-check to make sure you have the correct rings before proceeding.

5   If the gap is too small, it must be enlarged or the ring ends may come in contact with each other during engine operation, which can

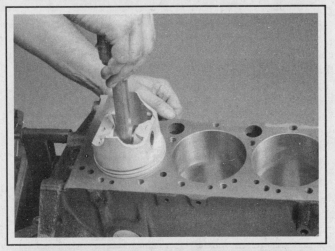

22.3 When checking piston ring end gap, the ring must be square in the cylinder bore (this is done by pushing the ring down with the top of a piston as shown)

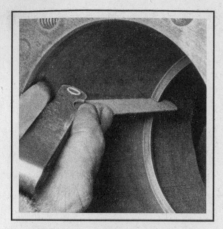

22.4 With the ring square in the cylinder, measure the end gap with a feeler gauge

22.5 If the end gap is too small, clamp a file in a vise and file the ring ends (from the outside in only) to enlarge the gap slightly

22.9a Installing the spacer/expander in the oil control ring groove

cause serious engine damage. The end gap can be increased by filing the ring ends very carefully with a fine file. Mount the file in a vise equipped with soft jaws, slip the ring over the file with the ends contacting the file teeth and slowly move the ring to remove material from the ends. When performing this operation, file only from the outside in (see illustration).

6   Excess end gap isn't critical unless it's greater than 0.040-inch. Again, double-check to make sure you have the correct rings for the engine.

7   Repeat the procedure for each ring that will be installed in the first cylinder and for each ring in the remaining cylinders. Remember to keep rings, pistons and cylinders matched up.

8   Once the ring end gaps have been checked/corrected, the rings can be installed on the pistons.

9   The oil control ring (lowest one on the piston) is usually installed first. It's composed of three separate components. Slip the spacer/expander into the groove (see illustration). If an anti-rotation tang is used, make sure it's inserted into the drilled hole in the ring groove. Next, install the lower side rail. Don't use a piston ring installation tool on the oil ring side rails, as they may be damaged. Instead, place one end of the side rail into the groove between the spacer/

expander and the ring land, hold it firmly in place and slide a finger around the piston while pushing the rail into the groove (see illustration). Next, install the upper side rail in the same manner.

10   After the three oil ring components have been installed, check to make sure both the upper and lower side rails can be turned smoothly in the ring groove.

11   The number two (middle) ring is installed next. It's usually stamped with a mark, which must face up, toward the top of the piston.

➡ **Note: Always follow the instructions printed on the ring package or box - different manufacturers may require different approaches. Don't mix up the top and middle rings, as they have different cross sections.**

12   Use a piston ring installation tool and make sure the identification mark is facing the top of the piston, then slip the ring into the middle groove on the piston (see illustration). Don't expand the ring any more than necessary to slide it over the piston.

13   Install the number one (top) ring in the same manner. Make sure the mark is facing up. Be careful not to confuse the number one and number two rings.

14   Repeat the procedure for the remaining pistons and rings.

22.9b DO NOT use a piston ring installation tool when installing the oil ring side rails

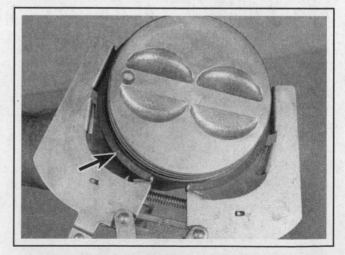

22.12 Installing the compression rings with a ring expander - the mark (arrow) must face up

23 Crankshaft - installation and main bearing oil clearance check

▶ **Refer to illustrations 23.11 and 23.15**

1 Crankshaft installation is the first step in engine reassembly. It's assumed at this point that the engine block and crankshaft have been cleaned, inspected and repaired or reconditioned.

2 Position the engine with the bottom facing up.

3 Remove the main bearing cap bolts and lift out the caps. Lay them out in the proper order to ensure correct installation.

4 If they're still in place, remove the original bearing inserts from the block and the main bearing caps. Wipe the bearing surfaces of the block and caps with a clean, lint-free cloth. They must be kept spotlessly clean.

MAIN BEARING OIL CLEARANCE CHECK

➡**Note: Don't touch the faces of the new bearing inserts with your fingers. Oil and acids from your skin can etch the bearings.**

5 Clean the back sides of the new main bearing inserts and lay one in each main bearing saddle in the block. If one of the bearing inserts from each set has a large groove in it, make sure the grooved insert is installed in the block. Lay the other bearing from each set in the corresponding main bearing cap. Make sure the tab on the bearing insert fits into the recess in the block or cap.

❖❖ CAUTION:

The oil holes in the block must line up with the oil holes in the bearing inserts. Do not hammer the bearing into place and don't nick or gouge the bearing faces. No lubrication should be used at this time.

6 The flanged thrust bearing must be installed in the number two cap and saddle (counting from the front of the engine).

7 Clean the faces of the bearings in the block and the crankshaft main bearing journals with a clean, lint-free cloth.

8 Check or clean the oil holes in the crankshaft, as any dirt here can go only one way - straight through the new bearings.

9 Once you're certain the crankshaft is clean, carefully lay it in position in the main bearings.

10 Before the crankshaft can be permanently installed, the main bearing oil clearance must be checked.

11 Cut several pieces of the appropriate size Plastigage (they should be slightly shorter than the width of the main bearings) and place one piece on each crankshaft main bearing journal, parallel with the journal axis (see illustration).

12 Clean the faces of the bearings in the caps and install the caps in their original locations (don't mix them up) with the arrows pointing toward the front of the engine. Don't disturb the Plastigage. Gently tap the caps into place with a soft-face hammer.

13 Starting with the center main and working out toward the ends, tighten the main bearing cap bolts to the torque figure listed in this Chapter's Specifications. Don't rotate the crankshaft at any time during this operation.

14 Remove the bolts and carefully lift off the main bearing caps. Keep them in order. Don't disturb the Plastigage or rotate the crankshaft. If any of the main bearing caps are difficult to remove, tap them gently from side-to-side with a soft-face hammer to loosen them.

15 Compare the width of the crushed Plastigage on each journal to the scale printed on the Plastigage envelope to obtain the main bearing oil clearance (see illustration). Check the Specifications to make sure it's correct.

16 If the clearance is not as specified, the bearing inserts may be the wrong size (which means different ones will be required). Before deciding different inserts are needed, make sure no dirt or oil was between the bearing inserts and the caps or block when the clearance was measured. If the Plastigage was wider at one end than the other, the journal may be tapered (refer to Section 20).

17 Carefully scrape all traces of the Plastigage material off the main bearing journals and/or the bearing faces. Use your fingernail or the edge of a credit card - don't nick or scratch the bearing faces.

FINAL CRANKSHAFT INSTALLATION

18 Carefully lift the crankshaft out of the engine.

19 Clean the bearing faces in the block, then apply a thin, uniform layer of moly-base grease or engine assembly lube to each of the bear-

23.11 Lay the Plastigage strips (arrow) on the main bearing journals, parallel to the crankshaft centerline

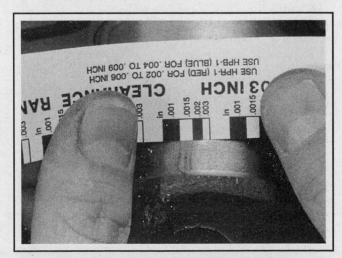

23.15 Measuring the width of the crushed Plastigage to determine the rod bearing oil clearance (be sure to use the correct scale - standard and metric ones are included)

GLOSSARY

B

Backlash - The amount of play between two parts. Usually refers to how much one gear can be moved back and forth without moving gear with which it's meshed.

Bearing Caps - The caps held in place by nuts or bolts which, in turn, hold the bearing surface. This space is for lubricating oil to enter.

Bearing clearance - The amount of space left between shaft and bearing surface. This space is for lubricating oil to enter.

Bearing crush - The additional height which is purposely manufactured into each bearing half to ensure complete contact of the bearing back with the housing bore when the engine is assembled.

Bearing knock - The noise created by movement of a part in a loose or worn bearing.

Blueprinting - Dismantling an engine and reassembling it to EXACT specifications.

Bore - An engine cylinder, or any cylindrical hole; also used to describe the process of enlarging or accurately refinishing a hole with a cutting tool, as to bore an engine cylinder. The bore size is the diameter of the hole.

Boring - Renewing the cylinders by cutting them out to a specified size. A boring bar is used to make the cut.

Bottom end - A term which refers collectively to the engine block, crankshaft, main bearings and the big ends of the connecting rods.

Break-in - The period of operation between installation of new or rebuilt parts and time in which parts are worn to the correct fit. Driving at reduced and varying speed for a specified mileage to permit parts to wear to the correct fit.

Bushing - A one-piece sleeve placed in a bore to serve as a bearing surface for shaft, piston pin, etc. Usually replaceable.

C

Camshaft - The shaft in the engine, on which a series of lobes are located for operating the valve mechanisms. The camshaft is driven by gears or sprockets and a timing chain. Usually referred to simply as the cam.

Carbon - Hard, or soft, black deposits found in combustion chamber, on plugs, under rings, on and under valve heads.

Cast iron - An alloy of iron and more than two percent carbon, used for engine blocks and heads because it's relatively inexpensive and easy to mold into complex shapes.

Chamfer - To bevel across (or a bevel on) the sharp edge of an object.

Chase - To repair damaged threads with a tap or die.

Combustion chamber - The space between the piston and the cylinder head, with the piston at top dead center, in which air-fuel mixture is burned.

Compression ratio - The relationship between cylinder volume (clearance volume) when the piston is at top dead center and cylinder volume when the piston is at bottom dead center.

Connecting rod - The rod that connects the crank on the crankshaft with the piston. Sometimes called a con rod.

Connecting rod cap - The part of the connecting rod assembly that attaches the rod to the crankpin.

Core plug - Soft metal plug used to plug the casting holes for the coolant passages in the block.

Crankcase - The lower part of the engine in which the crankshaft rotates; includes the lower section of the cylinder block and the oil pan.

Crank kit - A reground or reconditioned crankshaft and new main and connecting rod bearings.

Crankpin - The part of a crankshaft to which a connecting rod is attached.

Crankshaft - The main rotating member, or shaft, running the length of the crankcase, with offset throws to which the connecting rods are attached; changes the reciprocating motion of the pistons into rotating motion.

Cylinder sleeve - A replaceable sleeve, or liner, pressed into the cylinder block to form the cylinder bore.

D

Deburring - Removing the burrs (rough edges or areas) from a bearing.

Deglazer - A tool, rotated by an electric motor, used to remove glaze from cylinder walls so a new set of rings will seat.

E

Endplay - The amount of lengthwise movement between two parts. As applied to a crankshaft, the distance that the crankshaft can move forward and back in the cylinder block.

F

Face - A machinist's term that refers to removing metal from the end of a shaft or the face of a larger part, such as a flywheel.

Fatigue - A breakdown of material through a large number of loading and unloading cycles. The first signs are cracks followed shortly by breaks.

Feeler gauge - A thin strip of hardened steel, ground to an exact thickness, used to check clearances between parts.

Free height - The unloaded length or height of a spring.

Freeplay - The looseness in a linkage, or an assembly of parts, between the initial application of force and actual movement. Usually perceived as slop or slight delay.

Freeze plug - See Core plug.

G

Gallery - A large passage in the block that forms a reservoir for engine oil pressure.

Glaze - The very smooth, glassy finish that develops on cylinder walls while an engine is in service.

H

Heli-Coil - A rethreading device used when threads are worn or damaged. The device is installed in a retapped hole to reduce the thread size to the original size.

I

Installed height - The spring's measured length or height, as installed on the cylinder head. Installed height is measured from the spring seat to the underside of the spring retainer.

J

Journal - The surface of a rotating shaft which turns in a bearing.

K

Keeper - The split lock that holds the valve spring retainer in position on the valve stem.

Key - A small piece of metal inserted into matching grooves machined into two parts fitted together - such as a gear pressed onto a shaft - which prevents slippage between the two parts.

Knock - The heavy metallic engine sound, produced in the combustion chamber as a result of abnormal combustion - usually detonation. Knock is usually caused by a loose or worn bearing. Also referred to as detonation, pinging and spark knock. Connecting rod or main bearing knocks are created by too much oil clearance or insufficient lubrication.

L

Lands - The portions of metal between the piston ring grooves.

Lapping the valves - Grinding a valve face and its seat together with lapping compound.

Lash - The amount of free motion in a gear train, between gears, or in a mechanical assembly, that occurs before movement can begin. Usually refers to the lash in a valve train.

Lifter - The part that rides against the cam to transfer motion to the rest of the valve train.

M

Machining - The process of using a machine to remove metal from a metal part.

Main bearings - The plain, or babbitt, bearings that support the crankshaft.

Main bearing caps - The cast iron caps, bolted to the bottom of the block, that support the main bearings.

O

O.D. - Outside diameter.

Oil gallery - A pipe or drilled passageway in the engine used to carry engine oil from one area to another.

Oil ring - The lower ring, or rings, of a piston; designed to prevent excessive amounts of oil from working up the cylinder walls and into the combustion chamber. Also called an oil-control ring.

Oil seal - A seal which keeps oil from leaking out of a compartment. Usually refers to a dynamic seal around a rotating shaft or other moving part.

O-ring - A type of sealing ring made of a special rubberlike material; in use, the O-ring is compressed into a groove to provide the sealing action.

Overhaul - To completely disassemble a unit, clean and inspect all parts, reassemble it with the original or new parts and make all adjustments necessary for proper operation.

P

Pilot bearing - A small bearing installed in the center of the flywheel (or the rear end of the crankshaft) to support the front end of the input shaft of the transmission.

Pip mark - A little dot or indentation which indicates the top side of a compression ring.

Piston - The cylindrical part, attached to the connecting rod, that moves up and down in the cylinder as the crankshaft rotates. When the fuel charge is fired, the piston transfers the force of the explosion to the connecting rod, then to the crankshaft.

Piston pin (or wrist pin) - The cylindrical and usually hollow steel pin that passes through the piston. The piston pin fastens the piston to the upper end of the connecting rod.

Piston ring - The split ring fitted to the groove in a piston. The ring contacts the sides of the ring groove and also rubs against the cylinder wall, thus sealing space between piston and wall. There are two types of rings: Compression rings seal the compression pressure in the combustion chamber; oil rings scrape excessive oil off the cylinder wall.

Piston ring groove - The slots or grooves cut in piston heads to hold piston rings in position.

Piston skirt - The portion of the piston below the rings and the piston pin hole.

Plastigage - A thin strip of plastic thread, available in different sizes, used for measuring clearances. For example, a strip of plastigage is laid across a bearing journal and mashed as parts are assembled. Then parts are disassembled and the width of the strip is measured to determine clearance between journal and bearing. Commonly used to measure crankshaft main-bearing and connecting rod bearing clearances.

Press-fit - A tight fit between two parts that requires pressure to force the parts together. Also referred to as drive, or force, fit.

Prussian blue - A blue pigment; in solution, useful in determining the area of contact between two surfaces. Prussian blue is commonly used to determine the width and location of the contact area between the valve face and the valve seat.

R

Race (bearing) - The inner or outer ring that provides a contact surface for balls or rollers in bearing.

Ream - To size, enlarge or smooth a hole by using a round cutting tool with fluted edges.

Ring job - The process of reconditioning the cylinders and installing new rings.

Runout - Wobble. The amount a shaft rotates out-of-true.

S

Saddle - The upper main bearing seat.

Scored - Scratched or grooved, as a cylinder wall may be scored by abrasive particles moved up and down by the piston rings.

Scuffing - A type of wear in which there's a transfer of material between parts moving against each other; shows up as pits or grooves in the mating surfaces.

Seat - The surface upon which another part rests or seats. For example, the valve seat is the matched surface upon which the valve face rests. Also used to refer to wearing into a good fit; for example, piston rings seat after a few miles of driving.

Short block - An engine block complete with crankshaft and piston and, usually, camshaft assemblies.

Static balance - The balance of an object while it's stationary.

Step - The wear on the lower portion of a ring land caused by excessive side and back-clearance. The height of the step indicates the ring's extra side clearance and the length of the step projecting from the back wall of the groove represents the ring's back clearance.

Stroke - The distance the piston moves when traveling from top dead center to bottom dead center, or from bottom dead center to top dead center.

Stud - A metal rod with threads on both ends.

T

Tang - A lip on the end of a plain bearing used to align the bearing during assembly.

Tap - To cut threads in a hole. Also refers to the fluted tool used to cut threads.

Taper - A gradual reduction in the width of a shaft or hole; in an engine cylinder, taper usually takes the form of uneven wear, more pronounced at the top than at the bottom.

Throws - The offset portions of the crankshaft to which the connecting rods are affixed.

Thrust bearing - The main bearing that has thrust faces to prevent excessive endplay, or forward and backward movement of the crankshaft.

Thrust washer - A bronze or hardened steel washer placed between two moving parts. The washer prevents longitudinal movement and provides a bearing surface for thrust surfaces of parts.

Tolerance - The amount of variation permitted from an exact size of measurement. Actual amount from smallest acceptable dimension to largest acceptable dimension.

U

Umbrella - An oil deflector placed near the valve tip to throw oil from the valve stem area.

Undercut - A machined groove below the normal surface.

Undersize bearings - Smaller diameter bearings used with re-ground crankshaft journals.

V

Valve grinding - Refacing a valve in a valve-refacing machine.

Valve train - The valve-operating mechanism of an engine; includes all components from the camshaft to the valve.

Vibration damper - A cylindrical weight attached to the front of the crankshaft to minimize torsional vibration (the twist-untwist actions of the crankshaft caused by the cylinder firing impulses). Also called a harmonic balancer.

W

Water jacket - The spaces around the cylinders, between the inner and outer shells of the cylinder block or head, through which coolant circulates.

Web - A supporting structure across a cavity.

Woodruff key - A key with a radiused backside (viewed from the side).

ing surfaces. Be sure to coat the thrust faces as well as the journal face of the thrust bearing. Install the rear oil seal (1990 and earlier models only) (see Section 25).

20 Make sure the crankshaft journals are clean, then lay the crankshaft back in place in the block.

21 Clean the faces of the bearings in the caps, then apply lubricant to them.

22 Install the caps in their original locations with the arrows pointing toward the front of the engine. Tap the caps into place with a soft-face hammer.

23 Install the bolts.

24 Tighten all except the thrust bearing cap bolts to the specified torque (work from the center out and approach the final torque in three steps).

25 Tighten the thrust bearing cap bolts to 10-to-12 ft-lbs.

26 Tap the ends of the crankshaft forward and backward with a lead or brass hammer to line up the main bearing and crankshaft thrust surfaces.

27 Retighten all main bearing cap bolts to the torque specified in this Chapter, starting with the center main and working out toward the ends.

28 On 1995 VIN K and all 1996 and later engines, install the side bolts securing the main bearing caps to the sides of the cylinder block. Tighten to the torque listed in this Chapter's Specifications.

29 Rotate the crankshaft a number of times by hand to check for any obvious binding.

30 The final step is to check the crankshaft endplay with feeler gauges or a dial indicator as described in Section 14. The endplay should be correct if the crankshaft thrust faces aren't worn or damaged and new bearings have been installed.

24 Camshaft and balance shaft - installation

▶ **Refer to illustrations 24.1 and 24.6**

CAMSHAFT

1   Lubricate the camshaft bearing journals and cam lobes with moly-base grease or engine assembly lube (see illustration).

2   Slide the camshaft into the engine. Support the cam near the block and be careful not to scrape or nick the bearings. Install the thrust plate and tighten the bolts to the torque listed in the Part A Specifications.

3   Refer to Part A for further instructions and complete the installation of the timing chain, sprockets and lifters.

BALANCE SHAFT (3800 ENGINE ONLY)

4   If the balance shaft timing was disturbed during engine repair, turn the camshaft so that with the camshaft sprocket temporarily installed, the timing mark is straight down.

5   With the camshaft sprocket and the camshaft gear removed, turn the balance shaft so the timing mark on the gear points straight down.

6   Align the marks on the balance shaft gear and the camshaft gear (see illustration) by turning the balance shaft.

7   Install the timing chain and sprockets (see Part A).

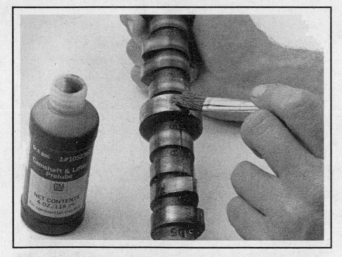

24.1 Be sure to prelube the camshaft bearing journals and lobes before installation

24.6 Align the marks on the gears as shown here (3800 engine only)

25 Crankshaft rear oil seal - installation

1985 THROUGH 1990 MODELS

▶ **Refer to illustrations 25.1, 25.2, 25.3a, 25.3b, 25.4 and 25.7**

➡ **Note: The crankshaft rear oil seal must be in place prior to crankshaft installation.**

1   Lay one seal section on edge in the seal groove in the block and push it into place with your thumbs. Both ends of the seal should extend out of the block slightly (see illustration).

2   Seat it in the groove by rolling a large socket or piece of bar stock along the entire length of the seal (see illustration). As an alternative, roll the seal very carefully into place with a wooden hammer handle.

3   Once the seal is completely seated in the groove, trim off the excess on the ends with a single edge razor blade or razor knife (see illustrations). The seal ends must be flush with the block-to-cap mat-

ing surfaces. Make sure the cut is clean so no seal fibers get caught between the block and cap.

4   Repeat the entire procedure to install the other half of the seal in the bearing cap. Apply a thin film of assembly lube or moly-base grease to the edge of the seal where it contacts the crankshaft (see illustration).

5   Soak the bearing cap side seals in kerosene or light oil prior to installation. They swell in the presence of oil and heat. The seals are slightly longer than the bearing cap grooves and must be cut to length.

6   Lightly coat the crankshaft journal with assembly lube or moly-base grease, then refer to Section 23 and install the crankshaft.

7   During final installation of the crankshaft (after the main bearing oil clearances have been checked with Plastigage), apply a thin, even coat of anaerobic gasket sealant to the parting surfaces of the bearing cap or block (see illustration). Don't get any sealant on the bearing or seal faces.

25.1 When correctly installed, the ends of the seal should extend out of the block slightly

25.2 Set the seal in the groove, but don't depress it below the bearing surface (the seal must contact the crankshaft journal)

25.3a Trim the seal ends flush with the block . . .

25.3b . . . but leave the inner edge (arrow) protruding slightly

25.4 Lubricate the seal with assembly lube or moly-base grease

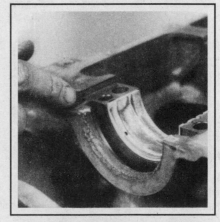

25.7 When applying sealant, be sure it gets into each corner and onto the vertical cap-to-block mating surfaces or oil leaks will result

8   Install the cap, bearing and oil seal assembly and tighten the cap bolts to the torque listed in this Chapter's specifications.

9   After the cap is bolted in place, insert the side seals into the grooves and make sure they're seated. As an alternative, you can inject RTV sealant into the grooves.

1991 AND LATER MODELS

10   These models use a one-piece, lip type seal. The crankshaft must be in place and the main bearing caps installed before the seal is installed. See Chapter 2, Part A for the replacement procedure. On later models, the seal is installed in the seal retainer (either on or off the engine) not in the engine block.

26 Pistons and connecting rods - installation and rod bearing oil clearance check

▶ **Refer to illustrations 26.5, 26.9, 26.11, 26.13 and 26.17**

1   Before installing the piston/connecting rod assemblies, the cylinder walls must be perfectly clean, the top edge of each cylinder must be chamfered, and the crankshaft must be in place.

2   Remove the cap from the end of the number one connecting rod (check the marks made during removal). Remove the original bearing inserts and wipe the bearing surfaces of the connecting rod and cap with a clean, lint-free cloth. They must be kept spotlessly clean.

CONNECTING ROD BEARING OIL CLEARANCE CHECK

➡**Note: Don't touch the faces of the new bearing inserts with your fingers. Oil and acids from your skin can etch the bearings.**

3   Clean the back side of the new upper bearing insert, then lay it in place in the connecting rod. Make sure the tab on the bearing fits into the recess in the rod. Don't hammer the bearing insert into place and be very careful not to nick or gouge the bearing face. Don't lubricate the bearing at this time.

4   Clean the back side of the other bearing insert and install it in the rod cap. Again, make sure the tab on the bearing fits into the recess in the cap, and don't apply any lubricant. It's critically important that the mating surfaces of the bearing and connecting rod are perfectly clean and oil free when they're assembled.

5   Stagger the piston ring gaps around the piston (see illustration).

6   Slip a section of plastic or rubber hose over each connecting rod cap bolt.

7   Lubricate the piston and rings with clean engine oil and attach a piston ring compressor to the piston. Leave the skirt protruding about

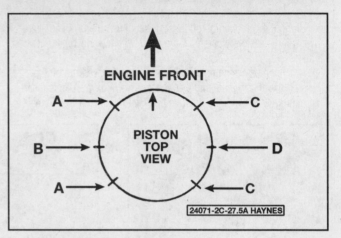

26.5 Ring end gap positions

A   Oil ring side rail gaps
B   Second compression ring gap
C   Oil ring spacer gap position in-between marks
D   Top compression ring gap

1/4-inch to guide the piston into the cylinder. The rings must be compressed until they're flush with the piston.

8   Rotate the crankshaft until the number one connecting rod journal is at BDC (bottom dead center) and apply a coat of engine oil to the cylinder walls.

9   With the mark or notch on top of the piston (see illustration) facing the front of the engine, gently insert the piston/connecting rod

26.9 The notch or arrow on each piston must face the front (timing chain) end of the engine as the pistons are installed

26.11 Drive the piston gently into the cylinder bore with the end of a wooden or plastic hammer handle

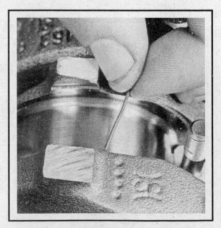

26.13 Lay the Plastigage strips on each rod bearing journal, parallel to the crankshaft centerline

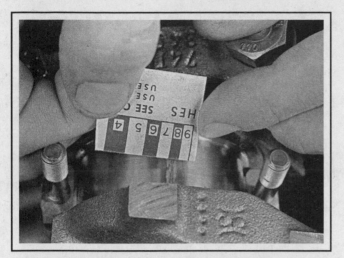

26.17 Measuring the width of the crushed Plastigage to determine the rod bearing oil clearance (be sure to use the correct scale - standard and metric ones are included)

assembly into the number one cylinder bore and rest the bottom edge of the ring compressor on the engine block.

10 Tap the top edge of the ring compressor to make sure it's contacting the block around its entire circumference.

11 Gently tap on the top of the piston with the end of a wooden or plastic hammer handle (see illustration) while guiding the end of the connecting rod into place on the crankshaft journal. The piston rings may try to pop out of the ring compressor just before entering the cylinder bore, so keep some downward pressure on the ring compressor. Work slowly, and if any resistance is felt as the piston enters the cylinder, stop immediately. Find out what's hanging up and fix it before proceeding. Do not, for any reason, force the piston into the cylinder - you might break a ring and/or the piston.

12 Once the piston/connecting rod assembly is installed, the connecting rod bearing oil clearance must be checked before the rod cap is permanently bolted in place.

13 Cut a piece of the appropriate size Plastigage slightly shorter than the width of the connecting rod bearing and lay it in place on the number one connecting rod journal, parallel with the journal axis (see illustration).

14 Clean the connecting rod cap bearing face, remove the protective hoses from the connecting rod bolts and install the rod cap. Make sure the mating mark on the cap is on the same side as the mark on the connecting rod.

15 Install the nuts and tighten them to the torque listed in this Chapter's Specifications. Work up to it in three steps.

➡**Note: Use a thin-wall socket to avoid erroneous torque readings that can result if the socket is wedged between the rod cap and nut. If the socket tends to wedge itself between the nut and the cap, lift up on it slightly until it no longer contacts the cap. Do not rotate the crankshaft at any time during this operation.**

16 Remove the nuts and detach the rod cap, being very careful not to disturb the Plastigage.

17 Compare the width of the crushed Plastigage to the scale printed on the Plastigage envelope to obtain the oil clearance (see illustration). Compare it to this Chapter's Specifications to make sure the clearance is correct.

18 If the clearance is not as specified, the bearing inserts may be the wrong size (which means different ones will be required). Before deciding different inserts are needed, make sure no dirt or oil was between the bearing inserts and the connecting rod or cap when the clearance was measured. Also, recheck the journal diameter. If the Plastigage was wider at one end than the other, the journal may be tapered (refer to Section 19).

FINAL CONNECTING ROD INSTALLATION

19 Carefully scrape all traces of the Plastigage material off the rod journal and/or bearing face. Be very careful not to scratch the bearing - use your fingernail or the edge of a credit card.

20 Make sure the bearing faces are perfectly clean, then apply a uniform layer of clean moly-base grease or engine assembly lube to both of them. You'll have to push the piston into the cylinder to expose the face of the bearing insert in the connecting rod - be sure to slip the protective hoses over the rod bolts first.

21 Slide the connecting rod back into place on the journal, remove the protective hoses from the rod cap bolts, install the rod cap and tighten the nuts to the torque specified in this Chapter. Again, work up to the torque in three steps.

22 Repeat the entire procedure for the remaining pistons/connecting rods.

23 The important points to remember are . . .

   a) *Keep the back sides of the bearing inserts and the insides of the connecting rods and caps perfectly clean when assembling them.*
   b) *Make sure you have the correct piston/rod assembly for each cylinder.*
   c) *The arrow or mark on the piston must face the front (timing chain end) of the engine.*
   d) *Lubricate the cylinder walls with clean oil.*
   e) *Lubricate the bearing faces when installing the rod caps after the oil clearance has been checked.*

24 After all the piston/connecting rod assemblies have been properly installed, rotate the crankshaft a number of times by hand to check for any obvious binding.

25 As a final step, the connecting rod endplay must be checked. Refer to Section 13 for this procedure.

26 Compare the measured endplay to this Chapter's Specifications to make sure it's correct. If it was correct before disassembly and the original crankshaft and rods were reinstalled, it should still be right. If new rods or a new crankshaft were installed, the endplay may be inadequate. If so, the rods will have to be removed and taken to an automotive machine shop for resizing.

27 Initial start-up and break-in after overhaul

✳ WARNING:

Have a fire extinguisher handy when starting the engine for the first time.

1   Once the engine has been installed in the vehicle, double-check the oil and coolant levels.

2   With the spark plugs out of the engine and the ECM fuse removed, crank the engine until oil pressure registers on the gauge or the light goes out.

3   Install the spark plugs, hook up the plug wires and install the ECM fuse.

4   Start the engine. It may take a few moments for the fuel system to build up pressure, but the engine should start without a great deal of effort.

➡ **Note: If the engine keeps backfiring, recheck the valve timing and spark plug wires.**

5   After the engine starts, it should be allowed to warm up to normal operating temperature. While the engine is warming up, make a thorough check for fuel, oil and coolant leaks.

6   Shut the engine off and recheck the engine oil and coolant levels.

7   Drive the vehicle to an area with minimum traffic, accelerate at full throttle from 30 to 50 mph, then allow the vehicle to slow to 30 mph with the throttle closed. Repeat the procedure 10 or 12 times. This will load the piston rings and cause them to seat properly against the cylinder walls. Check again for oil and coolant leaks.

8   Drive the vehicle gently for the first 500 miles (no sustained high speeds) and keep a constant check on the oil level. It's not unusual for an engine to use oil during the break-in period.

9   At approximately 500 to 600 miles, change the oil and filter.

10   For the next few hundred miles, drive the vehicle normally. Don't pamper it or abuse it.

11   After 2000 miles, change the oil and filter again and consider the engine broken in.

Specifications

General

Displacement

Engine VIN codes E or L (1987 and earlier)	181 cubic inches (3.0L)
Engine VIN code B and 3	231 cubic inches (3.8L)
Engine VIN code C*	231 cubic inches (3.8L)
Engine VIN code L* (1991 through 1994)	231 cubic inches (3800)
Engine VIN code K* (1995 and later)	231 cubic inches (3800)

* Equipped with balance shaft

Cylinder compression pressure	Lowest reading cylinder must be 70% of highest reading cylinder (100 psi minimum)
Oil pressure (minimum)	
1998 and earlier	37 psi @ 2,400 rpm
1999 and later	60 psi @ 1850 rpm

Cylinder head

Warpage limit	0.0035 inch
Maximum allowable cut	0.010 in

Valves and related components

Minimum recommended valve margin	0.025 in
Intake valve	
Seat angle	45-degrees
Stem diameter	
Models through 2004	0.3412 to 0.3401 in
2005 models	0.3129 to 0.3136 in

Stem-to-guide clearance	
Models through 2001	0.0015 to 0.0035 in
2002 and later	0.0012 to 0.0028 in
Exhaust valve	
Seat angle	45-degrees
Stem diameter	0.3412 to 0.3401 in
Stem-to-guide clearance	
Models through 2001	0.0015 to 0.0032 in
2002 and later	0.0014 to 0.0029 in
Valve spring installed height	1.690 to 1.727 in
Valve spring free length	1.960 to 1.981 inch

Cylinder bore

Diameter	3.800 in
Maximum taper/out-of-round	0.001 in

Crankshaft and connecting rods

Connecting rod journal	
Diameter	2.2487 to 2.2499 in
Bearing oil clearance	0.0005 to 0.0026 in
Maximum taper/out-of-round	0.0003 in
Connecting rod side clearance (end play)	
1997 and earlier	0.003 to 0.015 in
1998 and later	0.004 to 0.020 in
Main bearing journal	
Diameter	2.4988 to 2.4998 in
Bearing oil clearance	0.0009 to 0.0018 in
Maximum taper/out-of-round	0.0003 in
Crankshaft end play (at thrust bearing)	0.003 to 0.011 in

Pistons and rings

Piston-to-bore clearance (new)	0.0004 to 0.0022 in
Piston ring end gap	
Compression rings	
Top ring	0.010 to 0.020 in
Second ring	
1997 and earlier	0.010 to 0.020 in
1998 and later	0.023 to 0.033 in
Oil ring	
1997 and earlier	0.015 to 0.055 in
1998 and later	0.010 to 0.030 in
Piston ring side clearance	
Compression rings	
1997 and earlier	0.003 to 0.005 in
1998 and later	0.0013 to 0.0031 in
Oil ring	
1997 and earlier	0.0035 in
1998 and later	0.0009 to 0.0079 in

Camshaft

Camshaft lobe lift	
Intake	
1997 and earlier	0.250 in
Exhaust	
Models through 2001	0.255 in
2002 and later models	0.258 in
Bearing journal diameter	
1997 and earlier	1.785 to 1.786 in
1998 and later	1.8448 to 1.8462 in
Bearing oil clearance	
1997 and earlier	
No. 1 journal	0.0005 to 0.0025 in
All others	0.0005 to 0.0035 in
1998 and later	0.0016 to 0.0047 in

Torque specifications*   Ft-lbs (unless otherwise indicated)

Main bearing cap bolts	
1985 through 1987	100
1988 through 1990	90
1991 through 1995 (except VIN K)	
Step one	26
Step two	Turn an additional 50-degrees
1995 VIN K, all 1996 and later models	
Step 1	52
Step 2	Loosen one complete turn (360-degrees)
Step 3	15
Step 4	30
Step 5	Tighten in three steps, 35 + 35 + 40-degrees, for a total of 110 degrees
Main bearing side bolts (1995 VIN K, all 1996 and later models)	
Step one	132 in-lbs
Step two	Turn an additional 45-degrees
Connecting rod nuts or bolts	
1985 through 1991	43
1992 and later	
Step one	20
Step two	Turn an additional 50-degrees
Balance shaft gear bolt	
Models through 2000	
Step 1	14
Step 2	Rotate an additional 35-degrees
2001 and later models	
Step 1	16
Step 2	Rotate an additional 70-degrees
Balance shaft retainer bolts	
Models through 2004	26
2005 models	22

* Note: Refer to Part A for additional torque specifications.

Section

Reference to other Chapters

3

COOLING, HEATING AND AIR CONDITIONING SYSTEMS

1 General information

ENGINE COOLING SYSTEM

All vehicles covered by this manual employ a pressurized engine cooling system with thermostatically controlled coolant circulation. An impeller type water pump mounted on the engine block pumps coolant through the engine and radiator. The coolant flows around each cylinder and back to the radiator. Cast-in coolant passages direct coolant around the intake and exhaust ports, near the spark plug areas and the exhaust valve guides.

A wax pellet type thermostat is located in a housing connected to the upper radiator hose. During warm up the closed thermostat prevents coolant from circulating through the radiator. As the engine nears normal operating temperature, the thermostat opens and allows hot coolant to travel through the radiator, where it's cooled before returning to the engine.

The cooling system is sealed by a pressure-type radiator cap, which raises the boiling point of the coolant and increases the cooling efficiency of the radiator. If the system pressure exceeds the cap pressure relief value, the excess pressure in the system forces the spring-loaded valve inside the cap off its seat and allows the coolant to escape through a hose into a coolant reservoir. When the system cools, the excess coolant is automatically drawn from the reservoir back into the radiator.

The coolant reservoir serves as both the point at which fresh coolant is added to the cooling system to maintain the proper level and as a holding tank for expelled coolant.

This type of cooling system is known as a closed design because coolant that escapes past the pressure cap is saved and reused.

HEATING SYSTEM

The heating system consists of a blower fan and heater core located in the heater box, the hoses connecting the heater core to the engine cooling system and the heater/air conditioning control head on the dashboard. Hot engine coolant is circulated through the heater core. When the heater mode is activated, a trap door opens to expose the heater box to the passenger compartment. A fan switch on the control head activates the blower motor, which forces air through the core, heating the air.

AIR CONDITIONING SYSTEM

The air conditioning system consists of a condenser mounted in front of the radiator, an evaporator mounted adjacent to the heater core, a compressor mounted on the engine, a filter-drier (accumulator), which contains a high pressure relief valve, and the hoses and lines connecting all of the above components.

A blower fan forces the warmer air of the passenger compartment through the evaporator core (sort of a radiator-in-reverse), transferring the heat from the air to the refrigerant. The liquid refrigerant boils off into low pressure vapor, taking the heat with it when it leaves the evaporator.

2 Antifreeze - general information

The cooling system should be filled with a water/ethylene glycol based antifreeze solution, which will prevent freezing down to at least -20-degrees F, or lower if local climate requires it. It also provides protection against corrosion and increases the coolant boiling point.

The cooling system should be drained, flushed and refilled at the specified intervals (see Chapter 1). Old or contaminated antifreeze solutions are likely to cause damage and encourage the formation of rust and scale in the system. Use distilled water with the antifreeze.

Before adding antifreeze, check all hose connections, because antifreeze tends to leak through very minute openings. Engines don't normally consume coolant, so if the level goes down, find the cause of the leak and correct it.

The exact mixture of antifreeze-to-water which you should use depends on the relative weather conditions. The mixture should contain at least 50 percent antifreeze, but should never contain more than 70 percent antifreeze. Consult the mixture ratio chart on the antifreeze container before adding coolant. Hydrometers are available at most auto parts stores to test the coolant. Use antifreeze which meets the vehicle manufacturer's specifications.

3 Thermostat - check and replacement

CHECK

1   Before assuming the thermostat is to blame for a cooling system problem, check the coolant level, drivebelt tension (see Chapter 1) and temperature gauge (or light) operation.

2   If the engine seems to be taking a long time to warm up (based on heater output or temperature gauge operation), the thermostat is probably stuck open. Replace the thermostat with a new one.

3   If the engine runs hot, use your hand to check the temperature of the upper radiator hose. If the hose isn't hot, but the engine is, the thermostat is probably stuck closed, preventing the coolant inside the engine from escaping to the radiator. Replace the thermostat.

3.9 Location of the thermostat cover clamp bolt (3.8L engine)

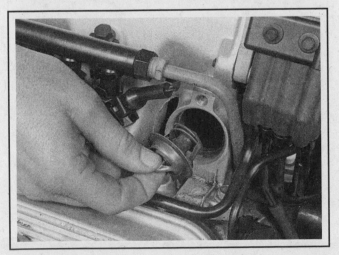

3.13 Install the thermostat with the spring end toward the engine

4   If the upper radiator hose is hot, it means the coolant is flowing and the thermostat is open. Consult the Troubleshooting section at the front of this manual for cooling system diagnosis.

REPLACEMENT

▶ **Refer to illustrations 3.9 and 3.13**

5   Disconnect the negative battery cable from the battery and drain the cooling system (see Chapter 1). If the coolant is relatively new or in good condition, save it and reuse it.

6   Follow the upper radiator hose to the engine to locate the thermostat housing.

7   Loosen the hose clamp, then detach the hose from the fitting. If the hose sticks, grasp it near the end with a pair of adjustable pliers and twist it to break the seal, then pull it off. If the hose is old or deteriorated, cut it off and install a new one.

➡ **Note: 3.0 liter engines have two hoses - remove both of them.**

8   If the outer surface of the large fitting that mates with the hose is deteriorated (corroded, pitted, etc.) it may be damaged further by hose removal. If it is, the thermostat cover will have to be replaced.

9   Remove the bolts and detach the thermostat cover (see illustration). If the cover is stuck, tap it with a soft-face hammer to jar it loose. Be prepared for some coolant to spill as the gasket seal is broken.

10 Note how it's installed (which end is facing up), then remove the thermostat.

11 Remove all traces of old gasket material and sealant from the housing and cover with a gasket scraper. Clean the gasket mating surfaces with lacquer thinner or acetone.

12 On 3800 engines, replace the gasket on the thermostat with the nubs facing up.

13 Install the new thermostat in the housing. Make sure the correct end faces up - the spring end is normally directed into the engine (see illustration).

14 On 3.0 liter engines, apply a thin, uniform layer of RTV sealant to both sides of the new gasket and position it on the housing. On 3.8 liter engines, install a new O-ring. On 3800 engines, install a new O-ring and gasket.

15 Install the cover and bolts. Tighten them to the torque figure listed in this Chapter's Specifications.

16 The remaining steps are the reverse of removal.

17 Refill the cooling system (see Chapter 1).

18 Start the engine and allow it to reach normal operating temperature, then check for leaks and proper thermostat operation (as described in Steps 2 through 4).

4   Engine cooling fans - check and replacement

CHECK

▶ **Refer to illustration 4.1**

1   To test the fan motor, unplug the electrical connector at the motor and use jumper wires to connect the fan directly to the battery (see illustration). If the fan still doesn't work, replace the motor.

2   If the motor tested OK, the fault lies in the coolant temperature switch, the relay or the wiring which connects the components. Carefully check all wiring and connections. If no obvious problems are found, further diagnosis should be done by a dealer service department or a repair shop.

REPLACEMENT

▶ **Refer to illustrations 4.6, 4.7 and 4.8**

3   Disconnect the negative battery cable from the battery.

※※※ **CAUTION:**

If the vehicle is equipped with a Delco Loc II or Theftlock audio system, make sure you have the correct activation code before disconnecting the battery. See the information at the front of this manual for the radio re-activation procedure.

4   Remove the fan guard and hose support.

5   Insert a small screwdriver into the connector to lift the lock tab and unplug the fan wire harness.

6   Unbolt the fan assembly (see illustration), then carefully lift it out of the engine compartment.

7   To detach the fan from the motor, remove the motor shaft nut (see illustration).

8   To remove the fan motor from the bracket, remove the mounting bolts/nuts (see illustration).

9   Installation is the reverse of removal.

4.1 The motor may be tested by running fused jumper wires directly from the battery to the connector (arrow)

4.6 Locations of the cooling fan mounting bolts (arrows)

4.7 Remove the nut (arrow) from the fan motor shaft, the pull the fan blade from the motor

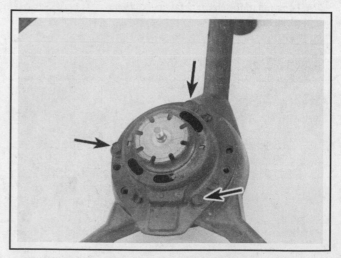

4.8 To detach the fan motor from the shroud, remove these three bolts (arrows)

5 Radiator - removal and installation

5.4a The transaxle cooler lines (arrows) are located in the upper . . .

5.4b . . . and lower corners of the radiator (viewed from below)

5.7 The coolant reservoir hose (arrow) connects to the right side of the radiator

▶ Refer to illustrations 5.4a, 5.4b, 5.7 and 5.9

❋❋ WARNING 1:

Wait until the engine is completely cool before beginning this procedure.

❋❋ WARNING 2:

Some models covered by this manual are equipped with an airbag system. Always disable the airbag system before working in the vicinity of the impact sensors, steering column or instrument panel to avoid the possibility of accidental deployment of the airbag(s), which could cause personal injury. See Chapter 12 for the airbag disarming procedure.

1 Disconnect the negative battery cable from the battery.

❋❋ CAUTION:

If the vehicle is equipped with a Delco Loc II or Theftlock audio system, make sure you have the correct activation code before disconnecting the battery. See the information at the front of this manual for the radio re-activation procedure.

2 Drain the cooling system (see Chapter 1). If the coolant is relatively new or in good condition, save it and reuse it.
3 Remove the air cleaner intake duct.
4 Disconnect the transaxle cooler lines from the radiator (see illustrations). Use a drip pan to catch spilled fluid.
5 Remove the engine cooling fan assemblies (see Section 4).
6 Loosen the hose clamps, then detach the radiator hoses from the fittings. If they're stuck, grasp each hose near the end with a pair of adjustable pliers and twist it to break the seal, then pull it off - be careful not to distort the radiator fittings! If the hoses are old or deteriorated, cut them off and install new ones.
7 Disconnect the reservoir hose from the radiator neck (see illustration).
8 Plug the lines and fittings.
9 Remove the upper radiator panel mounting bolts (see illustration).

Remove the bolts securing the radiator to the condensor if equipped. On 2001 and later models, remove the four bolts securing the tiebar, which is accessed once the support brackets (two bolts each) are removed.

10 Separate the radiator from the condensor if equipped and carefully lift out the radiator. Don't spill coolant on the vehicle or scratch the paint.
11 With the radiator removed, it can be inspected for leaks and damage. If it needs repair, have a radiator shop or dealer service department perform the work as special techniques are required.
12 Bugs and dirt can be removed from the radiator with compressed air and a soft brush. Don't bend the cooling fins as this is done.
13 Check the radiator mounts for deterioration and make sure there's nothing in them when the radiator is installed.
14 Installation is the reverse of the removal procedure.
15 After installation, fill the cooling system with the proper mixture of antifreeze and water. Refer to Chapter 1 if necessary.
16 Start the engine and check for leaks. Allow the engine to reach normal operating temperature, indicated by the upper radiator hose becoming hot. Recheck the coolant level and add more if required.
17 Check and add transmission fluid as needed.

5.9 Locations of the radiator mounting panel bolts (arrows)

6 Coolant reservoir - removal and installation

♦ **Refer to illustration 6.3**

1   Remove the battery (see Chapter 5).

✳✳ CAUTION:

If the vehicle is equipped with a Delco Loc II or Theftlock audio system, make sure you have the correct activation code before disconnecting the battery. See the information at the front of this manual for the radio re-activation procedure.

2   Disconnect the radiator overflow hose from the top of the radiator.

3   Remove the mounting bolts(s) and nuts (see illustration) and lift the coolant reservoir from the vehicle.

4   Installation is the reverse of removal.

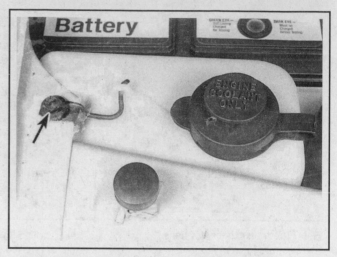

6.3 Remove the mounting bolt (arrow) to detach the coolant reservoir

7 Water pump - check

♦ **Refer to illustrations 7.4 and 7.5**

1   A failure in the water pump can cause serious engine damage due to overheating.

2   There are three ways to check the operation of the water pump while it's installed on the engine. If the pump is defective, it should be replaced with a new or rebuilt unit.

3   With the engine running at normal operating temperature, squeeze the upper radiator hose. If the water pump is working properly, a pressure surge should be felt as the hose is released.

✳✳ WARNING:

Keep your hands away from the fan blades!

4   Water pumps are equipped with weep or vent holes. If a failure occurs in the pump seal, coolant will leak from the hole (see illustration). In most cases you'll need a flashlight and mirror to find the hole on the under side of the water pump to check for leaks.

5   If the water pump shaft bearings fail there may be a howling sound coming from the drivebelt area while the engine is running. Shaft wear can be felt if the water pump pulley is rocked up-and-down (see illustration). Don't mistake drivebelt slippage, which causes a squealing sound, for water pump bearing failure.

7.4 If the pump is leaking, stains will form below the shaft (arrow) - pump removed for clarity

7.5 Rock the pulley back-and-forth to check for bearing play

8 Water pump - removal and installation

♦ **Refer to illustrations 8.4, 8.5 and 8.8**

✳✳ WARNING:

Wait until the engine is completely cool before beginning this procedure.

➥Note: It may be necessary to raise the engine to allow for pump removal (see Chapter 2A, Section 22).

1   Disconnect the negative battery cable from the battery.

✳✳ CAUTION:

If the vehicle is equipped with a Delco Loc II or Theftlock audio system, make sure you have the correct activation code before disconnecting the battery. See the information at the front of this manual for the radio re-activation procedure.

2   Drain the cooling system (see Chapter 1). If the coolant is relatively new or in good condition, save it and reuse it. Remove the windshield washer reservoir, if necessary.

3   If necessary, remove the power steering pump (without disconnecting the hoses) and position the pump aside. On models with an engine mount at the front of the engine, support the engine and remove the front engine mount (see Chapter 2A).

4   Loosen the water pump pulley mounting bolts (see illustration). Remove the drivebelt (see Chapter 1), then remove the water pump pulley.

5   Remove the bolts and detach the water pump from the engine (see illustration).

➥Note: On some models it may be necessary to remove a long bolt through an access hole in the frame rail.

6   Clean the fastener threads and any threaded holes in the engine to remove corrosion and sealant.

7   Compare the new pump to the old one to make sure they're identical.

8   Remove all traces of old gasket material from the engine with a gasket scraper (see illustration).

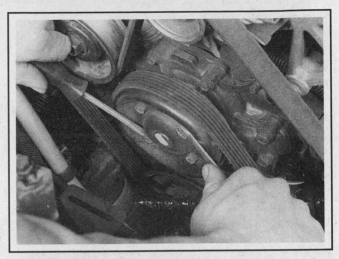

8.4 Loosen the water pump pulley bolts with the drivebelt in place

9   Clean the engine and water pump mating surfaces with lacquer thinner or acetone.

10 Apply a thin coat of RTV sealant to the engine side of the new gasket.

11 Apply a thin layer of RTV sealant to the gasket mating surface of the new pump, then carefully mate the gasket and the pump. Slip a couple of bolts through the pump mounting holes to hold the gasket in place.

12 Carefully attach the pump and gasket to the engine and start the bolts finger tight.

13 Tighten the bolts in 1/4-turn increments to the torque figure listed in this Chapter's Specifications. Start with all the short bolts, then tighten the longer bolts. Don't overtighten them or the pump may be distorted.

14 Reinstall all parts removed for access to the pump. Tighten the pulley bolts to the torque listed in this Chapter's Specifications.

15 Refill the cooling system (see Chapter 1). Run the engine and check for leaks.

8.5 Locations of the water pump mounting bolts (typical)

8.8 Remove all traces of old gasket material with a scraper

9 Coolant temperature sending unit - check and replacement

▶ Refer to illustration 9.1

❋❋ WARNING:

Wait until the engine is completely cool before beginning this procedure.

1   The coolant temperature indicator system is composed of a light or temperature gauge mounted in the instrument panel and a coolant temperature sending unit mounted on the engine (see illustration). Most vehicles have more than one sending unit, but the one used for the indicator system has only one wire. On 2002 and later models, the CTS is located at the side of the manifold, just below the thermostat housing.

2   If the light or gauge indicates the engine is overheating, check the coolant level in the system and then make sure the wiring between the light or gauge and the sending unit is secure and all fuses are intact.

3   When the ignition switch is turned on and the starter motor is turning, the indicator light (if equipped) should be on (overheated engine indication).

4   If the light isn't on, the bulb may be burned out, the ignition switch may be faulty or the circuit may be open. Test the circuit by grounding the wire to the sending unit while the ignition is on (engine not running for safety). If the gauge now deflects full scale or the light comes on, replace the sending unit.

5   As soon as the engine starts, the light should go out and remain out unless the engine overheats. Failure of the light to go out may be due to a grounded wire between the light and the sending unit, a defective sending unit or a faulty ignition switch. Check the coolant to make

9.1 On 3.0 and 3.8 liter engines, the coolant temperature sending unit is located at the drivebelt end of the intake manifold (arrow) (on 3800 engines, it's located on the other end)

sure it's the proper type. Plain water may have too low a boiling point to activate the sending unit.

6   If the sending unit must be replaced, simply unscrew it from the engine and install the replacement. Use a light coat of sealant on the threads. Make sure the engine is cool before removing the defective sending unit. There will be some coolant loss as the unit is removed, so be prepared to catch it. Check the level after the replacement has been installed.

10 Heater and air conditioner blower motor - removal and installation

▶ Refer to illustrations 10.2a, 10.2b, 10.3 and 10.6

1   Disconnect the cable from the negative battery terminal.

❋❋ CAUTION:

If the vehicle is equipped with a Delco Loc II or Theftlock audio system, make sure you have the correct activation code before disconnecting the battery. See the information at the front of this manual for the radio re-activation procedure.

1999 AND EARLIER

2   Working in the engine compartment, remove the plastic firewall cover (see illustrations).

3   Disconnect the electrical connector from the blower motor (see illustration).

4   Detach the blower motor cooling tube.

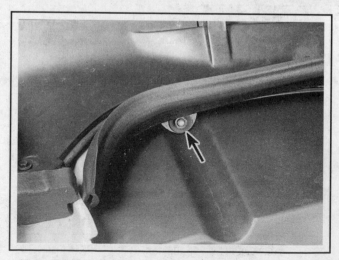

10.2a This nut . . .

10.2b . . . and this one secure the firewall cover

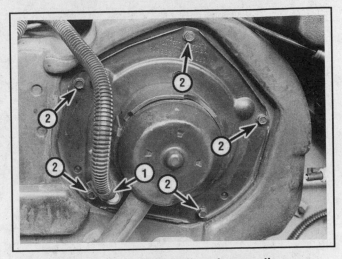

10.3 Detach the wiring harness clip and remove the mounting screws around the perimeter of the blower - later models have only 3 retaining screws

5　Remove the blower motor mounting bolts and separate the motor/fan assembly from the housing.

➡**Note: On some models it may be necessary to remove the rear rocker arm cover for access (see Chapter 2A).**

6　Remove the retaining nut and slide the fan off the motor shaft (see illustration).

7　Installation is the reverse of the removal procedure.

2000 AND LATER

8　Working in the passenger compartment, remove the right sound insulator panel from below the glove box, then remove the glove box (see Chapter 11).

9　Peel back the carpeting slightly and remove dash intregration module bracket retaining screws. On models through 2002, position the dash integration module aside without unplugging the electrical connectors to allow access to the blower motor. Position the dash integration module aside without unplugging the electrical connections to allow access the blower motor.

10　Disconnect the electrical connector from the blower motor.

11　Remove the blower motor mounting screws and separate the motor/fan assembly from the housing.

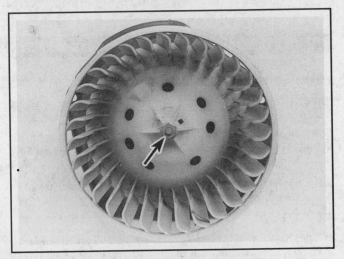

10.6 Remove the nut (arrow) and slip the fan off the motor shaft

11　Heater core - removal and installation

▸ **Refer to illustration 11.5**

❄❄ WARNING:

Some models covered by this manual are equipped with an airbag system. Always disable the airbag system before working in the vicinity of the impact sensors, steering column or instrument panel to avoid the possibility of accidental deployment of the airbag(s), which could cause personal injury. See Chapter 12 for the airbag disarming procedure.

1　Disconnect the battery cables (negative cable first, then the positive one).

❄❄ CAUTION:

If the vehicle is equipped with a Delco Loc II or Theftlock audio system, make sure you have the correct activation code before disconnecting the battery. See the information at the front of this manual for the radio re-activation procedure.

2　Raise the front of the vehicle and support it securely on jackstands. Apply the parking brake and block the rear wheels to keep the vehicle from rolling off the jackstands.

3　Drain the cooling system (see Chapter 1).

4　Remove the drain hose from the bottom of the heater case.

5   Working in the engine compartment, disconnect the heater hoses at the firewall (see illustration).

6   Remove the right sound insulator and steering column trim cover.

7   Disassemble the dash as necessary for access (see Chapter 11). Remove the heater outlet duct and glove box.

8   Remove the heater core cover. Pull straight back to avoid breaking the drain tube.

9   Remove the heater core clamps and remove the heater core.

10   Installation is the reverse of removal.

11.5 The heater hoses connect to the heater core in the right (passenger's) side of the engine compartment at the firewall - remove the plastic firewall cover for access

12   Heater and air conditioner control assembly - removal and installation

▶ Refer to illustrations 12.2, 12.3, 12.4 and 12.5

※※ WARNING:

Some models covered by this manual are equipped with an airbag system. Always disable the airbag system before working in the vicinity of the impact sensors, steering column or instrument panel to avoid the possibility of accidental deployment of the airbag(s), which could cause personal injury. See Chapter 12 for the airbag disarming procedure.

➡Note: Removal of the heater core is a difficult procedure for the home mechanic. Since it requires removal of the entire instrument panel and the cowl support structure behind the instrument panel. There are numerous electrical connectors and fasteners involved, some of which can be difficult to access, and we recommend that you have considerable mechanical experience before performing a heater core replacement.

1   Disconnect the negative battery cable from the battery.

※※ CAUTION:

If the vehicle is equipped with a Delco Loc II or Theftlock audio system, make sure you have the correct activation code before disconnecting the battery. See the information at the front of this manual for the radio re-activation procedure.

2   Remove the trim panel from around the heater and air conditioner control (see illustration).

3   Remove the two screws on each side of the heater and air conditioner control assembly (see illustration).

4   Pull the control out of the dash far enough to label the various wires and/or vacuum hoses.

➡Note: Do not disconnect the individual vacuum lines - disconnect the vacuum connector (if equipped) as a unit (see illustration).

5   On manually controlled units, carefully pry the cable housing and cable retaining clips off with a small screwdriver (see illustration).

6   Installation is the reverse of removal.

12.2 Remove the trim panel screws (arrows) and detach the panel (typical)

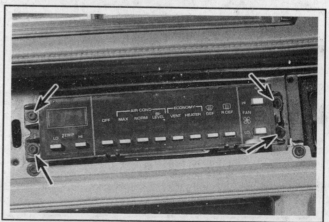

12.3 Remove the screws on both sides of the heater and air conditioner control (arrows) (typical)

12.4 Unplug the electrical and vacuum connectors from the back of the control panel - be sure to unplug the entire vacuum connector, not the individual hoses (typical manual control shown)

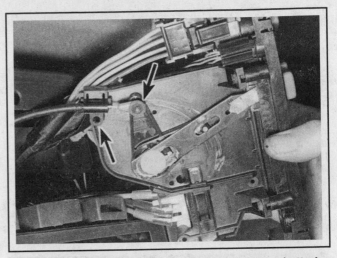

12.5 On manual systems, two mounting clips (arrows) attach the temperature control cable and cable housing to the heater and air conditioner control assembly (typical manual control shown)

13 Air conditioning system - check and maintenance

❊❊ WARNING:

The air conditioning system is under high pressure. DO NOT loosen any hose or line fittings or remove any components until after the system has been discharged by a dealer service department or service station. Always wear eye protection when disconnecting air conditioning system fittings.

➡ **Note 1:** The air conditioning system on 1994 and later models uses the non-ozone depleting refrigerant, referred to as R-134a. The R-134a refrigerant and its lubricating oil are not compatible with the R-12 system and under no circumstances should the two different types of refrigerant and lubricating oil be intermixed. If mixed, it could result in costly compressor failure due to improper lubrication.

➡ **Note 2:** Because of recent Federal regulations proposed by the Environmental Protection Agency, R-12 refrigerant is no longer available to the do-it-yourselfer.

➡ **Note 3:** Kits are available in many auto parts stores or at dealerships that can be used to convert your system from R-12 to R-134A refrigerant. The conversion is not difficult and usually includes all new O-rings, a new accumulator and fittings that go over your R-12 service ports to convert them to the R-134A type.

1 The following maintenance checks should be performed on a regular basis to ensure the air conditioner continues to operate at peak efficiency.

a) *Check the compressor drivebelt. If it's worn or deteriorated, replace it (see Chapter 1).*

b) *Check the system hoses. Look for cracks, bubbles, hard spots and deterioration. Inspect the hoses and all fittings for oil bubbles and seepage. If there's any evidence of wear, damage or leaks, replace the hose(s).*

c) *Inspect the condenser fins for leaves, bugs and other debris. Use a "fin comb" or compressed air to clean the condenser.*

d) *Make sure the system has the correct refrigerant charge.*

2 It's a good idea to operate the system for about 10 minutes at least once a month, particularly during the winter. Long term non-use can cause hardening, and subsequent failure, of the seals.

3 Because of the complexity of the air conditioning system and the special equipment necessary to service it, in-depth troubleshooting and repairs are not included in this manual (refer to the *Haynes Automotive Heating & Air Conditioning* manual). However, simple checks and component replacement procedures are provided in this Chapter.

4 The most common cause of poor cooling is simply a low system refrigerant charge. If a noticeable loss of cool air output occurs, one of the following quick checks may help you determine if the refrigerant level is low.

5 Warm the engine up to normal operating temperature.

6 Place the air conditioning temperature selector at the coldest setting and put the blower at the highest setting. Open the doors (to make sure the air conditioning system doesn't cycle off as soon as it cools the passenger compartment).

7 With the compressor engaged - the compressor clutch will make an audible click and the center of the clutch will rotate - feel the orifice tube located adjacent to the accumulator.

8 If a significant temperature drop is noticed, the refrigerant level is probably okay. Further inspection of the system is beyond the scope of the home mechanic and should be left to a professional.

9 If the inlet line has frost accumulation or feels cooler than the accumulator surface, the refrigerant charge is low and the system must be charged. Take the vehicle to a qualified shop.

14 Air conditioner accumulator - removal and installation

♦ Refer to illustrations 14.4 and 14.6

❋❋ WARNING:

The air conditioning system is under high pressure. DO NOT loosen any hose or line fittings or remove any components until after the system has been discharged by a dealer service department or service station. Always wear eye protection when disconnecting air conditioning system fittings.

1   Have the system discharged (see **Warning** above).
2   Disconnect the negative battery cable from the battery. On 2001 and later models, remove the nut securing the positive battery cable to the underhood fuse/relay box.

❋❋ CAUTION:

If the vehicle is equipped with a Delco Loc II or Theftlock audio system, make sure you have the correct activation code before disconnecting the battery. See the information at the front of this manual for the radio re-activation procedure.

3   Remove the accumulator cover if equipped.
4   Disconnect the refrigerant lines from the accumulator (see illustration). Use a back-up wrench to prevent twisting the tubing.
5   Plug the open fittings to prevent entry of dirt and moisture.
6   Loosen the mounting bracket bolt (see illustration) and remove the accumulator.
7   If a new accumulator is being installed, remove the Schrader valve and pour the existing oil out into a measuring cup, noting the amount. Add the correct type of fresh refrigerant oil to the new accumulator, equal to the amount removed from the old unit, plus the following amount:

a) *1985 through 1993 models: plus one ounce of fresh 525 viscosity refrigerant oil.*
b) *1994 and later models: plus one ounce of polyalkaline glycol (PAG) refrigerant oil.*

8   Installation is the reverse of removal.
9   Have the system evacuated, recharged and leak tested by the shop that discharged it.

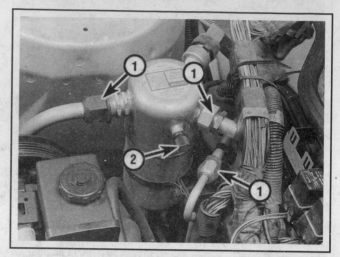

14.4 The accumulator is located on the passenger side of the engine compartment (1999 and earlier shown, 2000 and later models similar)

1   Refrigerant line fittings   *2   Charging port*

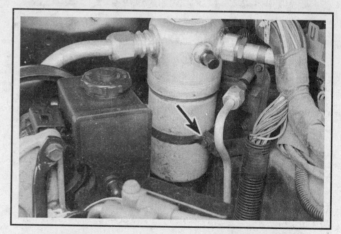

14.6 Location of the accumulator bracket bolt (arrow)

15 Air conditioner compressor - removal and installation

♦ Refer to illustration 15.6

❋❋ WARNING:

The air conditioning system is under high pressure. DO NOT loosen any hose or line fittings or remove any components until after the system has been discharged by a dealer service department or service station. Always wear eye protection when disconnecting air conditioning system fittings.

➡Note: The accumulator (see Section 14) should be replaced whenever the compressor is replaced.

1   Have the system discharged (see **Warning** above).
2   Disconnect the negative battery cable from the battery.

❋❋ CAUTION:

If the vehicle is equipped with a Delco Loc II or Theftlock audio system, make sure you have the correct activation code before disconnecting the battery. See the information at the front of this manual for the radio re-activation procedure.

3   Set the parking brake and block the rear wheels. Raise the front of the vehicle and support it securely on jackstands.
4   Remove the right lower splash shield. Also remove the inner fenderwell liner.
5   Refer to Chapter 1, and remove the drivebelt.
6   Disconnect the compressor clutch wiring (see illustration).
7   Disconnect the refrigerant lines from the rear of the compressor.

15.6 Disconnect the compressor clutch wiring (arrow)

Plug the open fittings to prevent entry of dirt and moisture.

8   Unbolt the compressor from the mounting brackets and lift it out of the vehicle.

9   If a new compressor is being installed, follow the directions with the compressor regarding the draining of excess oil prior to installation.

10 The clutch may have to be transferred from the original to the new compressor.

11 Installation is the reverse of removal. Replace all O-rings with new ones specifically made for A/C system use and lubricate them with refrigerant oil.

12 Have the system evacuated, recharged and leak tested by the shop that discharged it.

16 Air conditioner condenser - removal and installation

▶ Refer to illustrations 16.4 and 16.6

✳✳ WARNING 1:

The air conditioning system is under high pressure. DO NOT loosen any hose or line fittings or remove any components until after the system has been discharged by a dealer service department or service station. Always wear eye protection when disconnecting air conditioning system fittings.

✳✳ WARNING 2:

Some models covered by this manual are equipped with an airbag system. Always disable the airbag system before working in the vicinity of the impact sensors, steering column or instrument panel to avoid the possibility of accidental deployment of the airbag(s), which could cause personal injury. See Chapter 12 for the airbag disarming procedure.

➡Note: The accumulator (see Section 14) should be replaced whenever the condenser is replaced.

1   Have the system discharged (see **Warning** above).
2   On models through 2000, remove the battery (see Chapter 5).

✳✳ CAUTION:

If the vehicle is equipped with a Delco Loc II or Theftlock audio system, make sure you have the correct activation code before disconnecting the battery. See the information at the front of this manual for the radio re-activation procedure.

3   Remove the cooling fans (see Section 4).
4   Disconnect the refrigerant lines from the condenser (see illustration). Plug the lines to keep dirt and moisture out.
5   Remove the upper radiator panel from above the condenser.
6   Tilt the top of the radiator toward the engine and remove the condenser mounting bolts (see illustration).

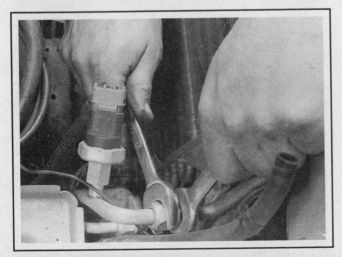

16.4 Use a back-up wrench to loosen/tighten the refrigerant lines

16.6 Remove the condenser mounting bolts (upper arrows), then carefully pull the condenser core back and lift it free of its rubber insulators (lower arrows)

7 Remove the condenser from the vehicle.

8 If the original condenser will be reinstalled, plug the line fittings to prevent oil from draining out.

9 If a new condenser is being installed, pour one ounce of refrigerant oil into it prior to installation.

10 Reinstall the components in the reverse order of removal. Be sure the rubber pads are in place under the condenser.

11 Have the system evacuated, recharged and leak tested by the service facility that discharged it.

2001 AND LATER MODELS

12 Remove the cover/seal over the top of the radiator.

13 Refer to Chapter 12 and remove the headlight housings.

14 Remove the bolts at the top of the radiator tiebar, and loosen the tiebar bolts in the headlight housing area and at the fascia to tiebar joints.

15 Remove the radiator upper brackets and disconnect the refrigerant hoses at the condenser.

16 The panel that mounts the headlight housings and front bumper fascia to the vehicle must be loosened enough to pull the panel forward for access to the bolts securing the condenser to the radiator.

17 Remove the condenser-to-radiator bolts (two) and pull the condenser up and out of the rubber grommets securing it at the bottom.

➡**Note: The two condenser-to-radiator bolts are a special length, and only these bolts should be used in this application.**

18 Installation is the reverse of the removal procedure.

Specifications

General

Cooling system capacity	See Chapter 1
Radiator cap pressure rating	See Chapter 1
Thermostat rating	195 degrees F
Drivebelt tension	See Chapter 1

Torque specifications

	Ft-lbs (unless otherwise indicated)
Thermostat cover bolts	15 to 20
Water pump mounting bolts	
Long bolts	22 to 29
Short bolts	96 to 132 in-lbs
Water pump pulley bolts	108 to 132 in-lbs

Section

Reference to other Chapters

4

FUEL AND EXHAUST SYSTEMS

1 General information

The fuel system consists of a fuel tank, an electric fuel pump, a fuel pump relay, an air cleaner assembly and a Port Fuel Injection (PFI) system. The PFI system is used on all V6 engines.

PORT FUEL INJECTION (PFI) SYSTEM

The PFI system utilizes six injectors, one for each cylinder. Each one has an electric solenoid that controls the amount of fuel injected. Fuel pressure at each injector is kept at a constant pressure by a pressure regulator, which returns excess fuel to the fuel tank. One injector is installed above each intake port. The throttle body serves only to control the amount of air passing into the system. Because each cylinder is equipped with an injector mounted immediately adjacent to the intake valve, much better control of the fuel/air mixture ratio is possible. Later models are equipped with an improved fuel injection system called SMPI, which stands for Sequential Multi-port Injection.

FUEL PUMP AND LINES

Fuel is circulated from the fuel tank to the fuel injection system, and back to the fuel tank, through a pair of metal lines running along the underside of the vehicle. An electric fuel pump is attached to the fuel sending unit inside the fuel tank. To reduce the likelihood of vapor lock, a vapor return system routes all vapors and hot fuel back to the fuel tank through a separate return line.

EXHAUST SYSTEM

The exhaust system includes an exhaust manifold fitted with an exhaust oxygen sensor, a catalytic converter, an exhaust pipe, and a tri-flow muffler. Later models have more than one oxygen sensor in the system.

The catalytic converter is an emission control device added to the exhaust system to reduce pollutants. A single-bed converter design is used in combination with a three-way (reduction) catalyst. Refer to Chapter 6 for more information regarding the catalytic converter.

2 Fuel pressure relief procedure

✖✖ WARNING:

Gasoline is extremely flammable, so take extra precautions when you work on any part of the fuel system. Don't smoke or allow open flames or bare light bulbs near the work area, and don't work in a garage where a gas-type appliance (such as a water heater or a clothes dryer) is present. Since gasoline is carcinogenic, wear latex gloves when there's a possibility of being exposed to fuel, and, if you spill any fuel on your skin, rinse it off immediately with soap and water. Mop up any spills immediately and do not store fuel-soaked rags where they could ignite. The fuel system is under constant pressure, so, if any fuel lines are to be disconnected, the fuel pressure in the system must be relieved first. When you perform any kind of work on the fuel system, wear safety glasses and have a Class B type fire extinguisher on hand.

→Note: After the fuel pressure has been relieved, it's a good idea to use a shop towel around any fuel connection to absorb any residual fuel that may spray when servicing the fuel system.

1 Before servicing any fuel system component, you must relieve the fuel pressure to minimize the risk of fire or personal injury.

2 Remove the cap from the gas tank - this will relieve any pressure built up in the tank.

3 On early models, remove the fuse marked FP (Fuel Pump) from the fuse block. On later models, the fuel pump fuse is located in the relay center on the right hand side of instrument panel (see Chapter 12).

4 Crank the engine over. It will start and run until the fuel supply remaining in the fuel lines is used. When the engine stops, engage the starter again for another three seconds to assure that any remaining pressure is dissipated.

5 With the ignition turned to Off, replace the fuel pump fuse. Disconnect the negative battery cable terminal to avoid any accidental fuel discharge if a attempt is made to start the engine.

6 Reconnect the negative battery cable.

3   Fuel pump/fuel pressure - testing

➡**Note: In order to perform the fuel pressure test, you will have to obtain a fuel pressure gauge and adapter set for the fuel injection system being tested.**

PRELIMINARY INSPECTION

1   If the fuel system fails to deliver the proper amount of fuel, or any fuel at all, to the fuel injection system, inspect it as follows.

2   Always make certain that there is fuel in the tank.

3   With the engine running, inspect for leaks at the threaded fittings at both ends of the fuel line (see Chapter 1). Tighten any loose connections. Inspect all hoses for flattening or kinks which would restrict the flow of fuel.

PRESSURE CHECK

▶ **Refer to illustration 3.5**

4   Relieve the fuel system pressure (see Section 2).

5   Install a fuel pressure gauge at the Schrader valve on the fuel rail (see illustration).

6   Turn the ignition switch on with the air conditioning off. The fuel pump should run for about two seconds. Note the reading on the pressure gauge. After the pump stops running the pressure should hold steady. It should be within the specified amount.

7   Start the engine and let it idle at normal operating temperature. The pressure should remain constant, although it might drop slightly from the previous reading. This is normal.

8   If the pressure reading was not within specification, disconnect the vacuum hose at the pressure regulator. Using a hand vacuum pump, apply 12-inches hg of vacuum to the pressure regulator. If the fuel pressure is now within specification, repair the vacuum source to the

regulator. If the regulator does not hold vacuum, replace the regulator.

9   If the fuel pressure is not within specification, check the following:

 a) If the pressure is higher than specified, check for a faulty regulator or a pinched or clogged fuel return hose or pipe.

 b) If the pressure is lower than specified:
 1) Inspect the fuel filter - make sure it's not clogged.
 2) Look for a pinched or clogged fuel hose between the fuel tank and the fuel rail.
 3) Check the pressure regulator for a malfunction.
 4) Look for leaks in the fuel line.
 5) Look for a pinched, broken or disconnected regulator vacuum hose.
 6) Check for leaking injectors.

10 After the testing is done, relieve the fuel pressure (see Section 2) and remove the fuel pressure gauge.

FUEL PUMP ELECTRICAL CIRCUIT CHECK

▶ **Refer to illustration 3.13**

➡**Note: Refer to Chapter 12 for additional wiring schematics that detail the fuel pump relay and circuit.**

11 If you suspect a problem with the fuel pump, verify the pump actually runs: Have an assistant turn the ignition switch to On - you should hear a brief whirring noise as the pump comes on and pressurizes the system If the pump does not turn on (makes no sound) with the ignition switch in the ON position, check the fuel pump fuse (see Chapter 12 for fuse locations). If the fuse is blown, replace the fuse and see if the pump works. If the pump now works, check for a short in the circuit between the fuel pump relay and the fuel pump.

12 If the pump still does not work, check the fuel pump relay circuit. With the help of an assistant cycle the ignition key ON and OFF (engine not running), while checking for battery voltage at the relay connector. There should be battery voltage at one terminal at all times and battery voltage at another terminal for several seconds after the ignition switch is turned On. If battery voltage does not exist, check the circuit from the PCM to the relay.

3.5 Connect the fuel pressure gauge to the Schrader valve on the fuel rail

3.13 On early models, the fuel pump relay is located on the firewall - on later models, the fuel pump relay is located in the engine compartment fuse box

➡Note 1: If oil pressure drops below the specified pressure level, the oil pressure switch will act as a fuel pressure cut-off device. Be sure to check the oil pressure switch and circuit in the event of a difficult problem diagnosing the fuel pump circuit (refer to the wiring diagrams at the end of Chapter 12).

➡Note 2: The theft deterrent system (if equipped) is equipped with a fuel enable circuit. If this system is malfunctioning it will not allow the PCM to signal the fuel pump relay or the engine to crank over. Be sure to check the theft deterrent system and circuit in the event of a difficult problem diagnosing the fuel pump circuit.

13 If battery voltage exists, check the relay (see Chapter 12) or replace the relay with a known good relay and retest (see illustration). If necessary, have the relay checked by a qualified automotive electrical specialist.

14 If the fuel pump does not activate, check for power to the fuel pump at the fuel tank. Access to the fuel pump is difficult, but it is possible to check for battery voltage at the electrical connector near the tank. If voltage is present at the fuel pump connector, replace the fuel pump.

4   Fuel lines and fittings - repair and replacement

▶ Refer to illustrations 4.12, 4.18 and 4.19

❋❋ WARNING:

Gasoline is extremely flammable, so extra precautions must be taken when working on any part of the fuel system. See the Warning in Section 2.

1   Always relieve the fuel pressure before servicing fuel lines or fittings (see Section 2).

2   The fuel feed and return lines extend from the in-tank fuel pump to the engine compartment. The lines are secured to the underbody with clip and screw assemblies. Both fuel feed lines must be occasionally inspected for leaks, kinks and dents.

3   If evidence of dirt is found in the system or fuel filter during disassembly, the line should be disconnected and blown out. Check the fuel strainer on the fuel gauge sending unit (see Section 7) for damage and deterioration.

STEEL TUBING

4   If replacement of a fuel line or emission line is called for, use welded steel tubing meeting the manufacturer's specifications.

5   Don't use copper or aluminum tubing to replace steel tubing. These materials cannot withstand normal vehicle vibration.

6   Because fuel lines used on fuel injected vehicles are under high pressure, they require special consideration.

7   Most fuel lines have threaded fittings with O-rings. Any time the fittings are loosened to service or replace components, make sure that:

   a) *A back-up wrench is used while loosening and tightening the fittings.*

   b) *Check all O-rings for cuts, cracks and deterioration. Replace any that appear worn or damaged.*

   c) *If the lines are replaced, always use original equipment parts, or parts that meet the manufacturer's standards.*

FLEXIBLE HOSE

8   When rubber hose is used to replace a metal line, use reinforced, fuel resistant hose with the word "Fluoroelastomer" imprinted on it (or hose meeting the same standards). Hose(s) not clearly marked like this

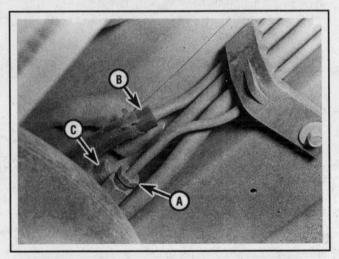

4.12 The fuel feed (A), fuel return (B) and vapor return (C) lines at the fuel tank

could fail prematurely and could fail to meet Federal emission standards. Hose inside diameter must match line outside diameter. Plastic hoses on later models should only be replaced with genuine factory parts.

9   Don't use flexible hose within four inches of any part of the exhaust system or within ten inches of the catalytic converter. Metal lines and flexible hoses must never be allowed to chafe against the frame. A minimum of 1/4-inch clearance must be maintained around a line or hose to prevent contact with the frame.

REMOVAL AND INSTALLATION

10 The following procedure and accompanying illustrations are typical for vehicles covered by this manual.

11 Relieve the fuel pressure, if not already done (see Section 2).

12 Disconnect the fuel feed, return or vapor line at the fuel tank (see illustration).

13 Detach the bracket from the rear crossmember.

14 Detach the bracket from the rear end of the left frame member, just in front of the left rear wheel.

15 Detach the three brackets from the left frame member.

4.18 Typical fuel feed and return line threaded fittings in the left side of the engine compartment

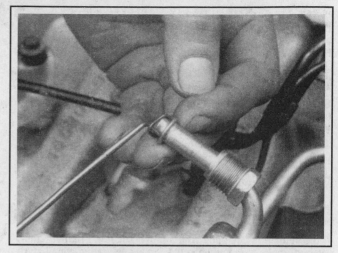

4.19 Always replace the fuel line O-rings (if equipped)

16 Detach the bracket from the left front end of the left frame member, just behind the left front wheel.

17 Detach the bracket from the lower left rear corner of the engine compartment.

18 Detach the threaded fitting(s) that attach the metal lines to the engine compartment fuel hoses (see illustration). Later models use quick-disconnect fittings, which are released by squeezing the plastic tabs and pulling the fitting apart.

19 Installation is the reverse of removal. Be sure to use new O-rings at the fittings (see illustration).

REPAIR

20 In repairable areas, cut a piece of fuel hose four inches longer than the portion of the line removed. If more than a six inch length of line is removed, use a combination of steel line and hose so hose lengths won't be more than ten inches. Always follow the same routing as the original line.

21 Cut the ends of the line with a tube cutter. Using the first step of a double flaring tool, form a bead on the end of both line sections. If the line is too corroded to withstand bead operation without damage, the line should be replaced.

22 Use a screw type hose clamp. Slide the clamp onto the line and push the hose on. Tighten the clamps on each side of the repair.

23 Secure the lines properly to the frame to prevent chafing.

Quick-disconnect fittings

▶ **Refer to illustration 4.25**

24 Twist the connector 1/4 turn in each direction to loosen any debris within the connector, then blow out the debris from the connector.

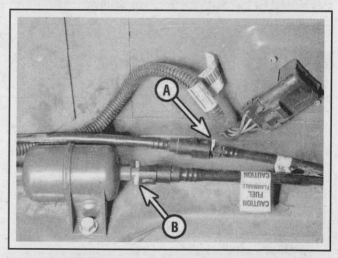

4.25 Some models are equipped with fuel lines that can be disconnected by pinching the tabs and separating each connector

A Fuel return line	B Fuel feed line

25 Squeeze the plastic tabs on the male end of the connector and pull it apart (see illustration).

26 Wipe off the male end of the connector, then apply a drop of clean engine oil onto the fuel line male connector where it is inserted into the female connector.

27 Insert the female connector onto the male connector and push it on until the plastic tabs snap and lock into place.

28 Pull on the female end of the connector to make sure it is securely locked into place.

5   Fuel tank - removal and installation

▶ **Refer to illustrations 5.6 and 5.8**

✳ WARNING:

Gasoline is extremely flammable, so extra precautions must be taken when working on any part of the fuel system. See the Warning in Section 2.

1   Relieve the fuel pressure (see Section 2).
2   Detach the cable from the negative terminal of the battery.

✳ CAUTION:

If the vehicle is equipped with a Delco Loc II or Theftlock audio system, make sure you have the correct activation code before disconnecting the battery. See the information at the front of this manual for the radio re-activation procedure.

3   If the tank is full or nearly full, either drain it into an approved gasoline container or drive the vehicle to use up the gas.
4   Raise the vehicle and place it securely on jackstands.
5   Loosen the exhaust heat shield at the rear, then remove the exhaust pipe hanger and lower the exhaust pipe down. On 2001 and later models, Refer to Chapter 10 and unbolt and remove the rear suspension support assembly to allow fuel tank removal.
6   Disconnect the fuel feed and return lines, the vapor return line and the filler neck and vent tubes (see illustration). Disconnect the electrical connector at the fuel pump/sending unit.
7   Support the fuel tank with a floor jack.
8   Disconnect both fuel tank retaining straps (see illustration) then lower the tank enough to disconnect the wires and ground strap from the fuel pump/fuel gauge sending unit.
9   If there's any fuel left in the tank, siphon it out at the fuel filler neck.
10   Remove the tank from the vehicle.
11   Installation is the reverse of removal.

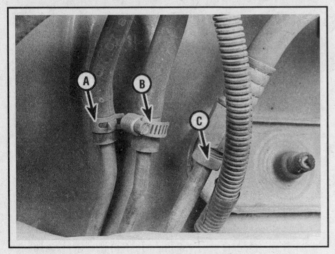

5.6 The vapor (A), return (B) and fuel feed (C) lines

5.8 Location of the fuel tank mounting strap bolts

6   Fuel tank cleaning and repair - general information

1   All repairs to the fuel tank or filler neck should be done by an experienced professional - the job is potentially dangerous! Even after cleaning and flushing of the fuel system, explosive fumes can remain and ignite during repair of the tank.

2   If the fuel tank is removed from the vehicle, it shouldn't be stored in an area where sparks or open flames could ignite the fumes coming out of it. Be especially careful inside a garage where a gas-type appliance is located, because it could cause an explosion.

7   Fuel pump - removal and installation

▶ **Refer to illustrations 7.5, 7.6a, 7.6b, 7.8, 7.9a and 7.9b**

❋❋ WARNING:

Gasoline is extremely flammable, so take extra precautions when you work on any part of the fuel system. See the Warning in Section 2.

1   Relieve the fuel system pressure (see Section 2).
2   Disconnect the cable from the negative battery terminal.

❋❋ CAUTION:

On models equipped with a Delco Loc II or Theftlock audio system, be sure the lockout feature is turned off before performing any procedure which requires disconnecting the battery.

3   The fuel pump/sending unit assembly is located inside the fuel tank on all models covered by this manual. On 1999 and earlier models, it will be necessary to remove the fuel tank from the vehicle to access the fuel pump (see Section 5). On 2000 and later models,

remove the finish panels from the trunk and detach the fuel pump access cover from the trunk floor. Also disconnect the electrical connectors and the fuel lines from the fuel pump module assembly.
4   The fuel pump/sending unit assembly is held in place by a cam lock ring mechanism consisting of an inner ring with three locking cams and a fixed outer ring with three retaining tangs.
5   To unlock the fuel pump/sending unit assembly, turn the inner ring counterclockwise until the locking cams are free of the retaining tangs (see illustration).
6   Lift the fuel pump/sending unit assembly out of the tank (see illustrations).

❋❋ CAUTION:

The fuel level float and sending unit are delicate. Don't bump them during removal or the accuracy of the sending unit may be affected.

7   Inspect the condition of the O-ring around the opening of the tank. If it is dried, cracked or deteriorated, replace it.
8   Remove the strainer from the lower end of the fuel pump (see illustration). If it's dirty, remove it, clean it with solvent and blow it out

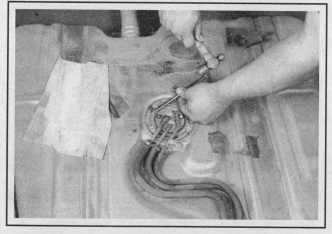

7.5 Use a brass punch to tap the lock-ring counterclockwise until the tabs align the recessed areas of the fuel tank

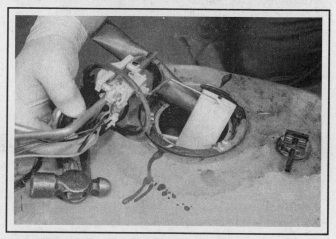

7.6a Carefully angle the fuel pump assembly out of the fuel tank

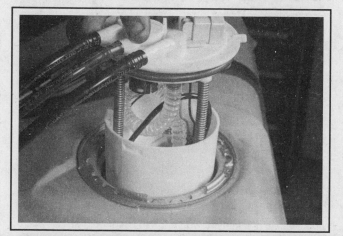

7.6b When removing the fuel pump module on 2000 and later models, have an approved container and shop rags available to catch fuel as the module is tilted out of the fuel tank

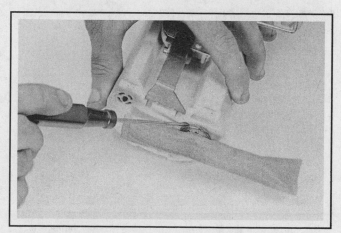

7.8 Inspect the fuel strainer for dirt; if it's too dirty to be cleaned, carefully pry the fuel strainer from the inlet pipe

7.9a On 1999 and earlier models, lift the tab to release the electrical connector from the fuel pump . . .

with compressed air. If it's too dirty to be cleaned, replace it.

9   On 1999 and earlier models (except 1997 and later Park Avenue), disconnect the electrical connector from the fuel pump and separate the fuel pump from the bracket (see illustrations). Installation of a new fuel pump is the reverse of removal.

10   On 1997 and later Park Avenue and all 200 and later models, the fuel pump is not serviced separately from the module assembly. In the event of failure, the complete module must be replaced. Transfer the fuel pressure sensor and the fuel level sending unit to the new fuel pump module assembly.

11   Position a new rubber O-ring around the opening in the fuel tank and guide the fuel pump/sending unit assembly into the tank.

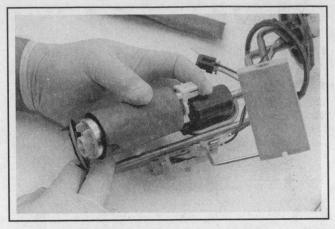

7.9b . . . then remove the fuel pump from the bracket

❄❄ **CAUTION:**

Do not fold or bend the fuel strainer during pump installation or fuel flow will be restricted.

12   Turn the inner lock ring clockwise until the locking cams are fully engaged by the retaining tangs.

➥**Note: Since you've installed a new O-ring, it may be necessary to push down on the inner lock ring until the locking cams slide under the retaining tangs.**

13   On 1999 and earlier models, install the fuel tank (see Section 5). On 2000 and later models, install the fuel pump access cover and the trunk finishing panels.

| **8   Air filter housing assembly - removal and installation** |

1985 THROUGH 1991 MODELS

▶ **Refer to illustrations 8.2 and 8.4**

1   Detach the cable from the negative terminal of the battery.

8.2 To remove the Mass Air Flow (MAF) sensor, unplug the electrical connectors (A), then loosen the hose clamps (B) (1985 through 1991 models)

❄❄ **CAUTION:**

If the vehicle is equipped with a Delco Loc II or Theftlock audio system, make sure you have the correct activation code before disconnecting the battery. See the information at the front of this manual for the radio re-activation procedure.

2   Unplug the electrical connector and unclamp the air duct from the Mass Air Flow (MAF) sensor, and remove the MAF sensor from the vehicle (see illustration).

3   Remove the air filter element (see Chapter 1).

4   Remove the air filter housing bolts and nuts (see illustration).

5   Installation is the reverse of removal.

1992 AND LATER MODELS

6   Detach the cable from the negative terminal of the battery.

❄❄ **CAUTION:**

If the vehicle is equipped with a Delco Loc II or Theftlock audio system, make sure you have the correct activation code before disconnecting the battery. See the information at the front of this manual for the radio re-activation procedure.

8.4 To remove the air cleaner housing, remove the nuts and bolts (arrows) and slide the assembly off the upper studs (1985 through 1991 models)

7 Remove the air filter element from the air cleaner housing (see Chapter 1).

8 Disconnect the electrical connector from the Intake Air Temperature (IAT) sensor.

➡**Note: On earlier models, the IAT sensor is mounted in the air cleaner housing. On later models the IAT sensor is mounted in the air intake duct.**

9 Detach the air intake duct from the throttle body, then remove upper air filter cover and the air intake duct as an assembly from the engine compartment.

10 On 1992 through 1999 models, remove the screws securing the lower air cleaner housing to the chassis and remove it from the engine compartment.

11 On 2000 and later models, release two clips and remove the PCM from the lower air cleaner housing without disconnecting the electrical connectors and position it aside. Pry the lower air cleaner housing out from the rubber retaining grommets and remove it from the engine compartment.

12 Installation is the reverse of removal.

9 Fuel injection system - general information

▶ **Refer to illustrations 9.4 and 9.5**

Electronic fuel injection provides optimum fuel/air mixture ratios at all stages of combustion and offers immediate throttle response characteristics. It also enables the engine to run at the leanest possible fuel/air mixture ratio, reducing exhaust gas emissions.

These vehicles are fitted with a Port Fuel Injection (PFI) system. This system is controlled by an Electronic Control Module (ECM), which monitors engine performance and adjusts the air/fuel mixture accordingly.

An electric fuel pump located in the fuel tank with the fuel gauge sending unit pumps fuel to the fuel injection system through the fuel feed line and an in-line fuel filter. A pressure regulator keeps fuel available at a constant pressure. Fuel in excess of injector needs is returned to the fuel tank by a separate line.

The throttle body (see illustration) has a throttle valve to control the amount of air delivered to the engine. The Throttle Position Sensor (TPS) and Idle Air Control (IAC) valves are located on the throttle body.

The fuel rail is mounted on the top of the engine. It distributes fuel to the individual injectors (see illustration).

Fuel is delivered to the input end of the rail by the fuel lines goes through the rail and then to the pressure regulator. The regulator keeps the pressure to the injectors at a constant level.

The remaining fuel is returned to the fuel tank.

9.4 Typical PFI throttle body assembly

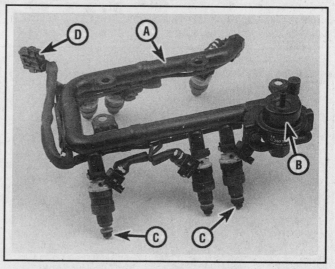

9.5 Typical PFI fuel rail assembly (VIN L shown)

A Fuel rail
B Fuel pressure regulator
C Fuel injectors
D Fuel injector harness electrical connector

10 Fuel injection system - check

❊❊ WARNING:

Gasoline is extremely flammable, so extra precautions must be taken when working on any part of the fuel system. See the Warning in Section 2.

➡**Note: The following procedure is based on the assumption that the fuel pump and fuel pressure are normal (see Section 3).**

1 Check the ground wire connections on the intake manifold for tightness. Check all wiring harness connectors that are related to the system. Loose connectors and poor grounds can cause many problems that resemble more serious malfunctions.

2 Check to see that the battery is fully charged, as the control unit and sensors depend on an accurate supply voltage in order to properly meter the fuel.

3 Check the air filter element - a dirty or partially blocked filter will severely impede performance and economy (see Chapter 1).

4 If a blown fuse is found, replace it and see if it blows again. If it does, search for a grounded wire in the harness to the fuel pump.

5 Check the air intake duct from the air mass sensor to the intake manifold for leaks, which will result in an excessively lean mixture. Also check the condition of all of the vacuum hoses connected to the intake manifold.

6 Remove the air intake duct from the throttle body and check for dirt, carbon or other residue build-up. If it's dirty, clean it with carburetor cleaner and a toothbrush.

7 With the engine running, place a screwdriver against each injector, one at a time, and listen through the handle for a clicking sound, indicating operation.

8 The remainder of the system checks should be left to a dealer service department or other qualified repair shop, as there is a chance that the control unit may be damaged if not performed properly.

11 Port Fuel Injection (PFI) - component removal and installation

❊❊ WARNING 1:

Gasoline is extremely flammable, so extra precautions must be taken when working on any part of the fuel system. See the Warning in Section 2.

❊❊ WARNING 2:

Before servicing an injector, fuel rail or pressure regulator, always relieve the fuel pressure in the fuel system to minimize the risk of fire and injury. After servicing the system, always cycle the ignition on and off several times (wait 10 seconds between cycles) and check the system for leaks.

IDLE AIR CONTROL (IAC) VALVE

▶ **Refer to illustrations 11.2, 11.3 and 11.7**

1 Detach the cable from the negative terminal of the battery.

❊❊ CAUTION:

If the vehicle is equipped with a Delco Loc II or Theftlock audio system, make sure you have the correct activation code before disconnecting the battery. See the information at the front of this manual for the radio re-activation procedure.

2 Unplug the electrical connector from the IAC valve assembly (see illustration).

3 On early models, unscrew the IAC valve assembly from the idle

11.2 There are two electrical connectors plugged into the throttle body - the IAC valve (A) and the TPS switch (B). The connector below the throttle body is for the knock sensor (C)

11.3 On early models, to remove the IAC valve from the throttle body, use a large adjustable wrench - if there is thread locking compound on the threads and you are going to install the same IAC, don't remove the thread compound

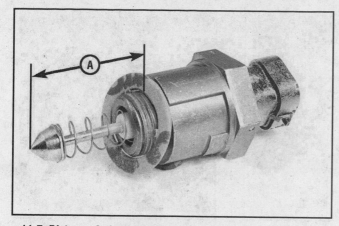

11.7 Distance A should be less than 1-1/8 inch - if it isn't, determine which kind of IAC valve you have and adjust it accordingly

air/vacuum signal housing assembly (see illustration). Remove the IAC valve assembly gasket and discard it.

> ❊ **CAUTION:**
>
> **Do not remove any thread locking compound from the threads.**

4   On later models, remove the two screws securing the IAC valve to the throttle body assembly. Remove the IAC valve assembly O-ring and discard it.

5   Clean the gasket mounting surface of the idle air/vacuum signal housing assembly to ensure a good seal.

> ❊ **CAUTION:**
>
> **The IAC valve assembly itself is an electrical component, and must not be soaked in any liquid cleaner or solvent, as damage may result.**

6   Before installing the IAC valve assembly, the position of the pintle must be checked. If the pintle is extended too far, damage to the assembly may occur.

7   Measure the distance from the gasket mounting surface of the IAC valve assembly to the tip of the pintle (see illustration).

8   If the distance is greater than 1-1/8-inch, reduce it as follows:

   a) If the IAC valve assembly has a collar around its electrical connector end, use firm hand pressure on the pintle to retract it (a slight side-to-side motion may help).

   b) If the IAC valve assembly has no collar, compress the pintle retaining spring toward the body of the IAC valve and try to turn the pintle clockwise. If the pintle will turn, continue turning it until the 1-1/8-inch dimension is reached. Return the spring to its original position with the straight part of the spring end lined up with the flat surface under the pintle head. If the pintle will not turn, use firm hand pressure to retract it.

9   Installation is the reverse of removal. On early models be sure to use a new gasket and tighten the IAC valve securely. On later models, install a new O-ring and tighten the screws securely.

➡**Note: No adjustment is made to the IAC valve assembly after reinstallation. IAC valve setting is controlled by the ECM when the vehicle is driven.**

11.14 To detach the throttle body from the plenum/intake manifold assembly, remove these two nuts (arrows)

THROTTLE BODY

◆ **Refer to illustrations 11.14 and 11.15**

10   Detach the cable from the negative terminal of the battery.

> ❊ **CAUTION:**
>
> **If the vehicle is equipped with a Delco Loc II or Theftlock audio system, make sure you have the correct activation code before disconnecting the battery. See the information at the front of this manual for the radio re-activation procedure.**

11   Remove the duct between the air cleaner housing and the throttle body (see Section 8).

12   Unplug the IAC, the TPS and the knock sensor electrical connectors (see illustration 11.2). On 1995 and later models, disconnect the vacuum hoses and unplug the MAF sensor electrical connector.

13   Detach the accelerator cable and, if equipped, the cruise control cables. Detach the cables from the bracket (see Section 12).

14   Remove the throttle body mounting nuts or bolts (see illustration) and separate the throttle body from the plenum/intake manifold assembly.

15   Turn the throttle body until its underside is exposed, loosen both hose clamps and detach the coolant hoses (models so equipped) (see illustration).

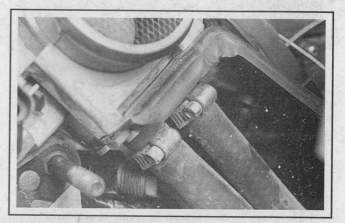

11.15 You can't see the two coolant hoses attached to the throttle body until you unbolt the throttle body from the plenum and turn it over

11.21 If you plan to install the same TPS switch, be sure to make an alignment between the plastic TPS switch body and the aluminum throttle body casting

11.22 On some models, you'll need a Torx bit or screwdriver to remove the TPS retaining screws

11.23 The TPS lever (A) must mate with the TPS drive lever (B) when you attach the TPS switch to the throttle body

16 Remove all old gasket material from the mating surfaces of the throttle body and the plenum.

17 Installation is the reverse of removal. Be sure to use a new gasket and tighten the throttle body mounting nuts or bolts to the specified torque.

THROTTLE POSITION SENSOR (TPS)

▶ **Refer to illustrations 11.21, 11.22 and 11.23**

18 Detach the cable from the negative terminal of the battery.

❋❋ CAUTION:

If the vehicle is equipped with a Delco Loc II or Theftlock audio system, make sure you have the correct activation code before disconnecting the battery. See the information at the front of this manual for the radio re-activation procedure.

19 Unplug the connector from the TPS switch (see illustration 11.2).

11.27 To remove the fuel pressure regulator from the fuel rail, detach the vacuum hose (A), unscrew the fuel return line threaded fitting (B) and remove the bracket mounting bolts (C) (bolt mounted)

20 If the TPS switch on your vehicle is located on the side of the throttle body, proceed to the next Step. If the switch is attached to the underside of the throttle body (like the one shown in this book), remove the throttle body (see Steps 10 through 16), then proceed to the next Step.

21 If you intend to install the same TPS switch, scribe or paint an alignment mark between the TPS and the throttle body (see illustration). If you are installing a new switch, you will have to take your vehicle to a dealer service department or automotive repair shop to make the final adjustment. A digital voltmeter or scan tool is needed to make these adjustments.

22 Using a Phillips or Torx screwdriver or bit (see illustration), remove the TPS switch.

23 Installation is the reverse of removal. Be sure to install the TPS switch onto the throttle body with the throttle valve in the closed position. Make sure that the TPS lever lines up with the TPS drive lever on the throttle shaft (see illustration).

24 If you had to remove the throttle body to get at the TPS, install the throttle body at this time (see Steps 10 through 17).

FUEL PRESSURE REGULATOR

➡ **Note: The pressure regulator is factory adjusted and is not serviceable. Do not attempt to remove the regulator from the fuel rail unless you intend to replace it with a new unit.**

Bolt mounted

▶ **Refer to illustrations 11.27 and 11.28**

25 Relieve the fuel pressure (see Section 2).
26 Detach the cable from the negative terminal of the battery.

❋❋ CAUTION:

If the vehicle is equipped with a Delco Loc II or Theftlock audio system, make sure you have the correct activation code before disconnecting the battery. See the information at the front of this manual for the radio re-activation procedure.

27 Detach the vacuum line from the top of the pressure regulator (see illustration).

11.28 Use a back-up wrench when disconnecting the fuel return line threaded fittings at the pressure regulator (bolt mounted)

11.35 On later models remove the snap ring (arrow) from the top of the fuel pressure regulator

28 Unscrew the fuel return line from the fuel pressure regulator. Be sure to use a back-up wrench (see illustration).

29 Remove the pressure regulator bracket mounting bolts and remove the regulator.

30 Installation is the reverse of removal.

Snap-ring mounted

▶ **Refer to illustration 11.35**

31 Relieve the fuel pressure (see Section 2)

32 Detach the cable from the negative terminal of the battery.

✳✳ CAUTION:

If the vehicle is equipped with a Delco Loc II or Theftlock audio system, make sure you have the correct activation code before disconnecting the battery. See the information at the front of this manual for the radio re-activation procedure.

33 Detach the vacuum line from the side of the pressure regulator.

34 Clean all dirt from the top of the regulator next to the snap ring or circlip.

35 On models secured with a snap-ring or circlip, use a scribe, a narrow screwdriver or a pair of snap ring pliers to remove the snap ring (see illustration).

36 Wrap a shop cloth around the regulator, carefully twist the regu-

lator and remove the regulator and O-rings.

37 Insert a clean shop cloth into the opening in the fuel rail to prevent the entry of foreign material.

38 Installation is the reverse of removal. Install new O-rings and apply a light coat of engine oil to them prior to installation.

FUEL RAIL ASSEMBLY

▶ **Refer to illustrations 11.40a, 11.40b, 11.41, 11.43 and 11.45**

➡**Note: There are three different configurations of fuel rails used among the various models and years. Although they visually appear different, removal and installation of the fuel rail assembly is similar for all models.**

39 Remove the acoustic engine cover/fuel injector sight shield from above the intake manifold if equipped (see Chapter 2A). Detach the vacuum hose from the pressure regulator (see illustration 11.27 and 11.35). Also position aside any spark plug wires that would interfere with the removal of the fuel rail.

40 On earlier models, it is easier to unplug the fuel rail wiring harness connector and remove the wiring harness with the fuel rail (see illustrations). On later models, label and unplug the electrical connectors at the fuel injectors (see illustration 11.50).

41 On earlier models, with hex fittings, use a back-up wrench and unscrew the fuel feed and return lines from the fuel rail (see illustration).

11.40a To detach the fuel injector wiring harness from the engine compartment wiring harness, slide the locking tab sideways with a small screwdriver . . .

11.40b . . . then release the locking tab on top of the plug and pull the two halves of the connector apart

11.41 Use a back-up wrench when detaching the fuel feed line from the fuel rail assembly (hex fitting only)

11.43 Fuel rail mounting bolts (B) (early models shown, later models similar) - on earlier models it will be necessary to remove the alternator bracket (A) before removing the fuel rail

42 On later models, with quick-connect fittings, use the special disconnect tools (see Section 4) to disconnect the fuel feed and return lines from the fuel rail

43 On earlier models, remove the alternator support bracket nut, if equipped (see illustration). Unscrew the alternator through bolt and remove the bracket.

44 Remove the four fuel rail assembly mounting bolts (see illustration 11.43).

45 Using a gentle rocking motion, remove the fuel rail and fuel injector assembly (see illustration).

46 Installation is the reverse of removal.

FUEL INJECTORS

▶ **Refer to illustrations 11.49, 11.50, 11.51, 11.52 and 11.53**

➡**Note: When purchasing new injectors, bring the old injector with you and match the part numbers and date code on the old one (stamped on the injector) to the replacement injectors.**

47 Relieve the fuel pressure (see Section 2).
48 Detach the cable from the negative terminal of the battery.

11.45 Use a gentle side-to-side rocking motion while pulling straight up to release the injectors from their bores in the intake manifold

✳✳ CAUTION:

If the vehicle is equipped with a Delco Loc II or Theftlock audio system, make sure you have the correct activation code before disconnecting the battery. See the information at the front of this manual for the radio re-activation procedure.

49 If you intend to service/replace more than one injector on a cylinder bank, it's a good idea to label the injectors to prevent mix-ups of the electrical leads at reassembly (see illustration).

50 To unplug the electrical connectors from the injectors you are going to service/replace, push in the connector wire retaining clip with your index finger (see illustration) and pull on the plug.

51 Pop off the fuel injector spring clip(s) with a small screwdriver (see illustration).

52 Using a gentle side-to-side wiggling motion, pull the injector from the fuel rail (see illustration). Use care when removing injectors to prevent damage to the electrical connector pins on the injector and the nozzle. Repeat the above Steps for each injector you wish to service or replace.

53 Even if you intend to reinstall the old injectors, be sure to remove

11.49 If you plan to remove more than one injector, it's a good idea to clearly label them to prevent mix-ups during reassembly

11.50 To unplug the electrical connectors from the injectors, push the wire retaining clip in and pull the connector straight up

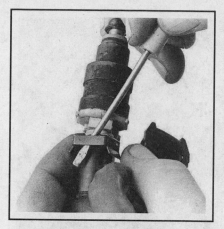

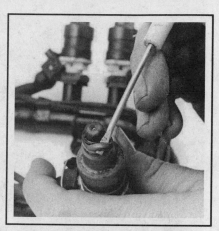

11.51 To detach an injector from the fuel rail assembly, pop off the spring clip with a small screwdriver . . .

11.52 . . . then pull on the injector while rocking it until the upper O-ring breaks loose from the fuel rail

11.53 If you plan to reuse the same injector, always remove the old O-rings with a small screwdriver and install new O-rings

the O-rings from both ends of each injector with a small screwdriver (see illustration) and install new ones. Coat the O-rings with a little engine oil and install them on the new injectors. If you are installing new injectors, they must also be fitted with new O-rings lubricated with a little engine oil.

54 Installation is the reverse of removal.

12 Accelerator cable - removal and installation

▶ **Refer to illustrations 12.3, 12.4, 12.5 and 12.7**

1   Disconnect the cable from the negative terminal of the battery.

❖❖ CAUTION:

If the vehicle is equipped with a Delco Loc II or Theftlock audio system, make sure you have the correct activation code before disconnecting the battery. See the information at the front of this manual for the radio re-activation procedure.

2   The accelerator cable will be retained to the throttle linkage by either a retaining clip or a plastic cable retainer.

3   To remove the retaining clip, use a small screwdriver to pop the retaining clip off the post on the throttle lever arm and remove the cable end (see illustration).

4   To remove the plastic cable retainer, push the cable end forward and lift up to detach the cable from the post on the throttle lever arm (see illustration).

5   To detach the accelerator cable and, if equipped, cruise control

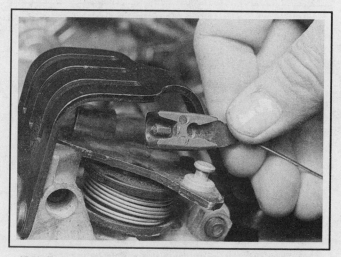

12.3 Use a small screwdriver to pop the clip off the post on the throttle arm

12.4 To remove the plastic cable retainer, push the cable end forward and lift up to detach the cable from the post

12.5 The throttle and cruise control cable retainer can be detached from the cable bracket by squeezing the locking tabs on the top and bottom of the retainer with a pair of needle-nose pliers

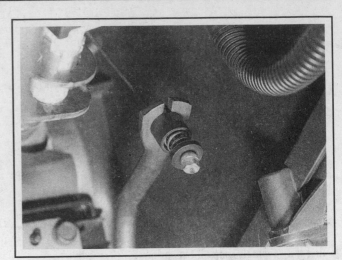

12.7 The accelerator cable at the pedal - once the cable is detached at the throttle body end, pull this end toward you and slide it out of the slot in the pedal

cables from the cable support bracket, use a pair of needle-nose pliers to squeeze the locking tabs on the top and bottom of the cable retainer (see illustration), then pull the cable assembly through the bracket.

6   Carefully study the routing of the cable before proceeding. On later models, remove the fasteners and push in on the tabs to remove the driver's side knee bolster (panel below the steering column) for access.

7   Working inside the vehicle, pull the cable toward you and detach it from the accelerator pedal (see illustrations).

8   Follow the cable with your hand until you locate the locking cable retainer on the passenger compartment side of the firewall. Again, with a pair of needle-nose pliers, squeeze the locking tabs of the retainer together and push the retainer through the firewall.

13 Exhaust system - removal and installation

▸ **Refer to illustrations 13.3 and 13.4**

✳✳ **CAUTION:**

DO NOT attempt to work on any part of the exhaust system until the entire system has completely cooled. Be especially careful around the catalytic converter, where the highest temperatures are generated.

1   Disconnect the cable from the negative terminal of the battery.

✳✳ **CAUTION:**

If the vehicle is equipped with a Delco Loc II or Theftlock audio system, make sure you have the correct activation code before disconnecting the battery. See the information at the front of this manual for the radio re-activation procedure.

2   Raise the vehicle and support it securely on jackstands.

3   After applying some penetrating oil and letting it soak in, disconnect the exhaust pipe from the exhaust manifold flange by removing the two bolts (see illustration).

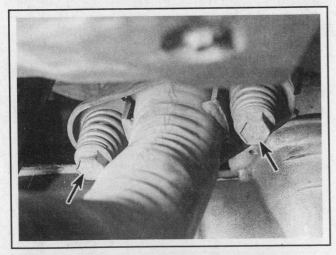

13.3 The exhaust pipe-to-exhaust manifold flange bolts - make sure you don't lose the springs when removing these bolts

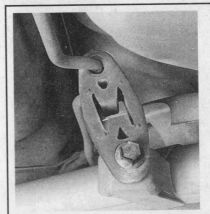

13.4 To detach this rubber block type hanger from the exhaust pipe, remove the large bolt

4   Remove the bolt from the rubber exhaust pipe hanger just in front of the fuel tank (see illustration).

5   Remove the bolt and detach the hanger from the forward end of the muffler.

6   Remove the bolt and detach the hanger from the rear end of the muffler.

7   Remove the exhaust pipe, catalytic converter and muffler as an assembly.

➡**Note: These components cannot be separated without a cutting torch. If you need to replace any of these parts, take the entire assembly to a dealer service department, service station or muffler shop for further service.**

Specifications

Fuel pressure - ignition on and engine off

1985 and 1986 (all)	37 to 43 psi
1987 through 1995	40 to 43 psi
1995 through 1997	40 to 47 psi
1998 and later	48 to 55 psi

Torque specifications   Ft-lbs (unless otherwise indicated)

PFI throttle body mounting nuts or bolts

1997 and earlier	120 to 180 in-lbs
1998 and later	89 in-lbs

Notes

Section

Reference to other Chapters

5

ENGINE ELECTRICAL SYSTEMS

1   Ignition system - general information

The 1985 models are equipped with a High Energy Ignition (HEI) system, consisting of an ignition switch, battery, coil, primary (low tension) and secondary (high tension) circuits, a distributor and spark plugs. The 1986 and later models are equipped with a distributorless ignition system (C3I).

HIGH ENERGY IGNITION (HEI) SYSTEM

HEI equipped vehicles use a special HEI distributor with Electronic Spark Timing (EST). The HEI distributor combines all the ignition components into one unit with the ignition coil integrated into the distributor cap. All spark timing changes are controlled by the Electronic Control Module (ECM), which monitors data from various engine sensors, computes the desired spark timing and signals the distributor to change timing accordingly. No vacuum or mechanical advance is used.

ELECTRONIC SPARK CONTROL (ESC)

Some engines are equipped with Electronic Spark Control (ESC). This system uses a knock sensor and a ESC module in conjunction with the ECM to control spark timing. Thus allowing the engine to maintain maximum spark advance without spark knock. Driveability and fuel economy are improved with this system.

COMPUTER CONTROLLED COIL IGNITION (C3I) SYSTEM

➡Note: Since 1992, the ignition system is no longer referred to as the C3I system by GM. All ignition systems and components operate in the same manner. The only change is the Electronic Control Module (ECM) is now being referred to as the Ignition Control Module (ICM).

This system, known as the computer controlled coil ignition (C3I) system, consists of the ECM, ignition module, ignition coils, a "Hall effect" cam, a crank sensor, a cam sensor and the connecting wires. On 3.0L V6 applications, the crank and cam sensor functions are combined into one dual sensor called a combination sensor, which is mounted at the vibration damper.

Two types of module/coil assemblies are commonly used. They can be distinguished by the configuration of their coil towers: Type I module/coil assemblies have evenly spaced towers, with three on either side; Type II's have all six towers (two per coil) placed on one side. The wiring harness and sensors, however, are interchangeable between either type.

The C3I system uses a "waste spark" method of spark distribution. Each cylinder is paired with its companion cylinder, i.e. 1-4, 5-2, 3-6. The spark occurs simultaneously in the cylinder coming up on compression and the cylinder coming up on exhaust.

The cylinder, on exhaust, requires very little of the available voltage to fire the spark plug. The remaining high voltage can then be used, as required, by the cylinder on compression.

The spark distribution is accomplished by a signal from the crank sensor, which is used by the ignition module to determine the proper time to trigger the next ignition coil. This signal is also processed by the C3I module into the reference signal used by the ECM.

The C3I system uses the electronic spark timing (EST) signal from the crankshaft to control spark timing.
Timing is controlled by the ECM using the following inputs:

Crankshaft position
Engine speed (rpm)
Engine temperature
Amount of air entering the intake (Mass air flow sensor)

2   Battery - removal and installation

▶ **Refer to illustration 2.3**

1   On 1999 and earlier models, the battery is located at the right or left front corner of the engine compartment. On 2000 and later models, the battery is located under the rear seat.

REMOVAL

2   On 1999 and earlier models, disconnect both cables from the battery terminals. On 2000 and later models, remove the rear seat (see Chapter 10), then disconnect both cables from the battery terminals.

3 Remove the hold-down clamp (see illustration) from the battery carrier. On 2001and later models, pull up and remove the battery vent panel and tube.

4 Carefully lift the battery out of the carrier.

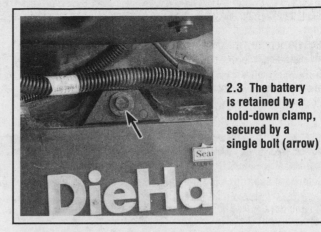

2.3 The battery is retained by a hold-down clamp, secured by a single bolt (arrow)

❄ **WARNING:**

Always keep the battery in an upright position to reduce the possibility of electrolyte spills. If you spill electrolyte on yourself or the vehicle, rinse it off immediately with plenty of water.

5 If you're installing a new battery, make sure you get one that's identical (same dimensions, amperage rating, cold cranking rating, etc.).

INSTALLATION

➡**Note: The battery carrier and hold-down clamp should be clean and free from corrosion before installing the battery. Make certain there are no parts in the carrier before installing the battery.**

6 Set the battery in position in the carrier. Don't tilt it.

7 Install the hold-down clamp and bolt. The bolt should be snug, but overtightening it may damage the battery case.

8 Install both battery cables - positive first, then negative.

➡**Note: The battery terminals and cable ends should be cleaned if necessary (see Chapter 1).**

3   Battery - emergency jump starting

Refer to the *Booster battery (jump) starting* procedure at the front of this manual.

4   Battery cables - check and replacement

◗ **Refer to illustration 4.1**

1 Periodically inspect the entire length of each battery cable for damage, cracked or burned insulation and corrosion (see illustration). Poor battery cable connections can cause starting problems and decreased engine performance.

2 Check the cable-to-terminal connections at the ends of the cables for cracks, loose wire strands and corrosion. The presence of white, fluffy deposits under the insulation at the cable terminal connection is a sign the cable is corroded and should be replaced. Check the terminals for distortion, missing mounting bolts or nuts and corrosion.

3 When removing the cables, always disconnect the negative cable first and hook it up last or the battery may be shorted by the tool used to loosen the cable clamps. Even if only the positive cable is being replaced, be sure to disconnect the negative cable from the battery first (see Chapter 1 for additional information related to battery cable removal).

❄ **CAUTION:**

If the vehicle is equipped with a Delco Loc II or Theftlock audio system, make sure you have the correct activation code before disconnecting the battery. See the information at the front of this manual for the radio re-activation procedure.

4 Disconnect the old cables from the battery, then trace each of them to their opposite ends and detach them from the starter solenoid and ground terminals. Note the routing of each cable to ensure correct installation.

5 If you're replacing either or both cables, take the old ones with

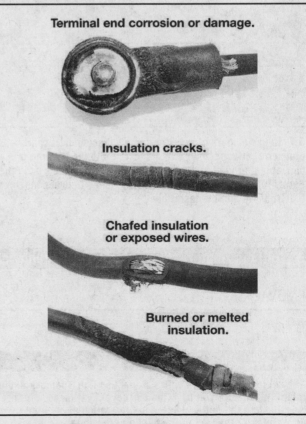

Terminal end corrosion or damage.

Insulation cracks.

Chafed insulation or exposed wires.

Burned or melted insulation.

4.1 Typical battery cable problems

you when buying the new ones - the replacements must be identical. Cables have characteristics that make them easy to identify: Positive cables are normally red and larger in diameter. Ground cables are normally black and smaller in diameter.

6   Clean the threads of the solenoid or ground connection with a wire brush to remove rust and corrosion. Apply a light coat of petro-

leum jelly to the threads to prevent future corrosion.

7   Attach the cable to the solenoid or ground connection and tighten the mounting nut/bolt securely.

8   Before connecting a new cable to the battery, make sure it reaches the battery post without having to be stretched.

9   Connect the positive cable first, followed by the negative cable.

5   Ignition system - check

♦ **Refer to illustrations 5.1 and 5.6**

✳ WARNING:

Because of the very high voltage generated by the ignition system, extreme care should be taken whenever an operation is performed involving ignition components. This not only includes the coil(s), control module and spark plug wires, but related items connected to the system as well, such as the plug connections, tachometer and any test equipment.

1   If the engine turns over but won't start, disconnect a spark plug

wire from any spark plug and attach a calibrated ignition tester (available at most auto parts stores) (see illustration).

2   Crank the engine and watch for a well defined bright blue spark at the tester. If spark occurs, sufficient voltage is reaching the plug to fire it. However, the plugs may be fouled, so remove them and check as described in Chapter 1. Repeat the check with the spark tester at the remaining plug wires to verify that the wires, distributor cap and rotor (if equipped), or coil(s) are OK.

3   On HEI systems, if no spark or intermittent spark occurs, remove the distributor cap and check the cap and rotor as described in Chapter 1.

4   On all models, if no spark occurs, check the spark plug wires and the wire connections at the coil(s). Check for voltage to the coil. Make any necessary repairs, then repeat the test again.

5.1 To use a spark plug tester, simply disconnect a spark plug wire, attach the wire to the tester, clip the tester to a convenient ground and operate the starter - if there's enough power to fire the plug, sparks will be visible between the electrode tip and the tester body

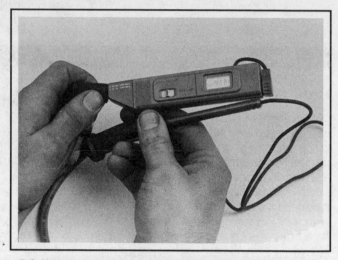

5.6 Use an ohmmeter to check the resistance of each plug wire - it should be less than 30,000 ohms

6   HEI distributor (1985 models only) - removal and installation

REMOVAL

1   Disconnect the cable from the negative battery terminal.

✳ CAUTION:

If the vehicle is equipped with a Delco Loc II or Theftlock audio system, make sure you have the correct activation code before disconnecting the battery. See the information at the front of this manual for the radio re-activation procedure.

2   Disconnect the electrical connectors at the distributor cap coil terminals.

3   Remove the distributor cap (see Chapter 1).

4   Mark the distributor housing at the point where the rotor tip is aligned. Also mark the relationship of the distributor to the engine block at the base of the distributor.

5   Remove the distributor hold-down bolt and clamp and lift the distributor straight up and out of the engine.

CAUTION:

Do not turn the crankshaft while the distributor is removed from the engine. If the crankshaft is turned, the alignment marks you made previously will be meaningless and the engine and distributor will have to be re-timed.

INSTALLATION

6   Clean the distributor-to-engine block mounting area and install a new gasket or O-ring.

7   If the crankshaft was not turned while the distributor was removed, insert it into the engine in exactly the same position in which it was removed. It will be necessary to start the rotor slightly off the mark when installing the distributor so that as the gears mesh it ends up properly aligned. Also, it may be necessary to slightly turn the oil pump driveshaft before the distributor is installed (this can be done with a long screwdriver) to enable the distributor shaft to properly engage the oil pump drive shaft when the distributor is seated.

8   If it was necessary to turn the crankshaft while the distributor was removed, position the engine with number one cylinder on TDC (see Chapter 2A). Install the distributor with the rotor aligned to the point where the number one spark plug wire is located (temporarily install the distributor cap to determine the number one plug wire location and mark the housing). Again, it's necessary to start the rotor slightly off the mark and adjust the oil pump driveshaft alignment as described in Step 7.

9   Once the distributor is installed, be sure it is properly seated against the block (no gap between the distributor housing flange and the block) and that the rotor is pointing exactly at your alignment mark. If not, remove the distributor and try again, meshing the gears in a different position.

10   The remainder of installation is the reverse of removal. Check and adjust the timing as described in Chapter 1.

---

7   HEI coil and ignition module (1985 models only) - removal and installation

COIL

▶ **Refer to illustration 7.3**

1   Remove the distributor cap (see Chapter 1).

2   Remove the screws and the coil cover on top of the distributor cap.

3   Remove the coil retaining screws. With a pair of needle-nose pliers pull the electrical connectors out of the cap and lift the coil off the cap (see illustration).

4   Installation is the reverse of removal. Install a new carbon button and gasket if the coil is being replaced.

7.3 To separate the coil from the distributor cap, clearly mark the wires, detach the coil ground wire and push the leads from the underside of the connectors

IGNITION MODULE

▶ **Refer to illustration 7.6**

5   Remove the distributor cap and rotor (see Chapter 1).

6   Remove the ignition module screws, disconnect the electrical connectors and lift the module from the distributor housing (see illustration).

7   Installation is the reverse of removal. Be sure to coat the base of the module with a thin layer of silicone grease (usually included with a new module) before installation.

➡**Note: This grease is necessary for proper operation of the module - if the grease is not applied, the module will overheat and fail.**

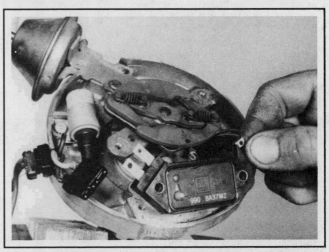

7.6 To detach the ignition module from the HEI distributor, remove the two mounting screws and unplug the connector from the terminals

8   Distributorless ignition coil and module (1986 and later models) - removal and installation

ALL MODELS

▶ **Refer to illustration 8.3**

1   Detach the cable from the negative terminal of the battery.

✳ CAUTION:

If the vehicle is equipped with a Delco Loc II or Theftlock audio system, make sure you have the correct activation code before disconnecting the battery. See the information at the front of this manual for the radio re-activation procedure.

2   Clearly label, then disconnect, all six spark plug wires from the coil.

8.3 Unbolt and unplug the connector from the C3I ignition coil/module assembly

8.5 The ignition coil/module mounting bracket has slots (arrows) that fit over three studs (1986 and 1987 models)

3   Unbolt and unplug the electrical connector at the module (see illustration).

1986 AND 1987 MODELS

▶ **Refer to illustration 8.5**

4   Detach the vacuum lines and the electrical connector from the EGR valve solenoid on the left end of the coil/module assembly (see Chapter 6 if necessary).

5   Remove the front bracket leg mounting nut from the stud on the intake manifold (next to the EGR valve) and the two rear bracket nuts from the studs protruding from the exhaust manifold side of the rear cylinder head (see illustration).

6   Remove the coil/module and support bracket assembly.

1988 AND LATER MODELS

7   Remove the mounting nuts from the bracket on the front of the valve cover and detach the assembly from the vehicle.

ALL MODELS

▶ **Refer to illustration 8.9**

8   Using a Torx screwdriver or bit, remove the coil-to-module screws.

9   Separate the coil pack from the module (see illustration). Label, then detach, the wires from the spade terminals on the underside of the coils.

10   Unbolt the module from the support bracket.

11   Installation is the reverse of removal. Be sure to attach the wires of the new module to the coil assembly spade terminals in exactly the same locations from which they were removed.

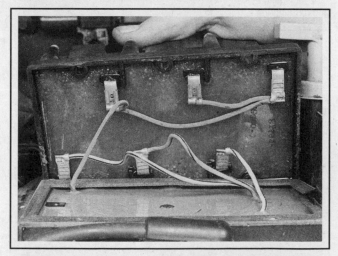

8.9 After removing the screws that hold the coil and module assemblies together, label and detach the wires from the module

9 Charging system - general information and precautions

➡**Note: On models equipped with the Delco Loc II audio system, be sure the lockout feature is turned off before performing any procedure which requires disconnecting the battery.**

The charging system consists of a belt-driven alternator with an integral voltage regulator and the battery. These components work together to supply electrical power for the ignition system, the lights and all accessories.

There are two types of alternators used. Earlier vehicles have the SI type and later models are equipped with the CS type. There are two types of CS alternators in use: The CS-130 and the CS-144. All types have a conventional pulley and fan.

To determine which type of alternator is installed on a vehicle, look at the fasteners employed to attach the two halves of the alternator housing. All CS models have rivets or special bolts instead of screws. CS alternators are rebuildable once the rivets are drilled out. However, we don't recommend this practice. For all intents and purposes, CS types should be considered non-serviceable and, if defective, exchanged as cores for new or rebuilt units.

The purpose of the voltage regulator is to limit the alternator's voltage to a preset value. This prevents power surges, circuit overloads, etc., during peak voltage output. On all models with which this manual is concerned, the voltage regulator is mounted inside the alternator housing.

The charging system doesn't ordinarily require periodic maintenance. However, the drivebelt, battery and wires and connections should be inspected at the intervals outlined in Chapter 1.

The dashboard warning light should come on when the ignition key is turned to Start, then go off immediately. If it stays on or comes on when the engine is running, a charging system problem has occurred (see Section 8).

Be very careful when making electrical circuit connections to a vehicle equipped with an alternator and note the following:

a) *When reconnecting wires to the alternator from the battery, be sure to note the polarity.*

b) *Before using arc welding equipment to repair any part of the vehicle, disconnect the wires from the alternator and the battery terminals.*

c) *Never start the engine with a battery charger connected.*

d) *Always disconnect both battery leads before using a battery charger.*

e) *The alternator is turned by an engine drivebelt which could cause serious injury if your hands, hair or clothes become entangled in it with the engine running.*

f) *Because the alternator is connected directly to the battery, it could arc or cause a fire if overloaded or shorted out.*

g) *Wrap a plastic bag over the alternator and secure it with rubber bands before steam cleaning the engine.*

10 Charging system - check

▶ **Refer to illustration 10.5**

1   If a malfunction occurs in the charging circuit, don't automatically assume the alternator is causing the problem. First check the following items:

a) *Check the drivebelt tension and condition (see Chapter 1). Replace it if it's worn or deteriorated.*

b) *Make sure the alternator mounting and adjustment bolts are tight.*

c) *Inspect the alternator wiring harness and the connectors at the alternator. They must be in good condition and tight.*

d) *Check the fusible link (if equipped) located between the starter solenoid and alternator. If it's burned, determine the cause, repair the circuit and replace the link (the engine won't start and/or the accessories won't work if the fusible link blows). Sometimes a fusible link may look good, but still be bad. If in doubt, remove it and check it for continuity.*

e) *Start the engine and check the alternator for abnormal noises (a shrieking or squealing sound indicates a bad bearing).*

f) *Check the specific gravity of the battery electrolyte. If it's low, charge the battery (doesn't apply to maintenance free batteries).*

g) *Make sure the battery is fully charged (one bad cell in a battery can cause overcharging by the alternator).*

h) *Disconnect the battery cables (negative first, then positive).*

Inspect the battery posts and cable clamps for corrosion. Clean them thoroughly if necessary (see Chapter 1). Reconnect the cable to the positive terminal.

i) *With the key off, connect a test light between the negative battery post and the disconnected negative cable clamp.*

1) *If the test light doesn't come on, reattach the clamp and proceed to the next Step.*

2) *If the test light comes on, there's a short (drain) in the electrical system of the vehicle. The short must be repaired before the charging system can be checked.*

3) *Disconnect the alternator wiring harness.*

(a) *If the light goes out, the alternator is bad.*

(b) *If the light stays on, pull each fuse until the light goes out (this will tell you which component is shorted).*

2   Using a voltmeter, check the battery voltage with the engine off. If should be approximately 12-volts.

3   Start the engine and check the battery voltage again. It should now be approximately 14-to-15 volts.

4   Locate the test hole in the back of the alternator.

➡**Note: If there is no test hole, the vehicle is equipped with a newer CS type alternator. Further testing of this type of alternator must be done by a service station, dealer service department or auto electric shop.**

5   Ground the tab inside the hole by inserting a screwdriver blade into the hole and touching the tab and the case at the same time (see illustration).

❈❈ CAUTION:

Don't run the engine with the tab grounded any longer than necessary to obtain a voltmeter reading. If the alternator is charging, it's running unregulated during the test. This condition may overload the electrical system and cause damage to the components.

6   The reading on the voltmeter should be 15-volts or higher with the tab grounded in the test hole.

7   If the voltmeter indicates low battery voltage, the alternator is faulty and should be replaced with a new one (see Section 11).

8   If the voltage reading is 15-volts or higher and a no charge condition exists, the regulator or field circuit is the problem. Remove the alternator (see Section 9) and have it checked by a service station, dealer service department or auto electric shop.

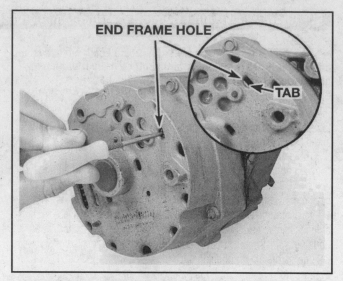

10.5 To full field the alternator, ground the tab located inside the test hole on the end frame (backside) of the alternator by inserting a screwdriver blade into the hole and touching the tab and the case at the same time

11 Alternator - removal and installation

◆ **Refer to illustration 11.2**

1   Detach the cable from the negative terminal of the battery.

❈❈ CAUTION:

If the vehicle is equipped with a Delco Loc II or Theftlock audio system, make sure you have the correct activation code before disconnecting the battery. See the information at the front of this manual for the radio re-activation procedure.

2   Detach the electrical connectors from the alternator (see illustration). On 2001 and later models, remove the bolt and nut securing the alternator brace to the alternator and the intake manifold.

3   Loosen the alternator mounting bolts and detach the drivebelt (see Chapter 1).

4   Remove the bolts and separate the alternator from the engine.

5   If you're replacing the alternator, take the old one with you when purchasing the new one. Make sure the new/rebuilt unit is identical to the old alternator. Look at the terminals - they should be the same in number, size and location as the terminals on the old alternator. Finally, look at the identification numbers - they'll be stamped into the housing or printed on a tag attached to the housing. Make sure the numbers are the same on both alternators.

6   Many new/rebuilt alternators DO NOT have a pulley installed, so

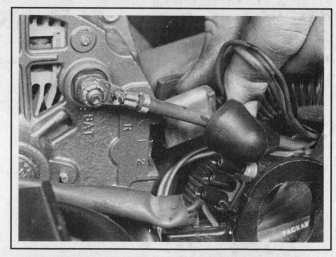

11.2 Remove the alternator electrical connectors

you may have to switch the pulley from the old one to the new/rebuilt one. When buying an alternator, find out the shop's policy regarding pulleys - some shops will perform this service free of charge.

7   Installation is the reverse of removal.

8   Check the charging voltage to verify proper operation of the alternator (see Section 8).

12 Alternator brushes - replacement

▶ **Refer to illustrations 12.3, 12.4, 12.5, 12.6, 12.7, 12.9a and 12.9b**

➡**Note: The following procedure applies only to SI type alternators. CS types have riveted housings or special bolts and shouldn't be disassembled.**

1 Remove the alternator from the vehicle (see Section 11).

2 Scribe or paint marks on the front and rear frame housings of the alternator to facilitate reassembly.

3 Remove the four through-bolts holding the front and rear end frames together, then separate the drive end frame from the rectifier end frame (see illustration).

4 Remove the bolts holding the stator to the rear end frame and separate the stator from the end frame (see illustration).

5 Remove the nuts attaching the diode trio to the rectifier bridge and remove the diode trio (see illustration).

6 Remove the screws attaching the resistor (not used on all models) and brush holder to the end frame and remove the brush holder (see illustration).

7 Remove the brushes from the brush holder by slipping the brush retainer off the brush holder (see illustration).

8 Remove the springs from the brush holder.

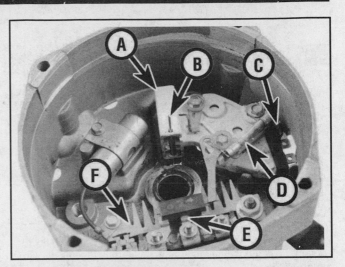

12.3 Inside a typical SI alternator

A	Brush holder	D	Resistor (not on all models)
B	Paper clip retaining brushes (for reassembly)	E	Diode trio
C	Regulator	F	Rectifier bridge

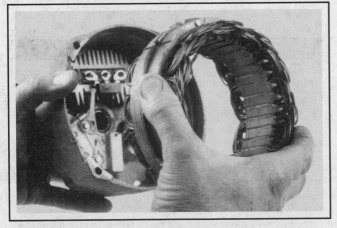

12.4 After removing the bolts holding the stator assembly to the end frame, remove the stator

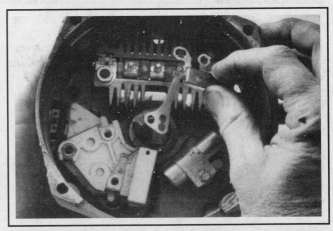

12.5 Remove the nuts attaching the diode trio to the rectifier bridge and remove the trio

12.6 After removing the screws that attach the brush holder and the resistor (if equipped) to the end frame, remove the brush holder

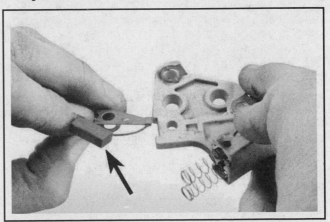

12.7 Slip the brush retainer off the brush holder and remove the brushes (arrow)

12.9a To hold the brushes in place during reassembly, insert a paper clip through the hole in the end frame nearest to the rotor shaft

12.9b With the paper clip in place, carefully lower the rectifier end frame over the drive end frame

9   Installation is the reverse of the removal procedure, noting the following: When installing the brushes in the brush holder, install the brush closest to the end frame first. Slip a paper clip through the rear of the end frame to hold both brushes in place while reassembly is completed (see illustrations). The paper clip should not be removed until the front and rear end frames have been bolted together.

13 Starting system - general information and precautions

➡**Note: On models equipped with the Delco Loc II audio system, be sure the lockout feature is turned off before performing any procedure which requires disconnecting the battery.**

The sole function of the starting system is to turn over the engine quickly enough to allow it to start.

The starting system consists of the battery, the starter motor, the starter solenoid and the wires connecting them. The solenoid is mounted directly on the starter motor. The solenoid/starter motor assembly is installed on the lower part of the engine, next to the transmission bellhousing.

When the ignition key is turned to the Start position, the starter solenoid is actuated through the starter control circuit. The starter solenoid then connects the battery to the starter. The battery supplies the electrical energy to the starter motor, which does the actual work of cranking the engine.

The starter on a vehicle equipped with an automatic transaxle can only be operated when the selector lever is in Park or Neutral.

Always observe the following precautions when working on the starting system:

a) *Excessive cranking of the starter motor can overheat it and cause serious damage. Never operate the starter motor for more than 15 seconds at a time without pausing to allow it to cool for at least two minutes.*

b) *The starter is connected directly to the battery and could arc or cause a fire if mishandled, overloaded or shorted out.*

c) *Always detach the cable from the negative terminal of the battery before working on the starting system.*

14 Starter motor - testing in vehicle

➡**Note: Before diagnosing starter problems, make sure the battery is fully charged.**

1   If the starter motor doesn't turn at all when the switch is operated, make sure the shift lever is in Neutral or Park.

2   Make sure the battery is charged and all cables, both at the battery and starter solenoid terminals, are clean and secure.

3   If the starter motor spins but the engine isn't cranking, the over-running clutch in the starter motor is slipping and the starter motor must be replaced.

4   If, when the switch is actuated, the starter motor doesn't operate at all but the solenoid clicks, then the problem lies with either the battery, the main solenoid contacts or the starter motor itself (or the engine is seized).

5   If the solenoid plunger can't be heard when the switch is actuated, the battery is bad, the fusible link is burned (the circuit is open) or the solenoid itself is defective.

6   To check the solenoid, connect a jumper lead between the battery (+) and the ignition switch wire terminal (the small terminal) on the

solenoid. If the starter motor now operates, the solenoid is okay and the problem is in the ignition switch, neutral start switch or the wiring.

7   If the starter motor still doesn't operate, remove the starter/solenoid assembly for disassembly, testing and repair.

8   If the starter motor cranks the engine at an abnormally slow speed, first make sure the battery is charged and all terminal connections are clean and tight. If the engine is partially seized, or has the wrong viscosity oil in it, it'll crank slowly.

9   Run the engine until normal operating temperature is reached, then disable the ignition system by unplugging the module (distributorless ignition models) or unplugging the primary wires from the distributor (models with a distributor).

10   Connect a voltmeter positive lead to the positive battery post and connect the negative lead to the negative post.

11   Crank the engine and take the voltmeter readings as soon as a steady figure is indicated. DO NOT allow the starter motor to turn for more than 15 seconds at a time. A reading of 9-volts or more, with the starter motor turning at normal cranking speed, is normal. If the reading is 9-volts or more but the cranking speed is slow, the motor is faulty. If the reading is less than 9-volts and the cranking speed is slow, the solenoid contacts are probably burned, the starter motor is bad, the battery is discharged or there's a bad connection.

15   Starter motor - removal and installation

▶ **Refer to illustrations 15.3 and 15.4**

➡**Note: On some vehicles, it may be necessary to remove the exhaust pipe(s) or frame crossmember to gain access to the starter motor. In extreme cases it may even be necessary to unbolt the mounts and raise the engine slightly to get the starter out.**

1   Detach the cable from the negative terminal of the battery.

CAUTION:

If the vehicle is equipped with a Delco Loc II or Theftlock audio system, make sure you have the correct activation code before disconnecting the battery. See the information at the front of this manual for the radio re-activation procedure.

2   Raise the front of the vehicle and support it securely on jackstands. Apply the parking brake and block the rear wheels to keep the vehicle from rolling off the jackstands. On 2001 and later models, remove the torque converter cover (see Chapter 7).

3   Working under the vehicle, clearly label, then disconnect the wires from the terminals on the starter motor and solenoid (see illustration).

4   Remove the mounting bolts and detach the starter (see illustration). Note the locations of the spacer shims (if used) - they must be reinstalled in the same positions.

5   Installation is the reverse of removal.

15.3   There are three terminals on the end of the typical starter solenoid

A   Battery terminal
B   Switch terminal (S)
C   Motor terminal (M)

15.4   Location of the starter mounting bolts (on some models it may be necessary to raise the engine slightly for access)

16 Starter solenoid - removal and installation

♦ **Refer to illustration 16.5**

1   Disconnect the cable from the negative terminal of the battery.

❊❊ CAUTION:

If the vehicle is equipped with a Delco Loc II or Theftlock audio system, make sure you have the correct activation code before disconnecting the battery. See the information at the front of this manual for the radio re-activation procedure.

2   Remove the starter motor (see Section 15).

3   Disconnect the strap from the solenoid to the starter motor terminal.

4   Remove the screws that secure the solenoid to the starter motor.

5   Twist the solenoid in a clockwise direction to disengage the flange from the starter body (see illustration).

6   Installation is the reverse of removal.

16.5 To remove the solenoid housing from the starter motor, remove the screws and turn it clockwise

Section

Reference to other Chapters

6

EMISSIONS AND ENGINE CONTROL SYSTEMS

1   General information

To prevent pollution of the atmosphere from incompletely burned and evaporating gases, and to maintain good driveability and fuel economy, a number of emission control devices are incorporated. They include the:

Fuel control system
Electronic Spark Timing (EST)
Electronic Spark Control (ESC) system
Exhaust Gas Recirculation (EGR) system
Evaporative Emission Control System (EECS)
Positive Crankcase Ventilation (PCV) system
Transaxle Converter Clutch (TCC)
Catalytic converter

All of these systems are linked, directly or indirectly, to the engine control/self-diagnosis system.

The Sections in this Chapter include general descriptions, checking procedures within the scope of the home mechanic and component replacement procedures (when possible) for each of the systems listed above.

Before assuming an emissions control system is malfunctioning, check the fuel and ignition systems carefully. The diagnosis of some emission control devices requires specialized tools, equipment and training. If checking and servicing become too difficult or if a procedure is beyond the scope of your skills, consult a dealer service department.

Remember, the most frequent cause of emissions problems is simply a loose or broken vacuum hose or wire, so always check the hose and wiring connections first.

This doesn't mean, however, that emission control systems are particularly difficult to maintain and repair. You can quickly and easily perform many checks and do most (if not all) of the regular maintenance at home with common tune-up and hand tools.

➡**Note: Because of a Federally mandated extended warranty which covers the emission control system components, check with your dealer about warranty coverage before working on any emissions-related systems. Once the warranty has expired, you may wish to perform some of the component checks and/or replacement procedures in this Chapter to save money.**

Pay close attention to any special precautions outlined in this Chapter. It should be noted that the illustrations of the various systems may not exactly match the system installed on your vehicle because of changes made by the manufacturer during production or from year-to-year.

A Vehicle Emissions Control Information label is located in the engine compartment. This label contains important emissions specifications and ignition timing procedures, as well as a vacuum hose schematic and emissions components identification guide. When servicing the engine or emissions systems, the VECI label in your particular vehicle should always be checked for up-to-date information.

2   Self-diagnosis system and trouble codes

➡**Note: 1994 and later models require a special SCAN tool to access the trouble codes - have the vehicle diagnosed by a dealer service department or other qualified automotive repair facility if the proper SCAN tool is not available. 1994 and later models utilize five-digit trouble codes. These "generic" trouble codes listed in the following table do not include the manufacturer's specific trouble codes. Consult a dealer service department or other qualified repair shop for additional information.**

DIAGNOSTIC TOOL INFORMATION

▸ **Refer to illustrations 2.1, 2.2 and 2.4**

1   A digital multi-meter is a necessary tool for checking fuel injection and emission related components (see illustration). A digital volt-ohmmeter is preferred over the older style analog multi-meter for several reasons. The analog multi-meter cannot display the volts-ohms or amps measurement in hundredths and thousandths increments. When working with electronic circuits which are often very low voltage, this accurate reading is most important. Another good reason for the digital multi-meter is its high impedance circuit. The digital multi-meter is equipped with a high resistance internal circuitry (10 million ohms). Because a voltmeter is hooked up in parallel with the circuit when testing, it is vital that none of the voltage being measured should be allowed to travel the parallel path set up by the meter itself. This dilemma does not show itself when measuring larger amounts of voltage (9 to 12 volt circuits) but if you are measuring a low voltage circuit such as the oxygen sensor signal voltage, a fraction of a volt may be a significant amount when diagnosing a problem.

2   Hand-held scanners are the most powerful and versatile tools for

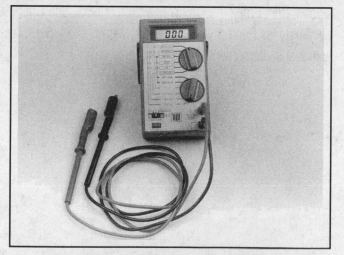

2.1 Digital multi-meters can be used for testing all types of circuits; because of their high impedance, they are much more accurate than analog meters for measuring millivolts in low-voltage computer circuits

analyzing engine management systems used on later model vehicles (see illustration). Early model scanners handle codes and some diagnostics for many OBD I systems. Each brand scan tool must be examined carefully to match the year, make and model of the vehicle you are working on. Often interchangeable cartridges are available to access the particular manufacturer; Ford, GM, Chrysler, etc.). Some manufacturers will specify by continent; Asia, Europe, USA, etc. Seek the advice of your local auto parts retailer.

2.4 Trouble code tools simplify the task of extracting the trouble codes

2.2 Scanners like the Actron Scantool and the AutoXray XP240 are powerful diagnostic aids - programmed with comprehensive diagnostic information, they can tell you just about anything you want to know about your engine management system

3   With the arrival of the federally mandated emission control system (OBD II), specially-designed scan tools were developed and are now available at tool and auto parts stores for home-mechanic use.

4   Another type of code reader is available at parts stores (see illustration). These tools simplify the procedure for extracting codes from the engine management computer by simply "plugging in" to the diagnostic connector on the vehicle wiring harness and are much less expensive (however, they will not work on models that require the use of a scan tool to extract codes).

GENERAL DESCRIPTION

5   The electronically controlled fuel and emissions system is linked with many other related engine management systems. It consists mainly of sensors, output actuators and an Electronic Control Module (ECM) or Powertrain Control Module (PCM) (see Section 1). Completing the system are various other components which respond to commands from the ECM/PCM.

6   In many ways, this system can be compared to the central nervous system in the human body. The sensors (nerve endings) constantly gather information and send this data to the ECM/PCM (brain), which processes the data and, if necessary, sends out a command for some type of vehicle change (limbs).

7   Here's a specific example of how one portion of this system operates: An oxygen sensor, mounted in the exhaust manifold and protruding into the exhaust gas stream, constantly monitors the oxygen content of the exhaust gas as it travels through the exhaust pipe. If the percent-

age of oxygen in the exhaust gas is incorrect, an electrical signal is sent to the ECM/PCM. The ECM/PCM takes this information, processes it and then sends a command to the fuel injectors, telling it to change the fuel/air mixture. To be effective, all this happens in a fraction of a second, and it goes on continuously while the engine is running. The end result is a fuel/air mixture which is constantly kept at a predetermined ratio, regardless of driving conditions.

OBTAINING TROUBLE CODES

▶ **Refer to illustrations 2.11a and 2.11b**

8   One might think that a system which uses exotic electrical sensors and is controlled by an on-board computer would be difficult to diagnose. This is not necessarily the case.

9   The On Board Diagnostic (OBD) system has a built-in self-diagnostic system, which indicates a problem by turning on a malfunction indicator light on the instrument panel when a fault has been detected.

➡**Note: Since some of the trouble codes do not set the CHECK ENGINE or SERVICE ENGINE SOON light, it is a good idea to access the OBD system and look for any trouble codes that may have been recorded and need tending.**

10 Perhaps more importantly, the ECM/PCM will recognize this fault, in a particular system monitored by one of the various information sensors, and store it in its memory in the form of a trouble code. Although the trouble code cannot reveal the exact cause of the malfunction, it greatly facilitates diagnosis as you or a dealer mechanic can "tap into" the ECM/PCM's memory and be directed to the problem area.

11 To retrieve this information from the ECM on 1993 and earlier models, you must use a short jumper wire to ground a diagnostic terminal. The terminal is part of an electrical connector called the Assembly Line Data Link (ALDL) (see illustrations). The ALDL is located under dashboard. On some models a small rectangular plate is used to cover

2.11a The 12-pin Assembly Line Data Link (ALDL) terminal identification

A   Ground
B   Diagnostic TEST terminal (if equipped)

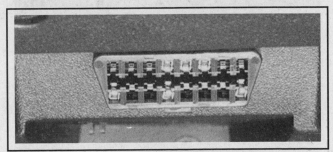

2.11b The 16 pin Data Link Connector (DLC) found on models equipped with OBD II

Typical component locations

1 Injectors

2 Idle Air Control (IAC) valve

3 Exhaust Gas Recirculation (EGR) valve

4 Evaporative emissions control canister

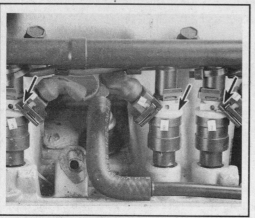

5 Oxygen sensor

6 Positive Crankcase Ventilation (PCV) valve

7 EGR solenoid (upper arrow)

16 Fuel pressure regulator (lower arrow)

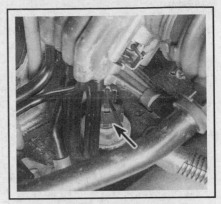

8 Electronic Spark Control (ESC) knock sensor

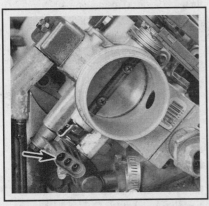

9 Throttle Position Sensor (TPS)

10 Camshaft position sensor

11 Crankshaft position sensor

12 Manifold Air Temperature (MAT) sensor

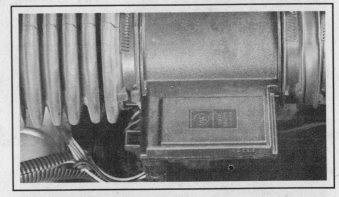

13 Mass Air Flow (MAF) sensor

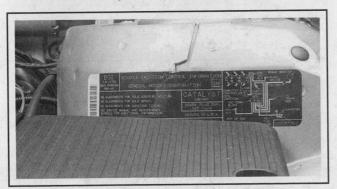

14 Vehicle Emission Control Information (VECI) label

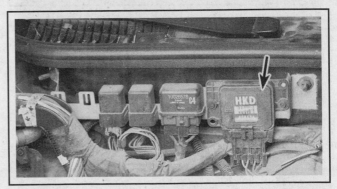

15 Electronic Spark Control (ESC) module

the connector and must be pried off to provide access to the terminals. With the electrical connector exposed, push one end of the jumper wire into the diagnostic TEST terminal and the other end into the GROUND terminal.

→Note: Some models with a 12-pin diagnostic connector may not have a terminal B present in the connector. On these models a scan tool is required to access trouble codes.

12 On 1994 and later models, a scan tool must be connected to the Assembly Line Data Link (ALDL). The scan tool is a hand held digital computer scanner that interfaces with the on-board computer. The scan tool is a very powerful tool; it not only reads the trouble codes but also displays the actual operating conditions of the sensors and actuators. Scan tools are expensive, but they are necessary to accurately diagnose a modern computerized fuel-injected engine. Scan tools are available from auto parts stores and specialty tool companies.

13 It should be noted that the self-diagnosis feature built into this system does not detect all possible faults. If you suspect a problem with the On Board Diagnostic (OBD) system, but the CHECK ENGINE or SERVICE ENGINE SOON light has not come on and no trouble codes have been stored, and performing the checks described in Section 4 doesn't pinpoint a problem, take the vehicle to a dealer service department or other qualified repair shop for diagnosis.

14 Furthermore, when diagnosing an engine performance, fuel economy or exhaust emissions problem (which is not accompanied by a CHECK ENGINE or SERVICE ENGINE SOON light) do not automatically assume the fault lies in this system. Perform all standard troubleshooting procedures, as indicated elsewhere in this manual, before turning to the On Board Diagnostic (OBD) system.

15 Finally, since this is an electronic system, you should have a basic knowledge of automotive electronics before attempting any diagnosis. Damage to the ECM/PCM, Programmable Read Only Memory (PROM) calibration unit or related components can easily occur if care is not exercised.

CLEARING TROUBLE CODES

16 To clear the trouble codes from the ECM's memory on a 1993 or earlier model, unplug the ECM electrical pigtail at the positive battery

cable and wait at least 30 seconds before plugging it back in. If the vehicle you are working on does not have this connector, disconnect the cable from the negative terminal of the battery for at least 30 seconds.

❊❊ CAUTION 1:

To prevent damage to the ECM, the ignition switch must be turned OFF when disconnecting or connecting power to the ECM.

❊❊ CAUTION 2:

If the vehicle is equipped with a Delco Loc II audio system, make sure you have the correct activation code before disconnecting the battery. See the information at the front of this manual for the radio reactivation procedure.

17 To clear the codes from the ECM/PCM memory on 1994 and later models, install the SCAN tool, scroll the menu for the function that describes "CLEARING CODES" and follow the prescribed method for that particular SCAN tool or momentarily remove the PCM/IGN fuse from the fuse box for 30 seconds. Clearing codes may also be accomplished by removing the fusible link (main power fuse) located near the battery positive terminal (see Chapter 12) or by disconnecting the cable from the positive terminal (+) of the battery.

❊❊ CAUTION:

If the vehicle is equipped with a Delco Loc II audio system, make sure you have the correct activation code before disconnecting the battery. See the information at the front of this manual for the radio reactivation procedure.

18 Disconnecting the power to the ECM/PCM to clear the memory can be an important diagnostic tool, especially on intermittent problems.

→Note: Not all codes apply to all models.

2-digit trouble codes

Code	Code definition
12	Diagnostic mode
13	Oxygen sensor or circuit
14	Coolant sensor or circuit/high temperature indicated
15	Coolant sensor or circuit/low temperature indicated
16	System voltage out of range
19	Crankshaft position sensor or circuit
21	Throttle Position Sensor (TPS) or circuit - voltage high
22	Throttle Position Sensor (TPS) or circuit - voltage low

Code	Code definition
23	Mixture Control (M/C) solenoid or circuit (carbureted models)
23	Manifold Air Temperature (MAT) sensor or circuit (1990 and earlier models)
23	Intake Air Temperature (IAT) sensor circuit (fuel-injected models)
24	Vehicle Speed Sensor (VSS) or circuit
25	Manifold Air Temperature (MAT) sensor or circuit - high temperature indicated (1990 and earlier models)
25	Intake Air Temperature (IAT) sensor or circuit - high temperature indicated (1991 and later models)

Code	Code definition
26	Quad Driver module circuit
27	Quad Driver module circuit
28	Quad Driver module circuit
29	Quad Driver module circuit
31	Park/Neutral Position (PNP) switch circuit
32	BARO sensor or circuit (carbureted models)
32	EGR circuit (fuel-injected models)
33	Manifold Absolute Pressure (MAP) sensor signal voltage high
33	Mass Air Flow (MAF) sensor or circuit - excessive airflow indicated
34	Manifold Absolute Pressure (MAP) sensor signal voltage low
34	Mass Air Flow (MAF) sensor signal - low airflow indicated
35	Idle Speed Control (ISC) switch or circuit (shorted) (carbureted models)
35	Idle Air Control (IAC) valve or circuit
38	Brake switch circuit
39	Torque Converter Clutch (TCC) circuit
41	No distributor signals to ECM, or faulty ignition module (carbureted models)
41	Cylinder select error - MEM-CAL or ECM problem (fuel-injected models)
41	Cam sensor circuit (3.8L engine)
42	Bypass or Electronic Spark Timing (EST) circuit
43	Low voltage at ECM terminal L (carbureted models)
43	Knock sensor circuit
44	Oxygen sensor or circuit - lean exhaust detected
45	Oxygen sensor or circuit - rich exhaust detected
46	Power steering pressure switch circuit
48	Misfire diagnosis
51	PROM, MEM-CAL or ECM problem
52	CALPAK or ECM problem
53	EGR fault (carbureted models only)
53	System over-voltage (ECM over 17.7 volts)
54	Mixture Control (M/C) solenoid or circuit (carbureted models)
54	Fuel pump circuit (1986 and later models)
55	Oxygen sensor circuit or ECM
55	Fuel lean monitor (2.2L engine)
61	Oxygen sensor signal faulty (possible contaminated sensor)
62	Transaxle gear switch signal circuits
63	Manifold Absolute Pressure (MAP) sensor voltage high (low vacuum detected)
64	Manifold Absolute Pressure (MAP) sensor voltage low (high vacuum detected)
66	Air conditioning pressure sensor or circuit

5-digit trouble codes

Code	Code definition
P0016	Crankshaft position/camshaft position, bank 1, sensor A - correlation
P0030	HO2S heater control circuit (bank 1, sensor 1)
P0036	HO2S heater control circuit (bank 1 sensor 2)
P0050	HO2S heater control circuit (bank 2, sensor 1)
P0101	Mass air flow or volume air flow circuit, range or performance problem
P0102	Mass air flow or volume air flow circuit, low input
P0103	Mass air flow or volume air flow circuit, high input
P0106	Manifold absolute pressure or barometric pressure circuit, range or performance problem
P0107	Manifold absolute pressure or barometric pressure circuit, low input
P0108	Manifold absolute pressure or barometric pressure circuit, high input
P0112	Intake air temperature circuit, low input
P0113	Intake air temperature circuit, high input
P0116	Engine coolant temperature circuit range/performance problem
Code	Code definition
P0117	Engine coolant temperature circuit, low input
P0118	Engine coolant temperature circuit, high input
P0121	Throttle position or pedal position sensor/switch circuit, range or performance problem
P0122	Throttle position or pedal position sensor/switch circuit, low input

5-digit trouble codes (continued)

Code	Code definition
P0123	Throttle position or pedal position sensor/switch circuit, high input
P0125	Insufficient coolant temperature for closed loop fuel control
P0128	Coolant thermostat (coolant temperature below thermostat regulating temperature)
P0130	O2 sensor circuit malfunction (bank 1, sensor 1)
P0131	O2 sensor circuit, low voltage (bank 1, sensor 1)
P0132	O2 sensor circuit, high voltage (bank 1, sensor 1)
P0133	O2 sensor circuit, slow response (bank 1, sensor 1)
P0134	O2 sensor circuit - no activity detected (bank 1, sensor 1)
P0135	O2 sensor heater circuit malfunction (bank 1, sensor 1)
P0136	O2 sensor circuit malfunction (bank 1, sensor 2)
P0137	O2 sensor circuit, low voltage (bank 1, sensor 2)
P0138	O2 sensor circuit, high voltage (bank 1, sensor 2)
P0139	O2 sensor circuit, slow response (bank 1, sensor 2)
P0140	O2 sensor circuit - no activity detected (bank 1, sensor 2)
P0141	O2 sensor heater circuit malfunction (bank 1, sensor 2)
P0150	O2 sensor circuit malfunction (bank 2, sensor 1)
P0151	O2 sensor circuit, low voltage (bank 2, sensor 1)
P0152	O2 sensor circuit, high voltage (bank 2, sensor 1)
P0153	O2 sensor circuit, slow response (bank 2, sensor 1)
P0154	O2 sensor circuit - no activity detected (bank 2, sensor 1)
P0155	O2 sensor heater circuit malfunction (bank 2, sensor 1)
P0171	System too lean (bank 1)
P0172	System too rich (bank 1)
P0173	Fuel trim malfunction (bank 2)
P0174	System too lean (bank 2)
P0175	System too rich (bank 2)
P0191	Fuel rail pressure sensor circuit, range or performance problem
P0192	Fuel rail pressure sensor circuit, low input
P0193	Fuel rail pressure sensor circuit, high input
P0201	Injector circuit malfunction - cylinder no. 1
P0202	Injector circuit malfunction - cylinder no. 2

Code	Code definition
P0203	Injector circuit malfunction - cylinder no. 3
P0204	Injector circuit malfunction - cylinder no. 4
P0205	Injector circuit malfunction - cylinder no. 5
P0206	Injector circuit malfunction - cylinder no. 6
P0207	Injector circuit malfunction - cylinder no. 7
P0208	Injector circuit malfunction - cylinder no. 8
P0218	Transmission overheating condition
P0230	Fuel pump primary circuit malfunction
P0243	Turbocharger wastegate solenoid A malfunction
P0300	Random/multiple cylinder misfire detected
P0301	Cylinder no. 1 misfire detected
P0302	Cylinder no. 2 misfire detected
P0303	Cylinder no. 3 misfire detected
P0304	Cylinder no. 4 misfire detected
P0305	Cylinder no. 5 misfire detected
P0306	Cylinder no. 6 misfire detected
P0315	Crankshaft position system - variation not learned
P0321	Crankshaft position (CKP) sensor/engine speed (RPM) sensor - range or performance problem
P0325	Knock sensor no. 1 circuit malfunction (bank 1 or single sensor)
P0326	Knock sensor no. 1 circuit, range or performance problem (bank 1 or single sensor)
P0327	Knock sensor no. 1 circuit, low input (bank 1 or single sensor)
P0332	Knock sensor no. 2 circuit, low input (bank 2)
P0335	Crankshaft position sensor A circuit malfunction
P0336	Crankshaft position sensor A circuit - range or performance problem
P0340	Camshaft position sensor "A", circuit malfunction (bank 1)
P0341	Camshaft position sensor "A", circuit - range or performance problem
P0342	Camshaft position sensor "A", circuit - low input
P0351	Ignition coil A primary or secondary circuit malfunction
P0352	Ignition coil B primary or secondary circuit malfunction
P0353	Ignition coil C primary or secondary circuit malfunction

Code	Code definition
P0354	Ignition coil D primary or secondary circuit malfunction
P0355	Ignition coil E primary or secondary circuit malfunction
Code	Code definition
P0356	Ignition coil F primary or secondary circuit malfunction
P0357	Ignition coil G primary or secondary circuit malfunction
P0358	Ignition coil H primary or secondary circuit malfunction
P0385	Crankshaft position sensor B circuit malfunction
P0386	Crankshaft position sensor B circuit, range or performance problem
P0401	Exhaust gas recirculation, insufficient flow detected
P0402	Exhaust gas recirculation, excessive flow detected
P0403	Exhaust gas recirculation circuit malfunction
P0404	Exhaust gas recirculation circuit, range or performance problem
P0105	Exhaust gas recirculation valve position sensor A circuit low
P0406	Exhaust gas recirculation valve position sensor A circuit high
P0410	Secondary air injection system malfunction
P0412	Secondary air injection system switching valve A, circuit malfunction
P0418	Secondary air injection system, relay A circuit malfunction
P0420	Catalyst system efficiency below threshold (bank 1)
P0440	Evaporative emission control system malfunction
P0441	Evaporative emission control system, incorrect purge flow
P0442	Evaporative emission control system, small leak detected
P0443	Evaporative emission control system, purge control valve circuit malfunction
P0446	Evaporative emission control system, vent control circuit malfunction
P0449	Evaporative emission control system, vent valve/solenoid circuit malfunction
P0452	Evaporative emission control system, pressure sensor low input
P0453	Evaporative emission control system, pressure sensor high input
P0454	Evaporative emission control system, pressure sensor intermittent

Code	Code definition
P0455	Evaporative emission (EVAP) control system leak detected (no purge flow or large leak)
P0461	Fuel level sensor circuit, range or performance problem
P0462	Fuel level sensor circuit, low input
P0463	Fuel level sensor circuit, high input
P0480	Cooling fan no. 1, control circuit malfunction
P0481	Cooling fan no. 2, control circuit malfunction
P0496	Evaporative emission system - high purge flow
P0502	Vehicle speed sensor circuit, low input
P0503	Vehicle speed sensor circuit, Intermittent, erratic or high input
P0506	Idle control system, rpm lower than expected
P0507	Idle control system, rpm higher than expected
P0520	Engine oil pressure sensor/switch circuit malfunction
P0522	Engine oil pressure sensor/switch circuit, low voltage
P0523	Engine oil pressure sensor/switch circuit, high voltage
P0530	A/C refrigerant pressure sensor, circuit malfunction
P0532	A/C refrigerant pressure sensor, low input
P0533	A/C refrigerant pressure sensor, high input
P0560	System voltage malfunction
P0561	System voltage unstable
P0562	System voltage low
P0563	System voltage high
P0574	Cruise control system - vehicle speed too high
P0600	Serial communication link malfunction
P0601	Internal control module, memory check sum error
P0602	Control module, programming error
P0603	Internal control module, keep alive memory (KAM) error
P0604	Internal control module, random access memory (RAM) error
P0605	Internal control module, read only memory (ROM) error
P0606	PCM processor fault
P0607	Control module performance
P0615	Starter relay - circuit malfunction
P0620	Generator control circuit malfunction
P0621	Generator lamp L, control circuit malfunction

5-digit trouble codes (continued)

Code	Code definition
P0622	Generator lamp F, control circuit malfunction
P0628	Fuel pump control - circuit low
P0629	Fuel pump control - circuit high
P0641	Sensor reference voltage A - circuit open
P0645	A/C clutch relay control circuit
Code	Code definition
P0650	Malfunction indicator lamp (MIL), control circuit malfunction
P0651	Sensor reference voltage B - circuit open
P0703	Torque converter/brake switch B, circuit malfunction
P0705	Transmission range sensor, circuit malfunction (PRNDL input)
P0706	Transmission range sensor circuit, range or performance problem
P0711	Transmission fluid temperature sensor circuit, range or performance problem
P0712	Transmission fluid temperature sensor circuit, low input
P0713	Transmission fluid temperature sensor circuit, high input
P0716	Input/turbine speed sensor circuit, range or performance problem
P0717	Input/turbine speed sensor circuit, no signal
P0719	Torque converter/brake switch B, circuit low

Code	Code definition
P0724	Torque converter/brake switch B circuit, high
P0730	Incorrect gear ratio
P0740	Torque converter clutch, circuit malfunction
P0741	Torque converter clutch, circuit performance or stuck in off position
P0742	Torque converter clutch circuit, stuck in on position
P0748	Pressure control solenoid, electrical problem
P0751	Shift solenoid A, performance problem or stuck in off position
P0752	Shift solenoid A, stuck in on position
P0753	Shift solenoid A, electrical problem
P0755	Shift solenoid B malfunction
P0756	Shift solenoid B, performance problem or stuck in off position
P0757	Shift solenoid B, stuck in on position
P0758	Shift solenoid B, electrical problem
P0856	Traction control input signal - malfunction
P0897	Transmission fluid deteriorated
P0973	Shift solenoid (SS) A - control circuit low
P0974	Shift solenoid (SS) A - control circuit high
P0976	Shift solenoid (SS) B - control circuit low
P0977	Shift solenoid (SS) B - control circuit high

3 Electronic Control Module (ECM)/Programmable Read Only Memory (PROM)/CALPAK/MEM-CAL/Powertrain Control Module

ECM REPLACEMENT

▶ **Refer to illustrations 3.1 and 3.5**

1   If you are replacing the ECM, check the service number on the new and old ECM to verify that they are the same (see illustration).

2   The Electronic Control Module (ECM) is located under the right hand side of the instrument panel.

3   Disconnect the negative battery cable from the battery.

✳ CAUTION:

If the vehicle is equipped with a Delco Loc II or Theftlock audio system, make sure you have the correct activation code before disconnecting the battery. See the information at the front of this manual for the radio re-activation procedure.

4   Remove the right sound insulator panel.

5   Remove the retaining bolts (see illustration) and carefully slide the ECM out far enough to unplug the electrical connector.

6   Unplug both electrical connectors from the ECM.

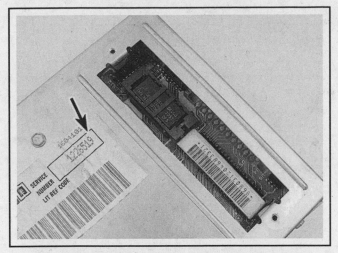

3.1 Make sure that the service numbers on the ECM and the PROM are the same - otherwise, depending on what you are replacing, you could end up with the wrong ECM or PROM

3.5 To remove the Electronic Control Module (ECM) from the vehicle, remove the mounting bolts and slide it out far enough to unplug the electrical connectors

❊❊ CAUTION:

The ignition switch must be turned off when pulling out or plugging in the connectors to prevent damage to the ECM.

7   Installation is the reverse of removal.

PROM

▶ **Refer to illustration 3.8**

8   To allow one model of ECM to be used for many different vehicles, a device called a PROM (Programmable Read Only Memory) is used (see illustration). The PROM is located inside the ECM and contains information on the vehicle's weight, engine, transaxle, axle ratio, etc. One ECM part number can be used by many GM vehicles but the PROM is very specific and must be used only in the vehicle for which it was designed. For this reason, it's essential to check the latest parts book and Service Bulletin information for the correct part number when replacing a PROM. An ECM purchased at a dealer doesn't come with a PROM. The PROM from the old ECM must be carefully removed and installed in the new ECM.

CALPAK

9   A device known as a CALPAK (see illustration 3.8) is used to allow fuel delivery if other parts of the ECM are damaged. The CAL-PAK has an access door in the ECM and replacement is the same as described for the PROM.

MEM-CAL (1985 through 1993 models)

10   The MEM-CAL contains the functions of the PROM, CALPAK and ESC module used on later models. Like the PROM, it contains the calibrations needed for a specific vehicle as well as the back-up fuel control circuitry required if the rest of the ECM is damaged or defective.

PROM, CALPAK OR MEM-CAL REPLACEMENT

▶ **Refer to illustrations 3.13, 3.14 and 3.15**

11   If you are replacing the PROM, CALPAK or MEM-CAL unit, check to make sure that the service numbers match on both the old and new parts.

12   Remove the access cover from the ECM/PCM.

13   Carefully grasp the PROM carrier at the ends (see illustration).

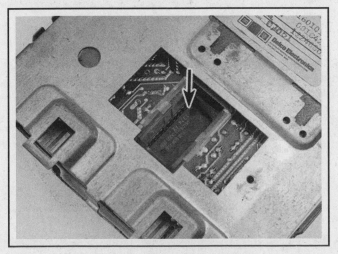

3.8 A typical PROM (arrow) installed in an ECM

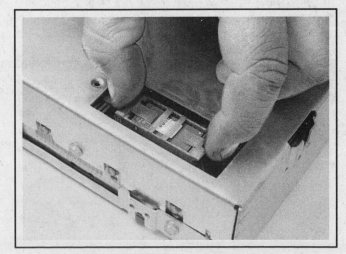

3.13 Grasp the PROM carrier at the ends and gently rock it and pull up until it is unplugged from the socket

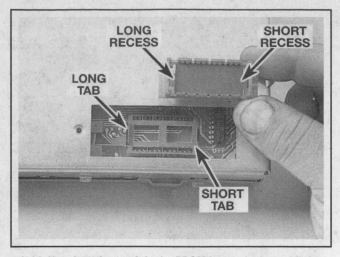

3.14 Note how the notch in the PROM is matched up with the smaller notch in the carrier

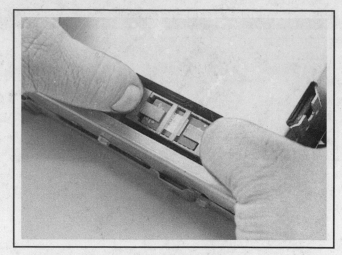

3.15 Press only on the ends of the PROM carrier - pressure on the area in between could result in bent or broken pins or damage to the PROM

Gently rock the carrier from end-to-end while pulling up firmly. The PROM carrier and PROM should lift off the PROM socket easily.

14 Note the reference end of the PROM carrier (see illustration) before setting it aside.

15 Position the carrier assembly squarely over the PROM socket with the small notched end of the carrier aligned with the small notch in the socket at the pin 1 end. Press on the carrier until it seats firmly in the socket (see illustration).

16 If the PROM is new, make sure the notch in the PROM is matched to the small notch in the carrier (see illustration 3.14).

※ **CAUTION:**

If the PROM is installed backwards and the ignition switch is turned on, the PROM will be destroyed.

---

17 Attach the access cover to the ECM and tighten the screw(s).

18 Install the ECM.

19 Start the engine and enter the diagnostic mode (see Section 2). If no trouble codes occur, the PROM is correctly installed.

POWERTRAIN CONTROL MODULE (PCM) REPLACEMENT (1996 AND LATER MODELS)

※ **CAUTION:**

The negative battery cable must be disconnected when pulling out or plugging in the connectors to prevent damage to the PCM.

➡ **Note 1:** If the PCM is going to be replaced on 1996 and 1997 models, remove the knock sensor module and reinstall it on the new PCM.

➡ **Note 2:** The powertrain control module on 1996 and later models performs basically the same function as the electronic control module used on previous model years, but it is not equipped with a replaceable PROM unit. If you replace the PCM on these models, the new unit will not be programmed. It must be programmed by the dealer for your specific year and model vehicle.

20 If you are replacing the PCM, check the service number on the new and old PCM to verify that they are the same.

21 The PCM is located in the front left-hand corner of the engine compartment.

22 Turn the ignition switch to the OFF position.

23 Detach the cable from the negative terminal of the battery.

※ **CAUTION:**

If the vehicle is equipped with a Delco Loc II or Theftlock audio system, make sure you have the correct activation code before disconnecting the battery. See the information at the front of this manual for the radio re-activation procedure.

---

24 Remove the air filter top cover and the air intake duct from the engine compartment.

25 On 1996 through 1999 models (except 1997 Park Avenue models), either unscrew the wing nuts or release the three hooks and remove the top cover from the PCM. Carefully pull the PCM unit up and out of the mounting bracket, then disconnect the electrical connectors. On 1997 Park Avenue and all 2000 and later models, release the two clips and carefully pull the PCM unit up and out of the lower air cleaner housing, then disconnect the electrical connectors.

26 Installation is the reverse of removal.

4   Information sensors

➡Note 1: See the component location illustrations at the front of this Chapter for the location of the following information sensors.

➡Note 2: After performing any checking procedure to any of the information sensors, be sure to clear the ECM of all trouble codes as described in Section 2.

❊❊ CAUTION:

If the vehicle is equipped with a Delco Loc II audio system, make sure you have the correct activation code before disconnecting the battery. See the information at the front of this manual for the radio re-activation procedure.

ENGINE COOLANT TEMPERATURE SENSOR

◆ Refer to illustrations 4.2 and 4.3

General description and check

1   The coolant sensor is a thermistor (a resistor which varies the value of its voltage output in accordance with temperature changes). A failure in the coolant sensor circuit should set a trouble code (see Section 2). This would indicate a failure in the coolant temperature circuit, so the appropriate solution to the problem will be either repair of a wire or replacement of the sensor. The sensor can also be checked with an ohmmeter, by measuring its resistance when cold, then warming up the engine and taking another measurement. If the difference in resistance readings is not approximately 500 ohms, the sensor is probably bad.

Replacement

2   To remove the sensor, release the locking tab (see illustration), unplug the electrical connector, then carefully unscrew the sensor.

❊❊ CAUTION:

Handle the coolant sensor with care. Damage to this sensor will affect the operation of the entire fuel injection system.

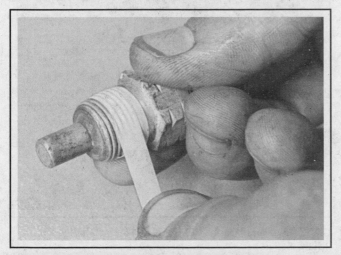

4.3 To prevent coolant leakage, be sure to wrap the temperature sensor threads with Teflon tape before installation

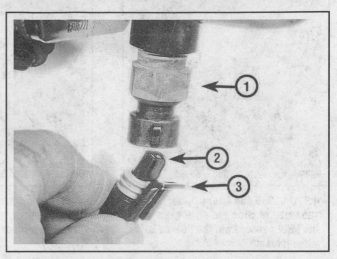

4.2 A typical engine coolant temperature sensor (1) electrical connector (2) has a locking tab (3) that must be released to unplug the connector

3   Before installing the new sensor, wrap the threads with Teflon sealing tape to prevent leakage and thread corrosion (see illustration).
4   Installation is the reverse of removal.

MANIFOLD ABSOLUTE PRESSURE (MAP) SENSOR

◆ Refer to illustrations 4.5 and 4.7

General description

5   The Manifold Absolute Pressure (MAP) sensor (see illustration) monitors the intake manifold pressure changes resulting from changes in engine load and speed and converts the information into a voltage output. The ECM uses the MAP sensor to control fuel delivery and ignition timing.

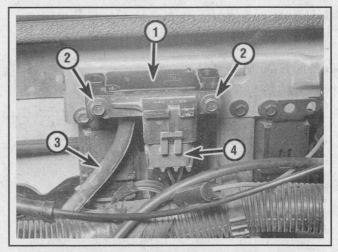

4.5 Typical MAP sensor installation details

1 MAP sensor	3 Vacuum line
2 Mounting screws	4 Electrical connector

4.7 On 1996 and later engines squeeze the clips and disengage the MAP sensor from the PCV valve housing

4.10 Removing an IAT sensor from the air cleaner housing (air cleaner removed from engine for clarity) (typical)

4.12a Typical Mass Airflow (MAF) sensor installation details (early models)

Check

6   A failure in the MAP sensor circuit should set a trouble code (see Section 2), but the operation of the sensor can also be checked using a high-impedance digital voltmeter. Unplug the electrical connector from the sensor and, using jumper wires, connect terminals A and C (the two outside terminals) to their corresponding terminals in the electrical connector. Connect the positive lead of the voltmeter to terminal B (the center terminal) of the sensor and the negative lead to ground. With the ignition On (engine not running) the voltage reading should be about 4.5 to 5 volts. Start the engine and let it warm up. The reading should now be different from the original reading, and should fluctuate with changes in engine rpm. If it doesn't, check the vacuum hose for breaks or blockage. If the hose is OK, the sensor is probably bad.

Replacement

7   To replace the sensor, detach the vacuum hose, unplug the electrical connector and remove the mounting screws (see illustration). Installation is the reverse of removal. Make sure the seal is in place and in good condition before reinstalling the MAP sensor.

MANIFOLD AIR TEMPERATURE (MAT) SENSOR OR INTAKE AIR TEMPERATURE (IAT) SENSOR

▶ **Refer to illustration 4.10**

General description

8   This sensor, located in the intake manifold air cleaner housing or air duct, is a thermistor (a resistor which changes the value of its voltage output as the temperature changes). The ECM uses the this signal to delay EGR until the manifold air temperature reaches 40-degrees F.

Check

9   A failure in the MAT/IAT sensor circuit should set either a trouble code (see Section 2). The sensor can also be checked with an ohmmeter, by measuring its resistance when cold, then warming it up (a hair dryer can be used for this) and taking another measurement. If the difference in resistance readings is not approximately 500 ohms, the sensor is probably bad.

Replacement

10   To remove a MAT/IAT sensor, unplug the electrical connector and remove the sensor with a wrench (see illustration).

11   Installation is the reverse of removal.

MASS AIR FLOW (MAF) SENSOR

▶ **Refer to illustration 4.12a and 4.12b**

General description

12   The Mass Air Flow (MAF) sensor, which is located in a housing between the air cleaner housing and the intake duct (see illustration), measures the amount of air entering the engine. On 2001 and later models, the MAF sensor is attached to the throttle body (see illustration). The ECM uses this information to control fuel delivery. A large quantity of air indicates acceleration, while a small quantity indicates deceleration or idle.

4.12b Typical Mass Airflow (MAF) sensor installation details (late models)

Check

13 If the sensor fails, a trouble code should set (see Section 2). A quick check of the sensor can also be made by tapping the flat portion of the sensor body with a screwdriver handle as the engine is running. If the engine stumbles or dies, the sensor is faulty.

Replacement

14 To replace the MAF sensor, unplug the electrical connector, loosen the clamps and detach the sensor from the air ducts.

15 Installation is the reverse of removal.

OXYGEN SENSOR

General description

16 The oxygen sensor is mounted in the exhaust system where it can monitor the oxygen content of the exhaust gas stream. By monitoring the voltage output of the oxygen sensor, the ECM will know what fuel mixture command to give the mixture control solenoid (carbureted models) or fuel injector(s). Later models have at least two oxygen sensors, one before the converter (in the exhaust manifold) and one after the converter. The PCM compares the two readings to determine converter efficiency.

17 The oxygen sensor produces no voltage when it's below its normal operating temperature of about 600-degrees F. During this initial period before warm-up, the ECM operates in open loop mode.

18 If the engine reaches normal operating temperature and/or has been running for two or more minutes, and if the oxygen sensor is producing a steady signal voltage between 0.35 and 0.55-volt, even though the TPS indicates the engine isn't at idle, the ECM will set a trouble code.

19 A delay of two minutes or more between engine start-up and normal operation or the sensor, followed by a low voltage signal or a short in the sensor circuit, will cause the ECM to set a code, indicating lean exhaust. If a high voltage signal occurs, the ECM will set a code indicating rich exhaust.

20 When any of the above codes occur, the ECM operates in the open loop mode - that is it controls fuel delivery in accordance with a programmed default value instead of feedback information from the oxygen sensor.

Check

21 An open in the oxygen sensor circuit should set a code indicating a faulty sensor or circuit. A low voltage in the circuit should set a code indicating lean exhaust. A high voltage in the circuit should set a code indicating rich exhaust. These codes may also be set as a result of fuel system problems.

22 The sensor can also be checked with a high-impedance digital voltmeter. Warm up the engine to normal operating temperature, then turn the engine off. Unplug the oxygen sensor electrical connector and connect the positive probe of the voltmeter to the sensor side of the connector.

> ✳✳ **CAUTION:**
>
> **Don't let the sensor wire or the voltmeter lead touch the exhaust pipe or manifold.**

Ground the negative probe of the meter, turn the meter to the millivolt setting and start the engine.

23 The reading on the voltmeter should fluctuate between 100 and 1,000 millivolts (0.1 and 1.0 volts). If the meter reading doesn't fluctuate, the sensor is probably bad (although a fuel system problem could be the cause).

Replacement

24 Refer to Section 5 for the oxygen sensor replacement procedure.

THROTTLE POSITION SENSOR (TPS)

25 The Throttle Position Sensor (TPS) is located on the TBI unit or throttle body.

26 By monitoring the output voltage from the TPS, the ECM can determine fuel delivery based on throttle valve angle (driver demand). A broken or loose TPS can cause intermittent bursts of fuel from the injector and an unstable idle because the ECM thinks the throttle is moving.

27 A problem in any of the TPS circuits will set a trouble code (see Section 2). Once a trouble code is set, the ECM will use an artificial default value for TPS and some vehicle performance will return.

28 Checking and replacement procedures for the TPS are contained in Chapter 4.

PARK/NEUTRAL (P/N) SWITCH

29 The Park/Neutral (P/N) switch, located on the rear upper part of the automatic transaxle, indicates to the ECM when the transaxle is in Park or Neutral. This information is used for Transaxle Converter Clutch (TCC), Exhaust Gas Recirculation (EGR) and Idle Air Control (IAC) valve operation.

> ✳✳ **CAUTION:**
>
> **The vehicle should not be driven with the Park/Neutral switch disconnected because idle quality will be adversely affected and a false trouble code may be set.**

30 For more information regarding the P/N switch, which is part of the Neutral start and back-up light switch assembly, see Chapter 7B.

AIR CONDITIONING (A/C) ON SIGNAL

31 This signal tells the ECM the A/C selector switch is in the On position and the high side low pressure switch is closed. The ECM uses this information to turn on the A/C and adjust the idle speed when the air conditioning system is working. If this signal isn't available to the ECM, idle may be rough, especially when the A/C compressor cycles.

32 If a trouble code related to the A/C On signal is set, check the circuit between the pressure switch and the ECM for an open or shorted condition.

VEHICLE SPEED SENSOR (VSS)

▶ **Refer to illustration 4.33**

General description and check

33 The Vehicle Speed Sensor (VSS) sends a pulsing voltage signal to the ECM, which the ECM converts to miles per hour. This sensor is used by the ECM to control the operation of the Torque Converter Clutch (TCC) system, as well as the speedometer/odometer. If a code

related to a faulty Vehicle Speed Sensor or circuit is set, remove the sensor and check for voltage fluctuations with a voltmeter while the drive gear is turned (see illustration). If a voltage fluctuation greater than 0.5-volts AC is not generated, replace the sensor.

Replacement

34 Disconnect the electrical connector.

35 Remove the VSS mounting bolt and remove the sensor from the transaxle case.

36 Installation is the reverse of removal, but be sure to use a new O-ring and lubricate it with clean automatic transmission fluid before installation.

CRANKSHAFT POSITION SENSOR

General description and check

37 The crankshaft position sensor sends a signal to the ECM to tell it both engine rpm and crankshaft position (see illustration 6.2a). A failure in this sensor or circuit should set a trouble code and usually will cause a no-start condition. Other symptoms include an occasional stumble at idle or stalling, usually after normal operating temperature has been reached. The vanes on the rear of the crankshaft balancer/vibration damper can also become damaged, which will affect the operation of the sensor (and in some cases may even damage the sensor physically).

Replacement

38 Refer to Chapter 2A and remove the vibration damper.

39 Unplug the electrical connector, remove the bolts and detach the sensor from the timing chain cover.

40 Position the crankshaft sensor on the timing chain cover and install the bolts, but don't tighten them yet. The sensor will have to be aligned before the bolts are tightened. A special tool is available for this purpose - check with your local auto parts store. If the tool is not available, temporarily install the vibration damper and adjust the position of the sensor so it straddles the vanes on the back of the damper, with equal spacing on either side of the vane, then tighten the bolts securely.

> ❊❊ **CAUTION:**
>
> **If this is not done, the sensor may contact the vanes on the damper, which will break the sensor and damage the vibration damper.**

CAMSHAFT POSITION SENSOR

General description and check

41 The camshaft position sensor is located in the timing chain cover, on early models it's above the CKS on the left side of the timing cover, and on 2001 and later models it's on the right side of the timing

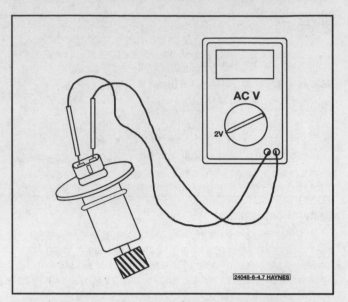

4.33 Connect the probes of a voltmeter directly to the VSS and observe A/C voltage fluctuations as the drive gear is rotated

cover (see illustration 6.2b). As the camshaft turns, a magnet on the camshaft activates the sensor as it passes by. This signal is used by the computer to synchronize the fuel injection system with the opening of the intake valves. In the event that the reference signal is lost or interrupted, the engine will usually not start. A trouble code should set in the event of a cam sensor or circuit malfunction. If a faulty sensor or circuit is indicated, always check the magnet on the camshaft sprocket before replacing the sensor or tracing the circuit for an open or shorted condition. This can be done by removing the sensor and turning the crankshaft with a large breaker bar and socket, looking through the sensor mounting hole for the presence of the magnet.

Replacement

42 Disconnect the cable from the negative terminal of the battery.

> ❊❊ **CAUTION:**
>
> **If the vehicle is equipped with a Delco Loc II audio system, make sure you have the correct activation code before disconnecting the battery. See the information at the front of this manual for the radio re-activation procedure.**

43 Unplug the electrical connector from the sensor, remove the sensor mounting bolt and pull the sensor out of the timing chain cover.

➡**Note: On 2001 and later models, remove the coolant reservoir first for access (see Chapter 3).**

44 Installation is the reverse of removal, but be sure to use a new O-ring and lubricate it with clean engine oil before installation.

5 Oxygen sensor - replacement

◆ **Refer to illustration 5.3**

➡**Note 1:** On 1994 and later models, the oxygen sensor is referred to as the Heated Oxygen Sensor (HO2S). The sensor is the same, only the name has changed.

➡**Note 2:** Because it's installed in the exhaust manifold or pipe, which contracts when cool, the oxygen sensor may be very difficult to loosen when the engine is cold. Rather than risk damage to the sensor (assuming you're planning to reuse it in another manifold or pipe), start and run the engine for a minute or two, then shut it off. Be careful not to burn yourself during the following procedure.

1 Disconnect the cable from the negative terminal of the battery.

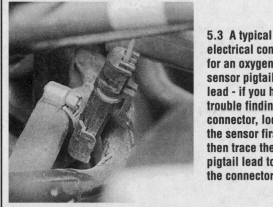

5.3 A typical electrical connector for an oxygen sensor pigtail lead - if you have trouble finding the connector, locate the sensor first, then trace the pigtail lead to the connector

❊ **CAUTION:**

If the vehicle is equipped with a Delco Loc II or Theftlock audio system, make sure you have the correct activation code before disconnecting the battery. See the information at the front of this manual for the radio re-activation procedure.

2 Raise the vehicle and support it securely on jackstands.
3 Carefully disconnect the electrical connector (see illustration).
4 Note the position of the silicone boot, if equipped, and carefully unscrew the sensor from the exhaust manifold. A special socket is available at auto parts store for removing oxygen sensors.

❊ **CAUTION:**

Excessive force may damage the threads.

5 Anti-seize compound must be used on the threads of the sensor to facilitate future removal. The threads of a new sensor will already be coated with it, but if an old sensor is removed and reinstalled, recoat the threads.

❊ **CAUTION:**

Do not get any anti-seize on the tip of the sensor or it will be ruined.

6 Installation is the reverse of removal.

6 Electronic Spark Timing (EST)

◆ **Refer to illustrations 6.2a and 6.2b**

1 To provide improved engine performance, fuel economy and control of exhaust emissions, the Electronic Control Module (ECM) controls spark advance (ignition timing) with the Electronic Spark Timing (EST) system.
2 The ECM receives a reference pulse from the crankshaft sensor, which indicates both engine rpm and crankshaft position (see illustration). The ECM also receives a reference pulse from the camshaft sensor (see illustration), which indicates the position of the number one piston on the power stroke. The ECM then determines the proper spark advance for the engine operating conditions and sends an EST pulse to the ignition module. A fault in the EST system will usually set a trouble code.

6.2a A typical crankshaft sensor (arrow)

6.2b A typical camshaft sensor, located near the water pump (arrow)

7 Electronic Spark Control (ESC) system

▶ **Refer to illustration 7.3**

GENERAL DESCRIPTION

1   Irregular octane levels in modern gasoline can cause detonation in an engine. Detonation is sometimes referred to as "spark knock."

2   The Electronic Spark Control (ESC) system is designed to retard spark timing up to 20-degrees to reduce spark knock in the engine. This allows the engine to use maximum spark advance to improve driveability and fuel economy.

3   The ESC knock sensor, which is located on the upper left end of the block under the throttle body (see Illustration), sends a voltage signal of 8 to 10-volts to the ECM when no spark knock is occurring and the ECM provides normal advance. When the knock sensor detects abnormal vibration (spark knock), the ESC module turns off the circuit to the ECM. The ECM then retards the timing until spark knock is eliminated.

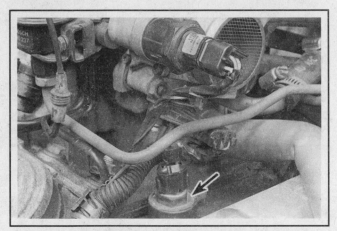

7.3 The Electronic Spark Control (ESC) knock sensor (arrow) is located on top of the block, underneath the throttle body

4   Failure of the ESC knock sensor signal or loss of ground at the ESC module will cause the signal to the ECM to remain high. This condition will result in the ECM controlling the EST as if no spark knock is occurring. Therefore, no retard will occur and spark knock may become severe under heavy engine load conditions. At this point, the ECM will set a trouble code (see Section 2).

5   Loss of the ESC signal to the ECM will cause the ECM to constantly retard EST. This will result in sluggish performance and also cause the ECM to set a trouble code.

KNOCK SENSOR REPLACEMENT

6   Detach the cable from the negative terminal of the battery.

✳✳ CAUTION:

If the vehicle is equipped with a Delco Loc II or Theftlock audio system, make sure you have the correct activation code before disconnecting the battery. See the information at the front of this manual for the radio re-activation procedure.

➡**Note: On 2001 and later models, there are two knock sensors located down low on the block, one on each side of the engine.**

7   Disconnect the wiring harness connector from the ESC sensor.

8   Remove the ESC sensor from the block. On models with block-mounted knock sensors, drain the cooling system before removal - the knock sensors are threaded into the block water jacket.

9   Installation is the reverse of the removal procedure.

✳✳ CAUTION:

Do not use Teflon tape or any kind of sealant on the threads of the knock sensors. The sensors already have a factory coating and any further coating will make the sensors less responsive.

8 Exhaust Gas Recirculation (EGR) system

GENERAL DESCRIPTION

1   The Exhaust Gas Recirculation (EGR) system is used to lower NOx (oxides of nitrogen) emission levels caused by high combustion temperatures. It does this by decreasing combustion temperature. The main element of the system is the EGR valve, which feeds small amounts of exhaust gas back into the combustion chamber.

2   The EGR valve is usually open during warm engine operation and anytime the engine is running above idle speed. The amount of gas recirculated is controlled by variations in vacuum and exhaust back pressure.

3   There are two types of EGR valves: a conventional EGR valve and a digital EGR valve. The conventional EGR valve is controlled by manifold vacuum, through an EGR solenoid valve, which is controlled by

the ECM. The digital EGR valve is electrically controlled directly by the ECM.

Digital EGR valve

▶ **Refer to illustration 8.5**

➡**Note: On 1994 and later models, the Digital EGR valve is replaced with a Linear EGR valve. The new type of valve operates similarly to the older version except the exhaust flow is controlled through a single orifice instead of three as on the digital EGR valve. Both types of valves perform the same function within the exhaust gas recirculation system.**

4   The digital EGR valve feeds small amounts of exhaust gas back into the intake manifold, which then flows into the combustion chamber.

8.5 Typical linear EGR valve assembly

A Electrical connector — *B Mounting bolts*

5 The digital EGR valve is designed to accurately supply exhaust gasses to the engine, independent of intake manifold vacuum. The valve controls EGR flow from the exhaust to the intake manifold through three orifices, which increment in size, to produce seven combinations. When a solenoid in the valve is energized, an armature, with an attached shaft and swivel pintle, is lifted, opening the orifice. The flow accuracy is dependent on metering orifice size only, which results in improved control (see illustration).

6 The digital EGR valve is opened by the ECM, grounding each solenoid circuit. This activates the solenoid, raises the pintle, and allows exhaust gas flow into the intake manifold. The exhaust gas then moves with the air/fuel mixture into the combustion chamber.

Conventional EGR valve

Refer to illustration 8.7

7 This valve is controlled by a flexible diaphragm which is spring loaded to hold the valve closed. Ported vacuum applied to the top side of the diaphragm overcomes the spring pressure and opens the valve in the exhaust gas port (see illustration).

8 The EGR vacuum control has a vacuum solenoid that uses pulse width modulation. This means the ECM turns the solenoid on and off many times a second and varies the amount of On time (the pulse width) to vary the amount of exhaust gas recirculated.

9 A diagnostic switch is part of the control and monitors vacuum to the EGR valve. This switch will trigger a Service Engine Soon light and set a trouble code in the event of a vacuum circuit failure.

CHECK

Conventional EGR valve

10 Disconnect the EGR solenoid vacuum harness. Rotate the harness and reinstall only the EGR valve side. Install a vacuum pump with a gauge on the manifold side of the EGR solenoid. Turn the ignition to On (engine stopped). Apply vacuum. Observe the EGR valve - it shouldn't move.

11 If the valve moves, disconnect the EGR solenoid electrical connector and repeat the test. If the valve still moves, replace the solenoid.

12 If the valve doesn't move, ground the diagnostic terminal and repeat the test. If the valve still doesn't move, replace the EGR valve.

8.7 Here's a typical conventional EGR valve, mounted on the firewall side of the intake manifold

13 If the valve moves, start the engine. Lift up on the EGR valve diaphragm and note the idle speed.

14 If there's no change in the idle, remove the EGR valve and check the passages for blockage. If the passages aren't plugged, replace the EGR valve.

15 Due to the complexity and the interrelationship with the ECM, any further checks should be left to a dealer service department.

Digital EGR valve

16 A special "scan" tool is needed to check this valve. Take the vehicle to a dealer service department.

COMPONENT REPLACEMENT

EGR valve

17 Disconnect the vacuum hose or electrical connector from the EGR valve.

18 Remove the fasteners securing the EGR valve to the intake manifold.

19 Separate the EGR valve from the engine.

20 In some cases, a conventional EGR valve and the passages in the intake manifold or adapter can be cleaned, and the old EGR valve can be returned to service (see below).

21 Using a scraper, remove all traces of old gasket material from the EGR valve (if it will be reused) and the manifold or adapter. Be careful not to gouge the delicate aluminum surfaces on the manifold or adapter.

22 Install the EGR valve, with a new gasket, on the intake manifold or adapter.

23 Installation is the reverse of removal.

EGR control solenoid (conventional systems only)

Refer to illustration 8.25

24 Disconnect the negative battery cable.

CAUTION:

If the vehicle is equipped with a Delco Loc II or Theftlock audio system, make sure you have the correct activation code before disconnecting the battery. See the information at the front of this manual for the radio re-activation procedure.

25 Unplug the solenoid electrical connector, then clearly label and remove the vacuum hoses (see illustration).

26 Remove the mounting nut and the solenoid.

27 Installation is the reverse of removal.

EGR VALVE CLEANING

> ❊❊ **CAUTION:**
>
> **On digital and linear EGR valves, never wash the valve in solvents or degreaser - both agents will permanently damage the solenoids in the armature assembly. Sandblasting is also not recommended because it will affect the sealing ability of the swivel pintle seals.**

28 Sometimes, a conventional EGR valve can be returned to service after cleaning it and the EGR passages in the manifold or adapter.

29 Inspect the valve pintle(s) for deposits.

30 Depress the valve diaphragm and check for deposits around the valve seat area.

31 Use a wire brush to carefully clean deposits from the pintle.

> ❊❊ **CAUTION:**
>
> **On 2001 and later models, do not use abrasives, wire brushes or scrapers on the pintle or the valve mounting surface. Wipe the pintle clean with a soft cloth only.**

8.25 A typical EGR control solenoid assembly

32 Remove any deposits from the valve outlet with a screwdriver.

33 If EGR passages in the intake manifold have an excessive build-up of deposits, the passages should be cleaned. Care should be taken to ensure that all loose particles are completely removed to prevent them from clogging the EGR valve or from being ingested into the engine.

➡ Note: It's a good idea to place a rag in the passage opening to keep debris from entering while cleaning the manifold.

9 Evaporative Emission Control System (EECS)

GENERAL DESCRIPTION

1   This system is designed to trap and store fuel vapors that evaporate from the fuel tank, throttle body and intake manifold.

2   The Evaporative Emission Control System (EECS) consists of a charcoal-filled canister and the lines connecting the canister to the fuel tank, ported vacuum and intake manifold vacuum.

3   Fuel vapors are transferred from the fuel tank, throttle body and intake manifold to a canister where they're stored when the engine isn't running. When the engine is running, the fuel vapors are purged from the canister by intake air flow and consumed in the normal combustion process.

4   On some engines, the ECM operates a solenoid valve (located on top of the canister) which controls vacuum to the purge valve in the charcoal canister. Under cold engine or idle conditions, the solenoid is turned on by the ECM, which closes the valve and blocks vacuum to the canister purge valve. The ECM turns off the solenoid valve and allows purge when the engine is warm.

CHECK

▶ **Refer to illustration 9.10**

5   Poor idle, stalling and poor driveability can be caused by an inoperative purge valve, a damaged canister, split or cracked hoses or hoses connected to the wrong tubes.

6   Evidence of fuel loss or fuel odor can be caused by liquid fuel leaking from fuel lines, a cracked or damaged canister, an inoperative purge valve or disconnected, misrouted, kinked, deteriorated or damaged vapor or control hoses.

7   Inspect each hose attached to the canister for kinks, leaks and cracks along its entire length. Repair or replace as necessary.

8   Inspect the canister. If it's cracked or damaged, replace it.

9   Look for fuel leaking from the bottom of the canister. If fuel is leaking, replace the canister and check the hoses and hose routing.

10   On models so equipped, check the filter at the bottom of the canister. If it's dirty, plugged or damaged, replace it (see illustration).

11   Any further testing should be left to a dealer service department.

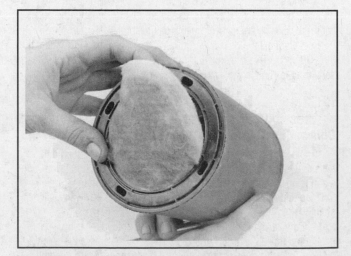

9.10 To replace the filter in the bottom of the canister, simply remove it and install another one

CHARCOAL CANISTER REPLACEMENT

♦ **Refer to illustration 9.13**

12 Detach the cable from the negative terminal of the battery.

> ⊗ **CAUTION:**
>
> **If the vehicle is equipped with a Delco Loc II or Theftlock audio system, make sure you have the correct activation code before disconnecting the battery. See the information at the front of this manual for the radio re-activation procedure.**

→**Note: On early models the canister is up front in the engine compartment, and on 2001 and later models it's at the rear of the vehicle, near the fuel tank.**

13 Unplug the solenoid electrical connector - if equipped (see illustration).

14 Clearly label, then detach the vacuum hoses.

15 Remove the canister mounting bolt and lift it out of the vehicle.

16 Installation is the reverse of removal.

9.13 A top view of the charcoal canister showing the solenoid electrical connector and the canister mounting bolt (arrows)

0 Positive Crankcase Ventilation (PCV) system

1 The Positive Crankcase Ventilation (PCV) system reduces hydrocarbon emissions by scavenging crankcase vapors. It does this by circulating fresh air from the air cleaner through the crankcase, where it mixes with blow-by gases and is then rerouted through a PCV valve to the intake manifold.

2 The main components of the PCV system are the PCV valve, a fresh air filtered inlet and the vacuum hoses connecting these two components with the engine and the EECS system.

3 To maintain idle quality, the PCV valve restricts the flow when the intake manifold vacuum is high. If abnormal operating conditions arise, the system is designed to allow excessive amounts of blow-by gases to flow back through the crankcase vent tube into the air cleaner to be consumed by normal combustion.

4 Checking and replacement of the PCV valve is covered in Chapter 1.

11 Transaxle Converter Clutch (TCC)

GENERAL DESCRIPTION

1 The Transaxle Converter Clutch (TCC) uses a solenoid-operated valve in the automatic transaxle to mechanically couple the engine flywheel to the output shaft of the transmission through the torque converter. This reduces the slippage losses in the converter, reducing emissions because engine rpm at any given speed is reduced. It also increases fuel economy.

2 For the converter clutch to operate properly, two conditions must be met:

a) *The engine must be warmed up before the clutch can apply. The engine coolant temperature sensor (see Section 4) tells the ECM when the engine is at normal operating temperature.*

b) *The vehicle must be traveling at the necessary minimum speed to raise the pressure to the level necessary to apply the valve. If the hydraulic pressure is correct, the ECM signals the solenoid to apply the converter clutch.*

3 After the converter clutch applies, the ECM uses the information from the TPS to release the clutch when the vehicle is accelerating or decelerating at a certain rate.

4 Another switch used in the TCC circuit is a brake switch, which opens the power supply to the TCC solenoid when the brake is applied.

5 A Third gear switch is placed in series on the battery side of the TCC solenoid to prevent TCC application until the transmission is in Third gear.

CHECKING

6 If the converter clutch is applied at all times, the engine will stall immediately, just like a manual transaxle with the clutch applied.

7 If the converter clutch doesn't apply, fuel economy may be lower than expected. If the Vehicle Speed Sensor (VSS) (see Section 4) fails, the TCC will not apply.

8 A TCC-equipped transaxle has different operating characteristics than an automatic transaxle without TCC. If you detect a "chuggle" or

"surge" condition, perform the following check.

9   Install a tachometer.

10   Drive the vehicle until normal operating temperature is reached, then maintain a 50 to 55 mph speed.

11   Lightly touch the brake pedal and check it for a slight bumpy sensation, indicating the TCC is releasing. A slight increase in rpm should also be noted.

12   Release the brake and check for reapplication of the converter clutch and a slight decrease in engine rpm.

13   If the TCC fails to perform satisfactorily during this test, take the vehicle to a dealer service department to have the TCC serviced.

12 Catalytic converter

GENERAL DESCRIPTION

1   The catalytic converter is an emission control device added to the exhaust system to reduce pollutants from the exhaust gas stream. A single-bed converter design is used in combination with a three-way (reduction) catalyst. The catalytic coating on the three-way catalyst contains platinum and rhodium, which lowers the levels of oxides of nitrogen (NOx) as well as hydrocarbons (HC) and carbon monoxide (CO).

CHECKING

2   The test equipment for a catalytic converter is expensive and highly sophisticated. If you suspect the converter is malfunctioning, take it to a dealer service department or authorized emissions inspection facility for diagnosis and repair. On models with upstream and downstream oxygen sensors, a faulty converter will set a diagnostic trouble code.

3   Whenever the vehicle is raised for servicing of underbody components, check the converter for leaks, corrosion and other damage. If damage is discovered, the converter should be replaced.

4   Because the converter is welded to the exhaust system, converter replacement requires removal of the exhaust pipe assembly (see Chapter 4). Take the vehicle, or the exhaust pipe system, to a dealer service department or a muffler shop.

5   Although catalytic converters don't break too often, they do become plugged. The easiest way to check for a restricted converter is to use a vacuum gauge to diagnose the effect of a blocked exhaust on intake vacuum.

   a) *Open the throttle until the engine speed is about 2000 RPM.*
   b) *Release the throttle quickly.*
   c) *If there is no restriction, the gauge will quickly drop to not more than 2 in Hg or more above its normal reading.*
   d) *If the gauge does not show 5 in Hg or more above its normal reading, or seems to momentarily hover around its highest reading for a moment before it returns, the exhaust system, or the converter, is plugged (or an exhaust pipe is bent or dented or the core inside the muffler has shifted).*

7 AUTOMATIC TRANSAXLE

Section

Reference to other Chapters

Automatic transaxle fluid and filter change - See Chapter 1
Automatic transaxle fluid level check - See Chapter 1
Neutral start switch check - See Chapter 1

1 General information

✳✳ CAUTION:

If the vehicle is equipped with a Delco Loc II or Theftlock audio system, make sure you have the correct activation code before disconnecting the battery. See the information at the front of this manual for the radio re-activation procedure.

Due to the complexity of the clutches and the hydraulic control system, and because of the special tools and expertise required to perform an automatic transaxle overhaul, it should not be undertaken by the home mechanic. Therefore, the procedures in this Chapter are limited to general diagnosis, routine maintenance, adjustment and transaxle removal and installation.

If the transaxle requires major repair work, it should be left to a dealer service department or an automotive or transmission repair shop. You can, however, remove and install the transaxle yourself and save the expense, even if the repair work is done by a transmission shop.

Replacement and adjustment procedures the home mechanic can perform include those involving the throttle valve (TV) cable (1985 through 1991 models) and the shift linkage.

✳✳ CAUTION:

Never tow a disabled vehicle with an automatic transaxle at speeds greater than 35 mph or distances over 50 miles.

2 Diagnosis - general

1 Automatic transaxle malfunctions may be caused by a number of conditions, such as poor engine performance, improper adjustments, hydraulic malfunctions and mechanical problems.

2 The first check should be of the transaxle fluid level and condition. Refer to Chapter 1 for more information. Unless the fluid and filter have been recently changed, drain the fluid and replace the filter (also in Chapter 1).

3 Road test the vehicle and drive in all the various selective ranges, noting discrepancies in operation.

4 Verify that the engine isn't at fault. If the engine hasn't had a tune-up recently, refer to Chapter 1 and make sure all engine components

are functioning properly.

5 On 1985 through 1991 models, check the adjustment of the throttle valve (TV) cable (see Section 3).

6 Check the condition of all vacuum and electrical lines and fittings at the transaxle, or leading to it.

7 Check for proper adjustment of the shift control cable (see Section 5). On 1995 and later models, use a scan tool to check for transmission diagnostic trouble codes.

8 If at this point a problem remains, there is one final check before the transaxle is removed for overhaul. The vehicle should be taken to a shop for a line pressure check.

Underside view of the automatic transaxle

1	Transaxle pan	3	Left driveaxle	5	Front mount (obscured by lower crossmember)
2	Right driveaxle	4	Rear mount		

3   Throttle valve (TV) cable (1985 through 1991 models) -   replacement and adjustment

➡Note: On 1992 and later models, the 4T60-E and 4T65-E electronic controlled transmission is not equipped with the Throttle Valve (TV) cable assembly. The 4T60 transmission functions that are controlled by the TV cable are controlled electronically in the 4T60-E and 4T65-E transmissions.

REPLACEMENT

◆ Refer to illustrations 3.2, 3.3, 3.5 and 3.6

1   Detach the cable from the negative battery terminal.

❋❋ CAUTION:

If the vehicle is equipped with a Delco Loc II or Theftlock audio system, make sure you have the correct activation code before disconnecting the battery. See the information at the front of this manual for the radio re-activation procedure.

2   Disconnect the TV cable from the throttle lever by grasping the connector, pulling it forward to disconnect it and then lifting up and off the lever pin (see illustration).

3   Disconnect the TV cable housing from the bracket by compressing the tangs and pushing the housing back through the bracket (see illustration).

4   Disconnect any clips or straps retaining the cable to the transaxle.

5   Remove the bolt retaining the cable to the transaxle (see illustration).

6   Pull up on the cover until the end of the cable can be seen, then disconnect it from the transaxle TV link (see illustration). Remove the cable from the vehicle.

7   To install the cable, connect it to the transaxle TV link and install the bolt. Tighten the bolt to the specified torque and push the cover securely over the cable. Route the cable to the top of the engine, push the housing through the bracket until it clicks into place, place the connector over the throttle lever pin and pull back to lock it. Secure the cable with any retaining clips or straps.

3.2   To detach the TV cable from the throttle lever pin, grasp it firmly, move it forward and lift up

3.3   Use needle-nose pliers to compress the TV cable tangs, then push the housing back through the bracket

3.5   Remove the TV cable bolt (arrow) and pull up on the cable until it's out of the transmission

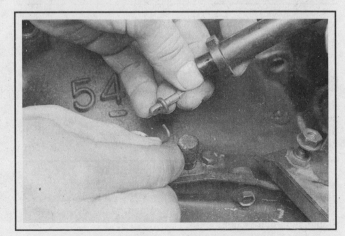

3.6   Hold the transaxle TV link and slide the cable link off the pin

3.9a To adjust the TV cable, press down on the re-adjust tab, move the slider back against the fitting until it stops, release the re-adjust tab and rotate the throttle lever toward the wide open position until you hear an audible click

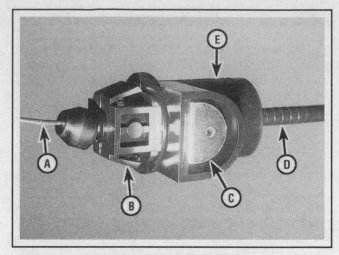

3.9b Details of the TV cable adjuster assembly

A	TV cable	D	Cable casing
B	Locking lugs	E	Slider
C	Release tab		

ADJUSTMENT

▶ **Refer to illustrations 3.9a and 3.9b**

8   The engine MUST NOT be running during this adjustment.

9   Depress the re-adjust tab and push the slider through the fitting (away from the throttle lever) as far as it will go (see illustrations).

10   Release the re-adjust tab.

11   Manually turn the throttle lever to the "wide open throttle" position until the re-adjust tab makes an audible click, then release the throttle lever. The cable is now adjusted.

➡Note: Don't use excessive force at the throttle lever to adjust the TV cable. If great effort is required to adjust the cable, disconnect the cable at the transaxle end and check for free operation. If it's still difficult, replace the cable. If it's now free, suspect a bent TV link in the transaxle or a problem with the throttle lever.

12   Reattach the cable to the negative battery terminal.

4	Neutral start switch - replacement and adjustment

REPLACEMENT

▶ **Refer to illustrations 4.2, 4.4 and 4.5**

1   Disconnect the negative cable from the battery.

❋ CAUTION:

If the vehicle is equipped with a Delco Loc II or Theftlock audio system, make sure you have the correct activation code before disconnecting the battery. See the information at the front of this manual for the radio re-activation procedure.

➡Note: On 2001 and later models, the neutral start function is performed by an internal mode switch (IMS). It is not service-able by the home mechanic, but a problem in the IMS will set a diagnostic trouble code.

2   Remove the cruise control servo assembly (see illustration).

3   Shift the transaxle into Neutral.

4   Disconnect the shift linkage (see illustration).

4.2 You'll need to remove the cruise control servo assembly to get at the neutral start switch: detach the vacuum hoses, remove the bracket nuts (arrows) and hang the cruise control servo assembly out of the way with a piece of wire

4.4 Detach the cable from the shift lever by prying it off with a screwdriver

5   Trace the wire harness from the neutral start switch to the connector (see illustration) and unplug it.

6   Remove the bolts (see illustration 4.9) and detach the switch.

7   To install the switch, line up the flats on the shift shaft with the flats in the switch and lower the switch onto the shaft.

8   Install the bolts. If the switch is new and the shaft hasn't been moved, tighten the bolts. If the switch requires adjustment, leave the bolts loose and follow the adjustment procedure below. The remainder of installation is the reverse of removal.

4.5 Unplug the electrical connector from the neutral start switch

ADJUSTMENT

9   Insert a 3/32-inch drill bit into the gauge hole in the side of the switch.

10 Rotate the switch until the drill bit can be felt dropping into the switch, indicating that it's now in the Neutral position. Tighten the switch bolts.

11 Install the cruise control assembly.

12 Connect the negative battery cable and verify that the engine will start only in Neutral or Park.

---

5   Automatic transaxle shift cable - replacement and adjustment

1   Disconnect the negative cable from the battery.

✳✳ CAUTION:

If the vehicle is equipped with a Delco Loc II or Theftlock audio system, make sure you have the correct activation code before disconnecting the battery. See the information at the front of this manual for the radio re-activation procedure.

REPLACEMENT

Floor shift

▶ **Refer to illustrations 5.4a, 5.4b and 5.6**

2   Working in the engine compartment, disconnect the shift cable from the transaxle lever (see illustration 4.4).

3   Remove the console between the seats (see Chapter 11).

4   Disconnect the shift cable from the shift lever and bracket (see illustrations).

5.4a To detach the cable at the shift control assembly, pull off the clip with needle-nose pliers . . .

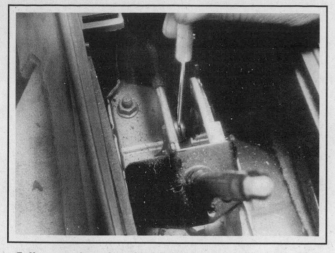

5.4b . . . and pry the cable off the lever pin with a screwdriver (single-post type floor shift shown, dual-type similar)

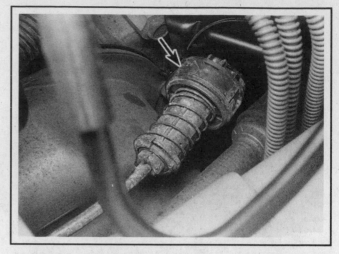

5.6 To detach the shift cable from the transaxle bracket, grasp the housing (arrow) firmly, cock it to one side and pull - though you can't see them in the photo, there's a tang on each side of the housing - the two tangs locate the housing in a square hole in the bracket, so when you rock the housing to one side, one of the tangs comes free (bracket design identical for column shift models)

5   Remove the right and left side sound insulators from the under dash portion of the console, then pull back the carpet for access to the cable (see Chapter 11).

6   Working in the engine compartment, detach the cable from the transaxle bracket (see illustration). On later models, disconnect the cable end by pulling up a clip at the bracket end.

7   Trace the cable up to the firewall grommet. Remove the screws from the grommet retainer, detach the grommet and retainer from the firewall and pull the cable assembly through the firewall.

8   Installation is the reverse of removal. After installation, adjust the cable as described below.

Column shift

9   Working in the engine compartment, disconnect the cable from the transaxle lever and bracket (see illustrations 4.4 and 5.6).

10   Working in the passenger compartment, remove the left sound

insulator located under the dash (see Chapter 12).

11   Disconnect the cable bracket on the steering column and detach the cable from the column shift lever.

12   Dislodge the grommet in the firewall and withdraw the cable from the vehicle.

13   Installation is the reverse of removal. After installation, adjust the cable as described below.

ADJUSTMENT

14   Place the shift lever and the transaxle lever in Neutral, then push the locking tab on the shift cable to automatically adjust the cable.

15   Reconnect the negative battery cable.

---

6   Automatic transaxle park/lock cable - removal and installation

➡Note: Some early column shift models don't have a park/lock cable.

REMOVAL

Floor shift models

1   Disconnect the negative cable from the battery.

❊❊ CAUTION:

If the vehicle is equipped with a Delco Loc II or Theftlock audio system, make sure you have the correct activation code before disconnecting the battery. See the information at the front of this manual for the radio re-activation procedure.

2   Remove the console (see Chapter 11).

3   Place the shift lever in Park and the ignition switch in the Run position.

4   Insert a screwdriver blade into the slot in the ignition switch inhibitor, depress the cable latch and detach the cable.

5   Push the cable connector lock button (located at the shift control base) to the up position and detach the cable from the park lock lever pin. Depress the two cable connector latches and remove the cable from the shift control base.

6   Remove the cable clips.

INSTALLATION

7   Make sure the cable lock button is in the up position and the shift lever is in Park. Snap the cable connector into the shift control base.

8   With the ignition key in the Run position (this is very important), snap the cable into the inhibitor housing. Do not attempt to insert the cable with the key in any other position.

9   Turn the ignition key to the Lock position.

10 Snap the end of the cable onto the shifter park/lock pin.

11 Push the nose of the cable connector forward to remove the slack.

12 With no load on the connector nose, snap down the cable connector lock button.

13 Check the operation of the park/lock cable as follows.

a) *With the shift lever in Park and the key in Lock, make sure the shift lever cannot be moved to another position and the key can be removed.*

b) *With the key in Run and the shift lever in Neutral, make sure the key cannot be turned to Lock.*

14 If it operates as described above, the park/lock cable system is properly adjusted.

15 If the park/lock system doesn't operate as described, return the cable connector lock to the up position and repeat the adjustment procedure. Push the cable connector down and recheck the operation.

Column shift models

16 Remove the driver's side lower sound panel from below the instrument panel, then remove the steering column covers (see Chapter 11).

17 Later models have an electronic shift lock actuator at the base of the steering column that is connected to the shift-lock system only with a wire harness, not a cable.

18 Disconnect the electrical connector at the base of the actuator. With a flat-bladed tool, pry the actuator from the steering column to remove it.

19 Installation of the shift lock actuator is the reverse of the removal procedure.

20 To adjust the actuator after installation, pull out the plastic tab on the side of the adjuster block.

21 By squeezing the sides of the adjuster block, you can slide it away from the shift lock solenoid.

22 Put the column shift lever into the Park position, then push the tab back into the adjuster block.

7   Transaxle differential seals - replacement

♦ **Refer to illustrations 7.3 and 7.6**

1   Raise the vehicle and support it securely on jackstands.

2   Remove the driveaxle(s) (see Chapter 8).

3   Use a hammer and chisel to pry up the outer lip of the seal to dislodge it so it can be pried out of the housing (see illustration).

4   Compare the new seal to the old one to make sure they're the same.

5   Coat the lips of the new seal with transmission fluid.

6   Place the new seal in position and tap it into the bore with a hammer and a large socket or a piece of pipe that's the same diameter as the outside edge of the seal (see illustration).

7   Reinstall the various components in the reverse order of removal.

7.3 Dislodge the metal-type differential seal by working around the outer edge with a chisel and hammer

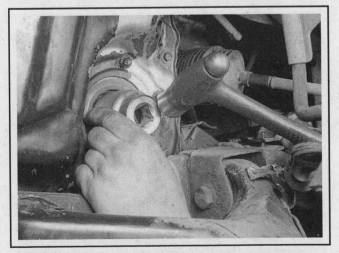

7.6 Tap the new seal into place with a large socket and hammer - use a socket with a diameter slightly smaller than the outside diameter of the seal itself

8 Automatic transaxle - removal and installation

♦ **Refer to illustrations 8.4, 8.6, 8.7, 8.8, 8.9, 8.10, 8.14, 8.15, 8.16, 8.18, 8.19, 8.20a, 8.20b, 8.21, 8.22, 8.23a, 8.23b, 8.23c, 8.23d, 8.24a, 8.24b, 8.25, 8.26, 8.27, 8.28 and 8.29**

REMOVAL

1   Detach the cable from the negative battery terminal.

2   Unplug the connector at the Mass Air Flow (MAF) sensor, loosen the hose clamps at the air intake duct and remove the duct and the MAF sensor as an assembly (see Chapter 4).

3   On 1985 through 1991 models, detach the TV cable from the throttle body linkage and the transaxle (see Section 3).

4   If the vehicle has cruise control, detach the cruise control cable from the throttle body, detach the vacuum hoses at the servo, remove the servo mounting bracket bolts (see illustration 4.2), unplug the servo electrical connector (see illustration) and remove the servo assembly.

5   Detach the shift control linkage from the mounting bracket on the transaxle and the lever at the manual shaft (see Section 5).

6   Unplug the wiring connectors at the neutral start

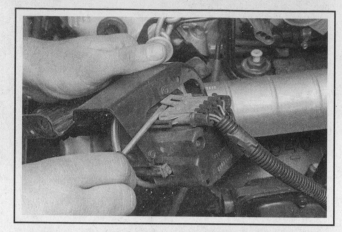

8.4 After you've detached the cruise control cable from the throttle body, detached the vacuum hoses from the cruise control servo and removed the servo mounting bracket bolts, unplug the electrical connector and remove the servo assembly

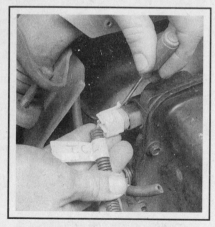

8.6 Unplug the wiring connector from the torque converter clutch

8.7 Unplug the vacuum hose at the vacuum modulator

8.8 Remove the dipstick tube bracket bolt and detach the dipstick tube assembly

8.9 Remove the upper transaxle-to-engine bolts (arrows)

8.10 Attach an engine support fixture or hoist - be sure to load the support fixture by raising the engine slightly to relieve tension on the frame and mounts

8.14 Unplug the vehicle speed sensor connector (arrow)

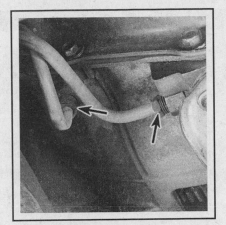

8.15 Unscrew and detach the oil cooler lines (arrows) from the transaxle

8.16 Detach this wire harness clip from the lower crossmember

switch (see illustration 4.5) and the torque converter clutch (see illustration). On later models, unclip the wiring harness from the bracket on the transmission.

7   Unplug the vacuum hose at the vacuum modulator on earlier models (see illustration).

8   Remove the dipstick tube bracket bolt (see illustration) and

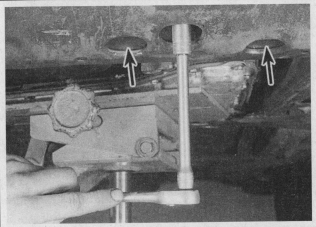

8.18 Remove the left front transaxle mount (arrows)

remove the dipstick tube assembly. Plug the hole to prevent leakage.

9   Remove the upper transaxle-to-engine bolts (see illustration).

10 Attach an engine support fixture or hoist (see illustration). Be sure to "load" the support fixture by raising the engine slightly to relieve tension on the frame and mounts.

11 Loosen the wheel lug nuts.

12 Raise the vehicle and place it securely on jackstands.

13 Remove both the front wheels.

14 Unplug the vehicle speed sensor (see illustration).

15 Detach the oil cooler lines from the transaxle and plug them (see illustration).

16 Detach the wire harness clip from the lower crossmember (see illustration).

17 Remove both driveaxles (see Chapter 8).

18 Remove the left front transaxle mount (see illustration).

19 Remove the right front engine mount-to-frame nuts (see illustration).

20 Remove the left rear transaxle mount-to-transaxle bolts (see illustration). Then remove the left rear transaxle mount-to-frame nuts (see illustration) and remove the mount.

21 Remove the right rear transaxle mount (see illustration). Unclip the power steering lines from the subframe.

22 Place a transaxle stand, or a floor jack, in position under the transaxle (see illustration). If you're using a floor jack, be sure to place a

8.19 Remove the right front engine mount-to-frame nuts (arrows)

8.20a Remove the left rear transaxle mount-to-transaxle bolts (arrows) . . .

8.20b . . . then remove the left rear transaxle mount-to-frame nuts (arrows)

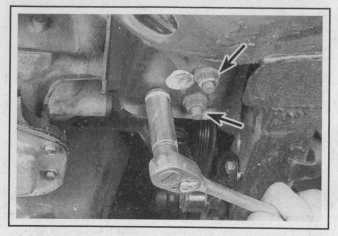

8.21 Remove the right rear transaxle mount (arrows)

wooden board between the jack and the transaxle to protect the transaxle pan.

➡**Note: Refer to Chapter 10 and loosen the steering-shaft pinch bolt, then unbolt the steering rack from the subframe and wire the rack to the vehicle. Disconnect the tie-rod ends at the steering knuckles and separate the lower balljoints.**

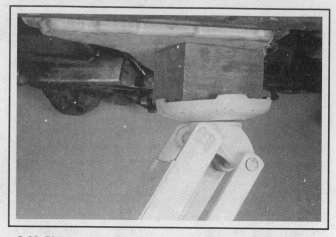

8.22 Place a transaxle stand or floor jack in position under the transaxle pan - if you're using a floor jack, be sure to put a board between the jack and the transaxle to protect the pan

23 Remove the engine subframe bolts (see illustrations). There are three bolts at the right front of the subframe and four bolts at the left rear (one of the four rear bolts is also a stabilizer bar clamp attaching bolts).

8.23a Remove the two right front engine subframe bolts . . .

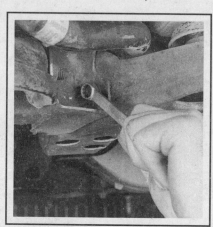

8.23b . . . then remove the four left rear subframe bolts - the lower left bolt is on the wheel well side . . .

8.23c . . . the lower right bolt is on the engine compartment side . . .

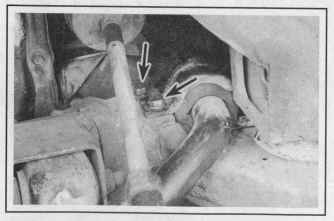

8.23d . . . and the two upper bolts, which can also be accessed through the wheel well, are on top of the corner of the subframe assembly (one of them is also a bushing clamp bolt for the stabilizer bar)

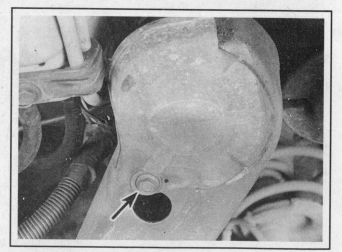

8.24a Remove the cover (arrow) . . .

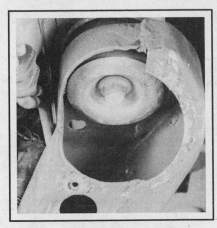

8.24b . . . then remove the front left frame-to-body bolt

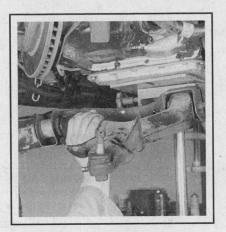

8.25 Remove the subframe assembly - this assembly is fairly heavy, so you may want assistance

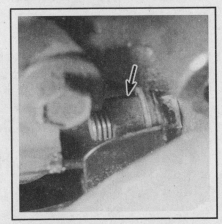

8.26 Remove the torque converter access cover

24 Remove the front left frame-to-body attaching bolt (see illustrations).

25 Remove the subframe assembly (see illustration).

26 Remove the torque converter access cover (see illustration).

27 Mark the converter-to-driveplate relationship to ensure proper reassembly (see illustration).

28 Remove the driveplate-to-converter bolts (see illustration).

29 Remove the two lower transaxle-to-engine fasteners (see illustration).

➡**Note: One of them is located between the transaxle case and the engine block and is installed in the opposite direction from the other fasteners.**

30 Make a final check that all wiring, cables, etc. which could interfere with transaxle removal are disconnected or moved out of the way.

31 To separate the transaxle from the engine, pry the bellhousing away very carefully with a large screwdriver or pry bar. Secure the torque converter to the transaxle so it won't fall out of the bellhousing during removal.

32 Carefully lower the transaxle assembly from the vehicle. Make sure you don't damage the driveaxle shaft seals or any hoses, lines or wiring, during removal.

INSTALLATION

33 Inspect the driveplate for missing weights, cracks, corrosion and damaged or broken starter gear teeth. Refer to Chapter 2 if the driveplate must be replaced.

34 If the engine rear main oil seal is leaking, refer to Chapter 2 for the replacement procedure.

35 Inspect all hoses, cables and wires that connect to the transaxle for damage and repair or replace them as necessary.

36 Whenever the transaxle is removed for overhaul or replacement of the torque converter, pump or case, make sure you flush the transaxle oil cooler lines prior to installation.

37 Make sure the torque converter hub is fully engaged in the oil pump.

38 Place the transaxle on a transaxle stand or a floor jack and raise it into position.

39 Install the lower transaxle-to-engine bolts and tighten them to the specified torque.

40 Install the rear transaxle mount bolts and tighten them securely.

41 Install the subframe assembly.

8.27 Mark the converter-to-driveplate relationship

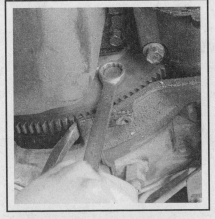

8.28 Remove the driveplate-to-converter bolts - wedge a large screwdriver or prybar between the bellhousing and the driveplate teeth to prevent the crankshaft from turning

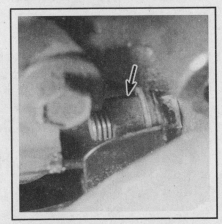

8.29 Remove the lower transaxle-to-engine stud nut (arrow) - a bolt (not visible in this photo) is located between the transaxle case and the engine block and is installed in the opposite direction

➡**Note: When installing the subframe assembly, be sure to align the studs for the front engine and transaxle mounts.**

42 Install the front transaxle mount and front engine mount bolts.

43 Install the torque converter bolts.

44 Install the torque converter access cover.

45 Attach the oil cooler lines to the transaxle and tighten them securely.

46 Attach the vehicle speed sensor connector.

47 Install the driveaxles (see Chapter 8).

48 Attach the balljoint to the steering knuckle (see Chapter 10).

49 Attach the stabilizer bar link bolt to the control arm (see Chapter 10).

50 Attach the outer tie-rods to the steering knuckles (see Chapter 10).

51 Install the wheels and hand tighten the lug nuts.

52 Lower the vehicle and tighten the lug nuts to the torque specified in Chapter 1.

53 Install the upper transaxle-to-engine bolts and tighten securely.

54 Remove the engine support fixture.

55 Install the dipstick tube assembly.

56 On 1985 through 1991 models, attach the TV cable to the transaxle (see Section 3).

57 Attach the shift cable to the shift lever (see Section 5).

58 Attach the starter safety switch and torque converter clutch electrical connectors.

59 Attach the vacuum modulator hose.

60 Install the cruise control servo assembly.

61 Install the air intake duct (see Chapter 4).

62 Attach the cable to the negative battery terminal.

63 Start the engine and check for leaks.

Specifications

Torque specifications

	Ft-lbs (unless otherwise indicated)
TV cable bolt	72 in-lbs
Transaxle-to-engine block bolts	
4T60-E	55
4T65E	
Models through 2000	83
2001 and later	55
Front left frame-to-body attachment bolt	74
Frame-to-body bolts, 2001 and later	141
Driveplate-to-converter bolts	46
Right rear transaxle mount	
Nuts	30
Bolts	40
Left transaxle mount bolts	
Models through 2000	30
2001 and later	
Bolts	46
Upper nut and lower bolt/nut	37
Right front transaxle mount-to-transaxle nuts	
Models through 2000	30
2001 and later	
Bracket bolts	42
Lower nut	52
Upper nut	59
Left front transaxle mount	
Nuts	30
Bolts	40
Strut bracket-to-transaxle bolt(s)	
4T60-E	46
4T65E	
Models through 2000	41
2001 and later	48

Section

Reference to other Chapters

8

DRIVEAXLES

1   General information

Power is transmitted from the transaxle to the front wheels by two driveaxles, which consist of splined axleshafts with constant velocity (CV) joints at each end. There are two types of inner CV joints used. On certain models a double-offset design using ball bearings with an inner and outer race is used to allow angular, as well as axial movement. The other CV joint used is a tri-pot design, with a spider bearing assembly and housing. To determine which CV joint is used on your vehicle, look at the housing - the tri-pot housing will have three major indentations in it. All outer CV joints are the double-offset type, and allow angular, but not axial, movement.

The CV joints are protected by rubber boots, which are retained by clamps so the joints are protected from water and dirt. The boots should be inspected periodically (see Chapter 1). Damaged CV joint boots must be replaced immediately or the joints can be damaged. Boot replacement involves removing the driveaxles (see Section 2). It's a good idea to disassemble, clean, inspect and repack the CV joint whenever replacing a CV joint boot to make sure the joint isn't contaminated with moisture or dirt, which would cause premature failure of the CV joint.

The most common symptom of worn or damaged CV joints, besides lubricant leaks, are a clicking noise in turns, a clunk when accelerating from a coasting condition or vibration at highway speeds.

2   Driveaxles - removal and installation

⬥ **Refer to illustrations 2.2, 2.5a, 2.5b, 2.6 and 2.8**

REMOVAL

1   Remove the wheel cover and loosen the hub nut. Loosen the driveaxle/hub nut 1/2-turn. Loosen the wheel lug nuts, raise the front of the vehicle and support it securely on jackstands. Apply the parking brake and block the rear wheels to keep the vehicle from rolling off the jackstands. Remove the front wheel.

2   Remove the driveaxle/hub nut. To prevent the hub from turning, insert a screwdriver through the caliper and into a rotor cooling vane, then remove the nut (see illustration).

3   Remove the brake caliper and disc and support the caliper out of the way with a piece of wire (see Chapter 9).

4   Remove the control arm-to-steering knuckle balljoint stud nut and separate the lower arm from the steering knuckle (see Chapter 10 if necessary).

5   Push the driveaxle out of the hub with a puller, then support the outer end of the driveaxle with a piece of wire to prevent damage to the inner CV joint (see illustrations).

6   Carefully pry the inner end of the driveaxle out of the transaxle,

2.2 A prybar will hold the hub stationary while loosening the driveaxle/hub nut

using a large prybar positioned between the transaxle housing and the CV joint housing (see illustration).

7   Support the CV joints and carefully remove the driveaxle from the vehicle.

2.5a A two-jaw puller works well for pushing the stub axle out of the hub

2.5b Support the driveaxle with a piece of wire after it has been freed from the hub - don't let it hang unsupported or the CV joint could be damaged

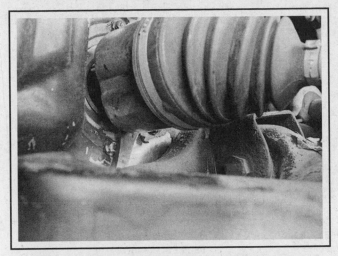

2.6 Use a large prybar positioned as shown to "pop" the inner CV joint out of the transaxle - it may be necessary to tap the prybar with a hammer if the driveaxle is stuck

INSTALLATION

8   Lubricate the differential seal with multi-purpose grease, raise tho driveaxle into position while supporting the CV joints and insert the splined end of the inner CV joint into the differential side gear. Seat the shaft in the side gear by positioning the end of a screwdriver in the groove in the CV joint and tapping it into position with a hammer (see illustration).

9   Apply a light coat of multi-purpose grease to the outer CV joint splines, pull out on the strut/steering knuckle assembly and install the stub axle in the hub.

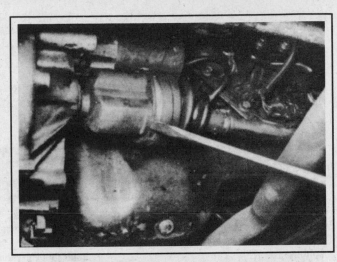

2.8 A large punch or screwdriver, positioned in the groove on the CV joint housing, can be used to seat the joint in the transaxle

10   Insert the control arm balljoint stud into the steering knuckle and tighten the nut. Be sure to use a new cotter pin (refer to Chapter 10).

11   Install the brake disc and caliper (see Chapter 9 if necessary).

12   Install the driveaxle/hub nut. Lock the disc so it can't turn, using a screwdriver or punch inserted through the caliper into a disc cooling vane, and tighten the driveaxle/hub nut securely.

13   Grasp the inner CV joint housing (not the driveaxle) and pull out to make sure the axle has seated securely in the transaxle.

14   Install the wheel and lug nuts and lower the vehicle.

15   Tighten the driveaxle/hub nut to the torque listed in this Chapter's Specifications. Tighten the wheel lug nuts to the torque listed in the Chapter 1 Specifications, then install the wheel cover.

---

3   Driveaxle boot - replacement

➡**Note: If the CV joint boots must be replaced, explore all options before beginning the job. Complete rebuilt driveaxles are available on an exchange basis, which eliminates much time and work. Whichever route you choose to take, check on the cost and availability of parts before disassembling the vehicle.**

1   Remove the driveaxle (see Section 2).

2   Place the driveaxle in a vise lined with rags to avoid damage to the axleshaft. Check the CV joint for excessive play in the radial direction, which indicates worn parts. Check for smooth operation throughout the full range of motion for each CV joint. If a boot is torn, disassemble the joint, clean the components and inspect for damage due to loss of lubrication and possible contamination by foreign matter.

INNER CV JOINT

▶ **Refer to illustrations 3.3a through 3.3w**

➡**Note: Some models use a "ball-and-cage" type inner CV joint instead of the usual tri-pot design. Aside from a wire retainer ring which must be removed before the ball-and-cage assembly can be removed from the CV joint housing, this unit is similar in construction to the outer CV joint, which is covered in the sequence beginning with illustration 3.4a. Also, 1993 through**

3.3a Cut off the boot retaining clamps, using wire cutters or a chisel and hammer

1996 models have a "bearing block" type spider assembly.

3   To replace the inner boot, refer to the accompanying illustrations (see illustrations 3.3a through 3.3w).

3.3b Slide the housing off the spider assembly

3.3c On models with a ball-and-cage inner joint, pry out the wire ring bearing retainer with a screwdriver

3.3d Slide the boot towards the center of the driveaxle

3.3e Spread the ends of the stop-ring apart and slide it towards the center of the shaft

3.3f Slide the spider (or ball-and-cage) assembly back to expose the retaining ring and pry off the ring

3.3g Carefully tap the spider (or ball-and-cage) off the axleshaft with a brass punch

3.3h When you slide the spider off the driveaxle, hold the bearings in place with your hand; even better, use tape or a cloth wrapped around the spider bearing assembly to retain them

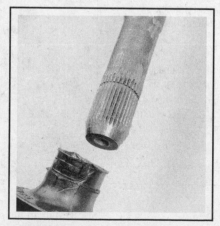

3.3i Slide the stop-ring and the boot off the axleshaft

3.3j Clean all of the old grease out of the housing and spider assembly, then remove each bearing, one at time

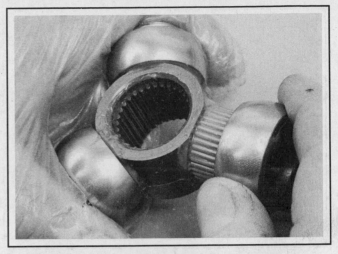

3.3k On models with needle bearings, carefully disassemble each section of the spider assembly, clean the needle bearings with solvent and inspect the rollers, spider cross, bearings and housing for scoring, pitting and other signs of abnormal wear

3.3l Apply CV joint grease to hold the needle bearings in place and slide the bearing over them - on 1993 through 1996 models, install the bearing blocks onto the posts of the spider, then turn them 90-degrees to lock them in place

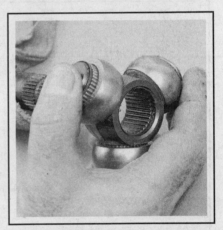

3.3m Wrap the axleshaft splines with tape to avoid damaging the boot, then slide the small clamp and boot onto the axleshaft

3.3n Remove the tape and slide the stop-ring onto the axleshaft, past the groove in which it seats

3.3o Install the spider assembly with the recess in the counterbore facing the end of the driveaxle

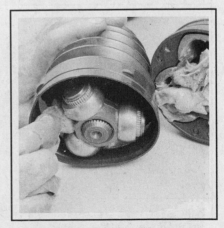

3.3p On ball-and-cage type inner joints, the small-diameter side of the cage must face the center of the axleshaft

3.3q Use a screwdriver to install the retaining ring, then slide the spider (or ball-and cage) assembly against it and install the stop-ring in its groove

3.3r Pack the housing with half of the grease furnished with the new boot and place the remainder in the boot

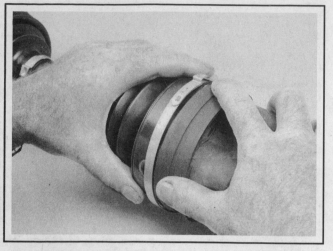

3.3s With the retaining clamps in place (but not tightened), install the joint housing; on 1993 through 1996 models, hold the bearing blocks in alignment so they can be inserted into the joint housing without cocking to the side

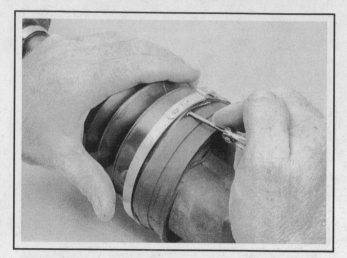

3.3t Seat the boot in the housing and axle seal grooves - a small screwdriver can make the job easier (make sure the boot isn't dimpled, stretched or out of shape)

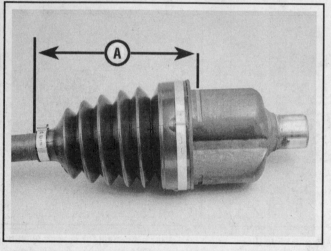

3.3u Adjust the length of the joint to the dimension listed in this Chapter's Specifications

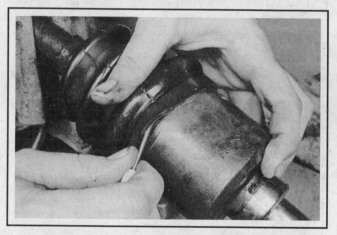

3.3v Equalize the pressure inside the boot by inserting a dull screwdriver between the boot and the outer race . . .

OUTER CV JOINT

▶ Refer to illustrations 3.4a through 3.4t

4   Refer to the accompanying illustrations and perform the outer CV joint boot replacement procedure (see illustrations 3.4a through 3.4t).

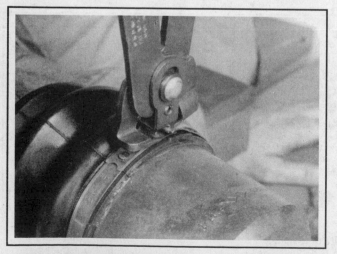

3.3w . . . then secure the boot clamps with special pliers (available at auto parts stores)

3.4a Cut off the band retaining the boot to the shaft, then slide the boot toward the center of the shaft

3.4b On models with a retaining ring, tap around the circumference of the retaining ring to remove it from the housing

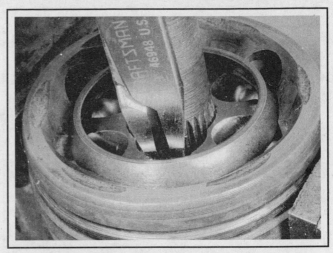

3.4c Remove the snap-ring, slide the joint off the shaft and remove the old boot

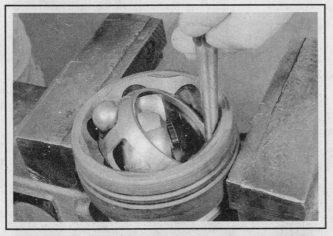

3.4d Press down on the inner race far enough to allow a ball bearing to be removed - if it's difficult to tilt, gently tap the cage and inner race with a brass punch and hammer

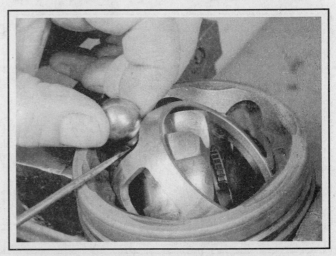

3.4e Pry the balls out of the cage, one at a time

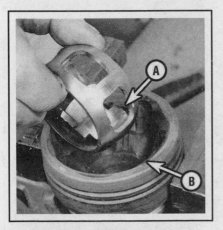

3.4f Tilt the inner race and cage 90-degrees, then align the windows in the cage (A) with the lands of the housing (B) and rotate the inner race up and out of the outer race

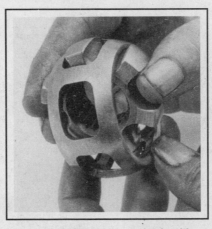

3.4g Align the inner race lands with the cage window and rotate the inner race out of the cage

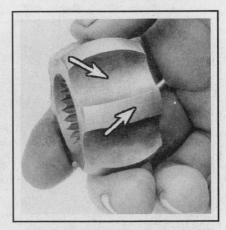

3.4h After cleaning the components with solvent, check the inner race lands and grooves for pitting and score marks

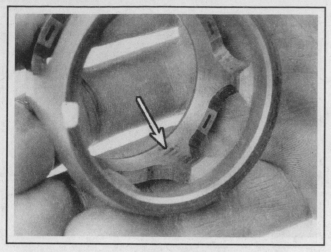

3.4i Check the cage for cracks, pitting and score marks - shiny spots are normal and don't affect operation

3.4j With the race and cage tilted at 90-degrees, lower the assembly into the housing

3.4k Rotate the assembly by gently tapping with a hammer and brass punch, then . . .

3.4l . . . press the balls into the cage windows, repeating until all of the balls are installed

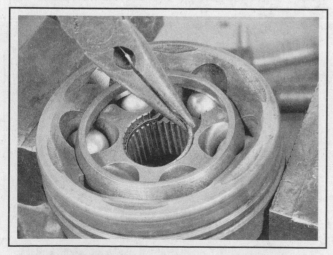

3.4m Use needle-nose pliers to lower a new snap-ring into the groove . . .

3.4n . . . then seat it into the groove with snap-ring pliers

3.4o Apply grease through the splined hole, then insert a wooden dowel (with a diameter slightly less than that of the axle) through the splined hole and push down - the dowel will force the grease into the joint - repeat until the bearing is completely packed

3.4p Install the small clamp and the boot on the driveaxle and apply grease to the inside of the axle boot until . . .

3.4q . . . the level is up to the end of the axle

3.4r Position the CV joint assembly on the driveaxle, aligning the splines, then use a soft-face hammer to drive the joint onto the driveaxle until the snap-ring is seated in the groove

3.4s Seat the inner end of the boot in the groove and install the retaining clamp, then do the same on the other end of the boot - equalize the pressure in the boot (see illustration 3.3v), then tighten the boot clamps with the special tool (see illustration 3.3w)

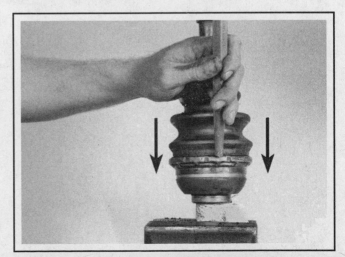

3.4t On models with a retaining ring, carefully tap around the circumference of the retaining ring to install it on the housing

Specifications

Driveaxles

Inner CV joint standard length
(dimension A in illustration 3.3u)

1990 and earlier models	5-1/16 inch
1991 and later models	4-29/32 inch

Torque specifications

	Ft-lbs
Driveaxle/hub nut	
1985 through 1990	180
1991	192
1992 through 1997	107
1998 and later	118
Wheel lug nuts	See Chapter 1

Section

Reference to other Chapters

9

BRAKES

1 General information

All vehicles covered by this manual are equipped with hydraulically operated front and rear brake systems. The front brakes are disc type, and the rear brakes are either drum or disc type. Two types of rear drum brakes are used - anchor plate (duo servo) and leading/trailing.

All brakes are self-adjusting. Disc brakes automatically compensate for pad wear, while the drum brakes incorporate an adjustment mechanism which is activated as the brakes are applied when the vehicle is driven in reverse (duo servo) or whenever the brake is actuated (leading/trailing).

The hydraulic system consists of two separate circuits, split diagonally. The master cylinder has separate reservoirs for the two circuits; in the event of a leak or failure in one hydraulic circuit, the other circuit will remain operative. A visual warning of circuit failure, induced by low fluid level, is given by a warning light activated by a fluid level switch in the master cylinder reservoir.

The parking brake mechanically operates the rear brakes only. It's activated by a pedal mounted under the driver's side of the dash.

The power brake booster, located in the engine compartment on the firewall, uses engine manifold vacuum and atmospheric pressure to provide assistance to the hydraulically operated brakes.

After completing any operation involving the disassembly of any part of the brake system, always test drive the vehicle to check for proper braking performance before resuming normal driving. Test the brakes while driving on a clean, dry, flat surface. Conditions other than these can lead to inaccurate test results. Test the brakes at various speeds with both light and heavy pedal pressure. The vehicle should stop evenly without pulling to one side or the other. Avoid locking the brakes because this slides the tires and diminishes braking efficiency and control.

Tires, vehicle load and front end alignment are factors which also affect braking performance.

2 Disc brake pads - replacement

❉❉ **WARNING:**

Disc brake pads must be replaced on both front or rear wheels at the same time - never replace the pads on only one wheel. Also, the dust created by the brake system is harmful to your health. Never blow it out with compressed air and don't inhale any of it. An approved filtering mask should be worn when working on the brakes. Do not, under any circumstances, use petroleum-based solvents to clean brake parts. Use brake cleaner or denatured alcohol only!

1999 AND EARLIER MODELS

Front

Removal

▶ **Refer to illustrations 2.6, 2.7, 2.9 and 2.10**

1 Remove and discard two-thirds of the brake fluid from the master cylinder.

2 Loosen the wheel lug nuts.

3 Raise the vehicle and support it securely on jackstands.

4 Remove the wheel and reinstall two lug nuts to hold the disc in position.

5 Clean the caliper, disc and other components with brake system cleaner before beginning any brake work. This should remove traces of potentially hazardous brake dust. Collect the drippings in a plastic container.

2.6 A large C-clamp can be used to compress the piston into the caliper for removal

6 Use a large C-clamp over the caliper to push the piston back into the body of the caliper (see illustration).

7 Remove the caliper mounting bolts (see illustration).

8 Pull the caliper straight up and off, but do not allow it to hang by the brake hose. Hang it by a length of wire if necessary.

9 Unclip the inner pad from the piston (see illustration).

10 Remove the outer pad from the caliper (see illustration). If necessary, use a screwdriver to pry up on the retaining spring of the outer pad while sliding it out of the caliper.

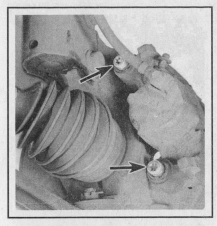

2.7 On most models, a Torx bit must be used to remove the two caliper mounting bolts (arrows)

2.9 Remove the inner pad by snapping it out of the piston in the direction shown (arrow)

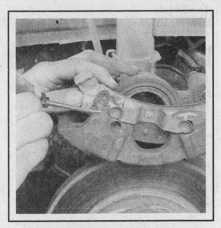

2.10 To remove the outer pad, pry off the retaining clip on the back of the pad with a screwdriver

Installation

▶ **Refer to illustrations 2.11a, 2.11b, 2.12a, 2.12b, 2.13, 2.14a, 2.14b and 2.15**

11 Check the caliper bolts and bushings for damage and, on 1996 and earlier models, the contact surfaces at the top and bottom edges of the caliper for corrosion. Also peel back the piston dust boot and check for fluid leakage and corrosion (see illustrations).

12 Coat the back of the new brake pads with an anti-squeal compound. Install the new inner and outer pads into the caliper (see illustrations).

13 If the boots for the caliper mounting bolts need replacing (1997

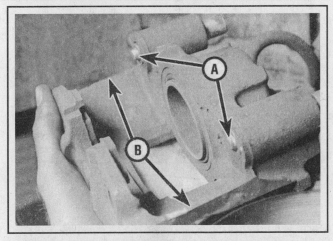

2.11a Inspect the caliper bolts and bushings (A) for damage and the contact surfaces (B) for corrosion

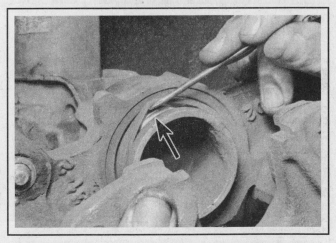

2.11b Carefully peel back the edge of the piston boot and check for corrosion and leaking fluid

2.12a Snap the inner pad retainer spring into the new pad in the direction shown (arrow) (on later models the clip is not detachable from the pad)

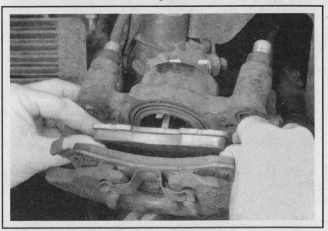

2.12b Place the pads in position and snap the inner pad into place in the piston

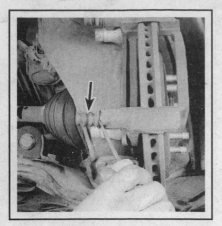

2.13 If the boots need replacing, pry them out (1997 and later models)

2.14a Lubricate the lower steering knuckle contact surface with a small amount of white lithium-base grease

2.14b Apply a light coat of white lithium-base grease to the upper steering knuckle-to-caliper contact surface

through 1999 models only), pry them out with a small screwdriver (see illustration).

14 On 1996 and earlier models, lubricate the upper and lower contact surfaces on the steering knuckle with high-temperature brake grease (see illustrations). Install the caliper over the disc. Lubricate the sleeves of the mounting bolts with high-temperature grease, install them and tighten to the torque listed in this Chapter's Specifications.

15 On 1997 through 1999 models, install the caliper over the disc and align the bolt holes. The rubber boots should be between the caliper and the mounting bracket. Lubricate the entire length of the mounting bolts with a thin film of high-temperature grease (see illustration). Install and tighten the caliper mounting bolts to the torque listed in this Chapter's Specifications. Lower the vehicle. Fill the reservoir with the recommended fluid (see Chapter 1) and pump the brake pedal several times to bring the pads into contact with discs. Road test the vehicle carefully before returning it to normal service.

Rear
Removal and installation

16 Removal and installation of the rear disc brake pads on 1999 and earlier models is the same procedure as described for 2000 models (see Step 17).

2000 AND LATER MODELS (FRONT OR REAR)

Removal

17 Remove and discard two-thirds of the brake fluid from the reservoir.

18 Loosen the wheel lug nuts, raise the vehicle and support it securely on jackstands. Remove the wheels.

19 Install two lug nuts to retain the disc when the caliper is removed.

20 On rear disc brakes, remove the bolt and washer attaching the cable support bracket to the caliper body, which allows enough cable slack to permit the caliper body to rotate enough for pad replacement. It isn't necessary to detach the cable or the brake hose.

21 On front or rear disc brakes, remove the lower caliper bolt, then pivot the caliper up. Use a piece of mechanic's wire to hold the caliper in the Up position so it doesn't put tension on the flexible brake hose.

22 Remove the pads and anti-rattle clips from the caliper mounting bracket.

Installation

▶ Refer to illustrations 2.23, 2.25a, 2.25b, 2.25c and 2.25d

23 On rear disc brakes, use a two-pin spanner or a pair of needle-

2.15 Lubricate the caliper mounting bolts before inserting and tightening them to Specifications

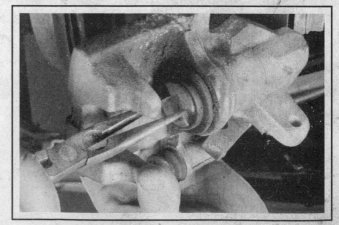

2.23 On rear disc brakes, use a pair of needle-nose pliers, with the tips engaged with the cutouts in the piston face, to turn the piston into the cylinder bore

2.25a Install the upper anti-rattle clip . . .

2.25b . . . and the lower anti-rattle clip in the caliper mounting bracket - make sure they're both fully seated

2.25c Install the inner pad . . .

2.25d . . . and the outer pad in the caliper mounting bracket, then swing the caliper down into place. Install the caliper mounting bolt and tighten it to the torque listed in this Chapter's Specifications

nose pliers, turn the piston to retract it into the caliper bore (see illustration). Turn the piston as necessary so that a line drawn through the slots in the piston face would be exactly perpendicular to a line drawn

through the caliper mounting bolt holes.

24 On front or rear disc brakes, use a small screwdriver, gently lift one edge of the piston boot to release any trapped air; the boot should lay flat.

25 Apply a coating of anti-squeal compound to the backing plates of the new pads. Install the anti-rattle clips and the pads in the caliper (see illustrations).

➡Note: The pad with the wear sensor is the outer pad.

26 Pivot the caliper assembly down over the disc. Lubricate the caliper mounting bolt with high-temperature grease and install it, tightening it to the torque listed in this Chapter's Specifications.

❊❊ CAUTION:

Make sure the projection on the backing plate of the inner pad fits into the cutout in the face of the piston.

27 The remainder of reassembly is the opposite of disassembly.

❊❊ WARNING:

Before driving the vehicle, pump the brakes several times to seat the pads against the disc. Road test the vehicle carefully before returning it to normal service.

---

3   Disc brake caliper - removal, overhaul and installation

▶ Refer to illustrations 3.9, 3.10, 3.11, 3.12, 3.15, 3.16, 3.17, 3.18 and 3.23

❊❊ WARNING:

Dust created by the brake system is harmful to your health. Never blow it out with compressed air and don't inhale any of it. An approved filtering mask should be worn when working on the brakes. Do not, under any circumstances, use petroleum-based solvents to clean brake parts. Use brake system cleaner only.

➡Note: If an overhaul is indicated (usually because of fluid leakage) explore all options before beginning the job. New and factory rebuilt calipers are available on an exchange basis, which makes this job quite easy. If you decide to rebuild the calipers, make sure rebuild kits are available before proceeding.

REMOVAL

1 Remove the cover from the brake fluid reservoir, siphon off two-thirds of the fluid into a container and discard it.

3.9 With the caliper padded to catch the piston, use compressed air to force the piston out of the bore - DO NOT position your hands or fingers between the piston and caliper!

3.10 Carefully pry the dust boot out of the housing, taking care not to scratch the bore surface

3.11 To avoid damage to the caliper bore or seal groove, remove the seal with a plastic or wooden tool (a pencil will do the job)

2   Loosen the wheel lug nuts, raise the front of the vehicle and support it securely on jackstands. Apply the parking brake and block the rear wheels to keep the vehicle from rolling off the jackstands. Remove the front wheels.

3   Reinstall two lug nuts on each rotor, flat side against the rotor, to hold them in place.

4   Bottom the piston in the caliper bore. This is accomplished by pushing on the caliper, although it may be necessary to carefully use a flat pry bar or a large C-clamp.

5   Remove the brake hose inlet fitting bolt and disconnect the fitting. Have a rag handy to catch spilled fluid and wrap a plastic bag tightly around the end of the hose to prevent fluid loss and contamination.

6   On 1999 and earlier models, use a no. 50 Torx bit, remove the two mounting bolts and detach the caliper from the vehicle (refer to Section 2 if necessary). On 2000 and later models, use the proper size wrench to loosen the caliper mounting bolts and detach the caliper from the vehicle.

OVERHAUL

Front (1999 and earlier models)

➡Note: This procedure applies to front calipers on 1999 and earlier models only. Overhauling a rear caliper is extremely difficult without the right tools. Furthermore, by the time a rear caliper requires overhaul, the parking brake actuator screw has probably become corroded (which would necessitate replacement of the caliper). If the caliper is leaking or malfunctioning, we recommend that you replace it with a new or rebuilt unit.

7   Refer to Section 2 and remove the brake pads from the caliper.
8   Clean the exterior of the caliper with brake system cleaner.

❊❊ WARNING:

DO NOT use gasoline, kerosene or petroleum-based cleaning solvents. Place the caliper on a clean workbench.

9   Position a wood block or several shop rags in the caliper as a cushion, then use compressed air to remove the piston from the bore (see illustration). Use only enough air pressure to ease the piston out. If it's blown out, even with the cushion in place, it may be damaged.

❊❊ WARNING:

Never place your fingers in front of the piston in an attempt to catch or protect it when applying compressed air - serious injury could result.

10   Carefully pry the dust boot out of the caliper bore (see illustration).

11   Using a wood or plastic tool, remove the piston seal from the

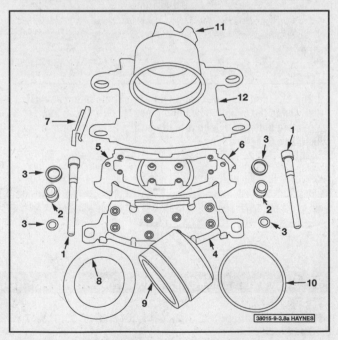

3.12 Exploded view of the disc brake caliper

1	Mounting bolt	7	Pad retainer (later models only)
2	Sleeve		
3	Bushing	8	Dust boot
4	Inner pad	9	Piston
5	Outer pad	10	Piston seal
6	Wear sensor	11	Bleeder screw
		12	Caliper housing

3.15 Position the seal in the caliper bore - make sure it isn't twisted

3.16 Install the new dust boot in the piston groove with the folds toward the open end of the piston

3.17 Push the piston squarely into the caliper bore

groove in the caliper bore (see illustration). Metal tools may damage the bore.

12 Remove the caliper bleeder valve, then remove and discard the sleeves and bushings from the caliper ears. Discard all rubber parts (see illustration).

13 Clean the remaining parts with brake cleaner or denatured alcohol, then blow them dry with compressed air.

14 Carefully examine the piston for scratches, nicks, burrs and loss of plating. If surface defects are noted, a new piston will be needed. Check the caliper bore in a similar way. Light polishing with crocus cloth is permissible to remove light corrosion and stains. Discard the mounting bolts if they're corroded or damaged.

15 When reassembling the caliper, lubricate the piston bore and seal with clean brake fluid. Position the seal in the caliper bore groove (see illustration).

16 Lubricate the piston with clean brake fluid, then install a new boot in the piston groove with the fold toward the open end of the piston (see illustration).

17 Insert the piston squarely into the caliper bore, then apply force to bottom the piston in the bore (see illustration).

18 Position the dust boot in the caliper counterbore, then use a

punch to drive it into position (see illustration). Make sure the boot is evenly installed below the caliper face.

19 Install the bleeder valve.

20 Install new bushings in the mounting bolt holes and fill the area between the bushings with silicone grease (supplied with the rebuild kit). Push the sleeves into the mounting bolt holes.

INSTALLATION

21 Inspect the mounting bolts for corrosion. Use new ones if the originals are pitted.

22 Place the caliper in position over the rotor and mounting bracket, install the bolts and tighten them to the torque figure listed in this Chapter's Specifications.

23 Check to make sure the clearance between the caliper and the bracket stops is between 0.005 and 0.012-inch (see illustration).

24 Install the brake hose-to-caliper bolt, using new copper washers, then tighten the bolt to the torque listed in this Chapter's Specifications. Bleed the brakes (see Section 9).

25 Install the wheels and lower the vehicle.

26 After the job has been completed, firmly depress the brake pedal a few times to bring the pads into contact with the rotor.

3.18 Use a seal driver to seal the boot in the caliper housing counterbore - if a seal driver isn't available, carefully tap around the outer edge of the boot with a punch until it's seated

3.23 Measure the clearance between the caliper and bracket stops at the top and bottom

4 Brake rotor (disc) - inspection, removal and installation

▶ **Refer to illustrations 4.2, 4.3, 4.4a, 4.4b, 4.5a and 4.5b**

INSPECTION

1   Loosen the wheel lug nuts, raise the front of the vehicle and support it securely on jackstands. Apply the parking brake and block the rear wheels to keep the vehicle from rolling off the jackstands. Remove the wheel and install two lug nuts to hold the rotor in place.

2   Remove the brake caliper as outlined in Section 3. You don't have to disconnect the brake hose. After removing the caliper bolts, suspend the caliper out of the way with a piece of wire - DO NOT let it hang by the hose (see illustration).

3   Visually inspect the rotor surface for score marks and other damage. Light scratches and shallow grooves are normal and may not be detrimental to brake operation, but deep score marks - over 0.015-inch (0.38 mm) deep - require rotor removal and refinishing by an automotive machine shop. Be sure to check both sides of the rotor (see illustration). If pulsating has been felt during application of the brakes, suspect excessive rotor runout.

4   To check rotor runout, mount a dial indicator with the stem resting at a point about 1/2-inch from the outer edge of the rotor (see illustration). Set the indicator to zero and turn the rotor. The indicator reading should not exceed the specified allowable runout limit. If it does, the rotor should be refinished by an automotive machine shop.

➡**Note: The rotors should be resurfaced, regardless of the dial indicator reading, to impart a smooth finish and ensure perfectly flat brake pad surfaces (which will eliminate pedal pulsations. At the very least, if you don't have the rotors resurfaced, remove the glaze with sandpaper or emery cloth using a swirling motion (see illustration).**

5   Never machine the rotor to a thickness less than the specified minimum allowable refinish thickness. The minimum wear (or discard) thickness is cast into the inside of the rotor (see illustration). The rotor thickness can be checked with a micrometer (see illustration).

4.2 Suspend the caliper with a piece of wire whenever it's necessary to reposition it - DO NOT let it hang by the brake hose!

4.3 The brake pads on this vehicle were obviously neglected as they wore down completely and cut deep grooves into the rotor - wear this severe will require replacement of the rotor

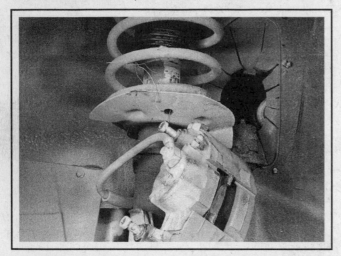

4.4a Check for runout with a dial indicator - mount it with the indicator needle about 1/2-inch from the outer edge

4.4b If you don't have the rotors machined, at the very least be sure to break the glaze on the rotor surface with sandpaper or emery cloth

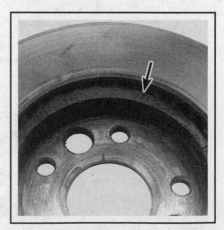

4.5a The minimum wear (or discard) thickness (arrow) is cast into the inside of the rotor

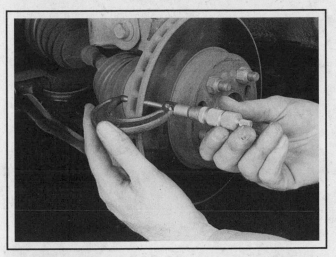

4.5b Measure the thickness of the rotor at several points with a micrometer

REMOVAL

6 Remove the two lug nuts that were put on to hold the rotor in place and remove the rotor from the hub.

INSTALLATION

7 Place the rotor in position over the threaded studs.

8 Install the caliper and brake pad assembly over the rotor and position it on the steering knuckle (refer to Section 3 for the caliper installation procedure, if necessary). Tighten the caliper bolts to the torque listed in this Chapter's Specifications.

9 Install the wheel, then lower the vehicle to the ground. Depress the brake pedal a few times to bring the brake pads into contact with the rotor. Bleeding of the system won't be necessary unless the brake hose was disconnected from the caliper. Check the operation of the brakes carefully before driving the vehicle in traffic.

---

5 Rear brake shoes - inspection and replacement

✳✳ WARNING:

Drum brake shoes must be replaced on both rear wheels at the same time - never replace the shoes on only one wheel. Also, the dust created by the brake system is harmful to your health. Never blow it out with compressed air and don't inhale any of it. An approved filtering mask should be worn when working on the brakes. Do not, under any circumstances, use petroleum-based solvents to clean brake parts. Use brake cleaner or denatured alcohol only!

✳✳ CAUTION:

Whenever the brake shoes are replaced, the return and hold-down springs should also be replaced. Due to the continuous heating/ cooling cycle the springs are subjected to, they lose tension over a period of time and may allow the shoes to drag on the drum and wear at a much faster rate than normal.

1 Loosen the wheel lug nuts, raise the rear of the vehicle and support it securely on jackstands. Block the front wheels to keep the vehicle from rolling off the jackstands.

2 Release the parking brake.

3 Remove the wheel.

➥Note: All four rear shoes must be replaced at the same time, but to avoid mixing up parts, work on only one brake assembly at a time.

ANCHOR PLATE (DUO-SERVO) TYPE

◆ **Refer to illustrations 5.4a through 5.4z**

4 Remove the brake drum. Refer to the accompanying photographs and perform the brake shoe inspection and, if necessary, the replacement procedure. Start with illustration 5.4a and be sure to read each caption.

5.4a If the brake drum won't come off, it may be necessary to remove the plug with a hammer and chisel, then turn the adjuster screw to move the brake shoes away from the drum

5.4b Before removing anything, clean the brake assembly with brake system cleaner - DO NOT use compressed air to blow the dust out of the brake assembly!

5.4c Anchor plate (duo-servo) type drum brake components - right side shown

1	Return spring	9	Primary shoe
2	Secondary shoe	10	Actuator link
3	Hold-down spring	11	Return spring
4	Adjuster screw spring	12	Parking brake lever
5	Adjuster screw assembly	13	Parking brake strut
6	Return spring	14	Wheel cylinder
7	Actuator lever	15	Strut spring
8	Hold-down spring		

➡Note: If the brake drum is stuck, make sure the parking brake is completely released, then apply some penetrating oil to the hub-to-brake drum joint. Allow the oil to soak in, then try to pull the drum off. If the drum still won't come off, the brake shoes will have to be retracted. This is done by removing the plug from the brake drum with a hammer and chisel (see illustration 5.4a). With the plug removed, pull the lever off the adjusting star wheel with a hooked tool while turning the star wheel with a small screwdriver or brake adjuster tool, moving the shoes away from the drum. The drum should now come off.

5.4d Remove the return springs with a brake spring tool

5.4e Remove the hold-down springs and pins by pushing the retainer in and turning it 90-degrees - the tool shown here is available at most auto parts stores

5.4f Lift up on the actuator lever and remove the actuating link from the anchor pin pivot along with the actuator lever and return spring (arrows)

5.4g Spread the shoes apart a the top and remove the parking brake strut

5.4h With the shoe assembly spread to clear the hub flange, lift it away from the backing plate

5.4I Disconnect the parking brake lever from the cable and remove the shoe assembly

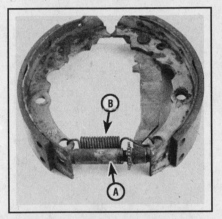

5.4j Remove the adjusting screw (A) and spring (B) from the shoe assembly - be sure to note how they're positioned

5.4k Remove the parking brake lever by prying off the C-clip

5.4l Install the parking brake lever on the new brake shoe and press the C-clip into place with needle-nose pliers

LEADING/TRAILING TYPE

▶ **Refer to illustrations 5.5a, 5.5b and 5.6a through 5.6z**

5 Remove the brake drum. If it's difficult to remove, back off the parking brake cable, remove the access hole plug from the backing plate (see illustration), insert a screwdriver through the hole and press in to push the parking brake lever off its stop (see illustration). This will allow the brake shoes to retract slightly. Insert a punch through the hole at the bottom of the splash shield and tap gently on the punch to loosen the drum. Use a rubber mallet to tap gently on the outer rim of the drum and/or around the inner drum diameter by the spindle. Avoid using

5.5m Lubricate the contact surfaces of the backing plate with high-temperature brake grease

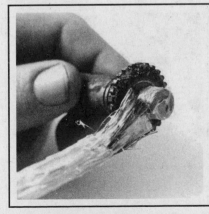

5.4n Lubricate the adjuster screw with high-temperature grease prior to installation

5.4o Connect the parking brake lever to the cable

5.4p Spread the brake assembly apart sufficiently to clear the hub flange and raise it into position

5.4q Install the parking brake strut and spring

5.4r Make sure the parking brake strut is positioned in the shoes properly (arrows)

5.4s Install the primary brake shoe hold-down pin and spring

5.4t Attach the actuator link and lever to the secondary brake shoe

5.4u Install the actuator lever return spring

excessive force.

6   Clean the brake assembly with brake system cleaner (see illustration 5.4b) - DO NOT use compressed air to blow the dust out of the brake assembly. Refer to the accompanying photographs for the actual shoe replacement procedure. If you're working on an early model with coil return springs, follow illustrations 5.5a, 5.5b and 5.6a through 5.6z. If you're working on a later model equipped with a single horseshoe shaped return spring, begin with illustration 5.6aa. Be sure to stay in order and read the caption under each illustration.

5.4v Install the secondary brake shoe hold-down pin and spring

5.4w Install the return springs

5.4x Center the brake shoe assembly so the drum will slide over it

5.4y Turn the adjusting screw so the drum fits snugly over the shoes

5.4z Remove the glaze from the drum with sandpaper or emery cloth - use a swirling motion

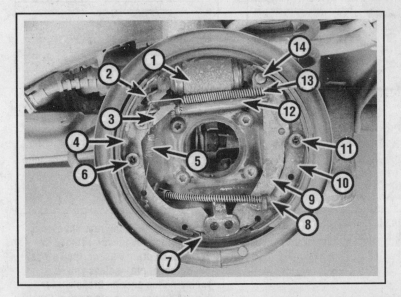

5.5a Leading/trailing type rear drum brake components

1   Wheel cylinder
2   Spring connecting link
3   Adjuster actuator
4   Leading brake shoe
5   Actuator spring
6   Hold-down spring, pin and retainer
7   Lower return spring
8   Parking brake cable
9   Parking brake lever
10 Trailing brake shoe
11 Hold-down spring, pin and retainer
12 Adjuster screw assembly
13 Upper return spring
14 Parking brake lever-to-shoe pin

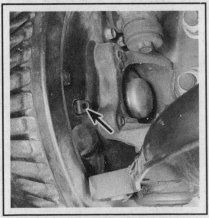

5.5b If you have trouble removing the drum, remove the access hole plug (arrow), insert a screwdriver through the hole and press in to push the parking brake lever off its stop - this will allow the brake shoes to retract slightly

5.6a Unhook the upper return spring with a pair of pliers . . .

5.6b . . . and remove it (to ensure proper reassembly, be sure to note the relationship of the spring to the front shoe and the parking brake lever)

5.6c Remove the hold-down spring and pin from the leading shoe (the tool shown here is available at most auto parts stores)

5.6d To remove the front shoe, pull it down . . .

5.6e . . . and detach the lower spring

5.6f Remove the adjuster (be sure to note the how the adjuster socket and the parking brake lever relate)

5.6g Remove the rear shoe hold-down spring, then pull down the rear shoe and parking brake assembly

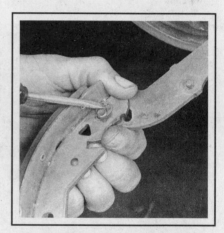

5.6h To separate the rear shoe from the parking brake lever, remove the C-clip and spring washer, then press out the lever pin - unless you're replacing the parking brake cable, it's not necessary to disconnect the cable and lever

5.6i Place the front shoe assembly on a workbench and note the relationship of the actuator spring, the spring connecting link and the adjuster actuator . . .

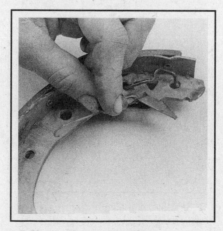

5.6j . . . then remove the actuator spring . . .

5.6k . . . the spring connecting link . . .

5.6l . . . and the adjuster actuator - transfer these parts to the new front shoe (see illustration 5.6z)

5.6m Lubricate the raised contact surfaces of the backing plate with white lithium-base grease

5.6n Install the parking brake lever on the new rear brake shoe, push the lever pin into place, install the spring washer and pop the C-clip into place with a pair of pliers (make sure the concave side of the spring washer faces toward the parking brake lever)

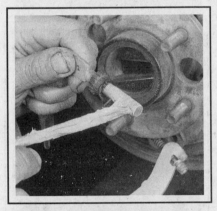

5.6o Clean the adjuster screw with brake cleaner or denatured alcohol, check the threads for smooth rotation over their full length, then lubricate the adjuster threads, the inside surface of the socket and the socket face with white lithium-base grease

5.6p Reattach the parking brake cable, if you disconnected it, place the rear brake shoe in position . . .

5.6q . . . and install the hold-down spring

5.6r Attach the lower return spring to the brake shoes as shown - make sure the spring runs behind the anchor plate, not in front of it

5.6s Place the front brake shoe in position as shown. . .

5.6t . . . and install the hold-down spring

5.6u Lubricate the adjuster screw with high-temperature grease prior to installation

5.6v Install the adjuster assembly - make sure the parking brake lever is sealed properly against the adjuster socket (see illustration 5.5a)

5.6w . . . and the adjuster actuator is properly seated against the adjuster screw (see illustration 5.6z)

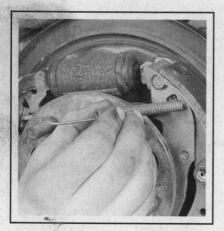

5.6x Install the upper return spring - be sure the rear end of the spring (the angled hook end) is installed through the parking brake lever and the rear shoe

5.6y . . . then grasp the forward end (the long straight section) with a pair of pliers and hook it over the crook in the spring connecting link (see illustration 5.6z)

5.6z Proper installation of the adjuster actuator, actuator spring and connecting link

5.6aa Drum brake components - leading/trailing type with a single return spring

1   Actuator spring
2   Trailing shoe
3   Return spring
4   Leading shoe
5   Adjusting screw
6   Actuator lever
7   Wheel cylinder
8   Backing plate

5.6ab Use a pair of needle-nose pliers to remove the actuator spring

5.6ac Wedge a flat-bladed screwdriver under the return spring and pry it out of the leading brake shoe, then remove the shoe, adjusting screw and actuator lever

5.6ad Lift the retractor spring from the trailing brake shoe and swing the shoe out of the hub area to gain access to the parking brake cable

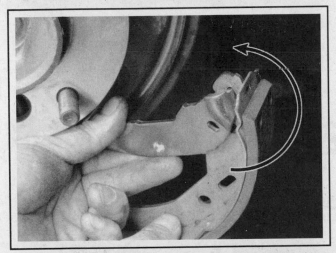

5.6ae Rotate the brake shoe to release the parking brake lever from the shoe

5.6af Use a screwdriver to pry the return spring over the alignment peg (arrow)

5.6ag Lubricate the contact surfaces of the backing plate with high-temperature grease

5.6ah Lubricate the adjuster screw with high-temperature grease prior to installation - to install new shoes on these models, reverse the removal steps, center the brake shoe assembly so the drum will slide over it, adjust the star wheel so the shoes will barely drag on the drum, then back-off the adjustment until they don't drag

BOTH TYPES

▶ **Refer to illustration 5.7**

7   Before reinstalling the drum, check it for cracks, score marks, deep scratches and hard spots, which will appear as blue discolored areas. If the hard spots can't be removed with fine emery cloth or if any of the other conditions listed above exist, the drum must be taken to an automotive machine shop to have it turned.

➡**Note: The drums should be resurfaced, regardless of the surface appearance, to impart a smooth finish and ensure a perfectly round drum (which will eliminate brake pedal pulsations related to out-of-round drums). At the very least, if you don't have the drums resurfaced, remove the glaze from the surface with medium-grit emery cloth using a swirling motion. If the drum won't "clean up" before the maximum service limit is reached in the machining operation, install a new one. The maximum wear diameter is cast into each brake drum (see illustration).**

8   Install the brake drum on the axle flange.

9   Mount the wheel, install the lug nuts, then lower the vehicle. Tighten the lug nuts to the torque listed in this Chapter's Specifications.

10   If the vehicle is equipped with anchor plate (duo servo) type brakes, make a number of forward and reverse stops to adjust the brakes until satisfactory pedal feel is obtained. If it's equipped with

5.7 The drum has a maximum permissible diameter cast into it (arrow) which is a wear dimension, not a refinish dimension

leading/trailing type brakes, apply and release the brake pedal 30 to 35 times using normal pedal force. Pause about one second between pedal applications. After adjustment, make sure that both wheels turn freely.

6   Rear wheel cylinder - removal, overhaul and installation

▶ **Refer to illustrations 6.4, 6.5a, 6.5b, 6.7 and 6.13**

➡**Note: If an overhaul is indicated (usually because of fluid leakage or sticking brakes) explore all options before beginning the job. New wheel cylinders are available, which makes this job quite easy. If you do rebuild the wheel cylinder, make sure rebuild kits are available before proceeding.**

REMOVAL

1   Raise the rear of the vehicle and support it securely on jackstands. Block the front wheels to keep the vehicle from rolling off the jackstands.

2   Remove the brake shoe assembly (Section 5).

3   Carefully clean the area around the wheel cylinder on both sides of the backing plate.

4   Unscrew the brake line fitting (see illustration), but don't pull the line away from the wheel cylinder.

5   Remove the wheel cylinder retaining bolts or clip (see illustrations).

6   Remove the wheel cylinder from the brake backing plate and place it on a clean workbench. Immediately plug the brake line to prevent fluid loss and contamination.

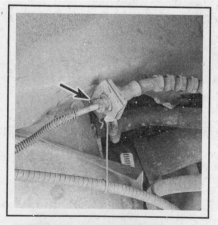

6.4 A flare nut wrench should be used to disconnect the brake line (arrow)

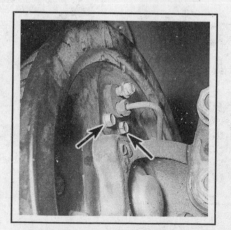

6.5a To remove the wheel cylinder, remove these two bolts (arrows) (strut removed for clarity)

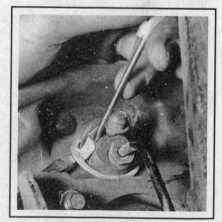

6.5b If equipped with a retaining clip instead of bolts, carefully pry the retaining clip off the wheel cylinder

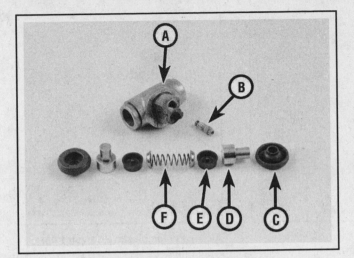

6.7 Wheel cylinder components - exploded view

A	Wheel cylinder body	D	Piston
B	Bleeder screw	E	Seal
C	Boot	F	Spring

OVERHAUL

7   Remove the bleeder valve, seals, pistons, boots and spring assembly from the wheel cylinder body (see illustration).

8   Clean the wheel cylinder with brake system cleaner.

✴✴ WARNING:

Do not, under any circumstances, use petroleum-based solvents to clean brake parts.

9   If available, use filtered, unlubricated compressed air to dry the wheel cylinder and blow out the passages.

6.13 A wood block (arrow) should be used to hold the wheel cylinder in position

10   Check the bore for corrosion and score marks. Crocus cloth may be used to remove light corrosion and stains, but the cylinder must be replaced with a new one if the defects can't be removed easily, or if the bore is scored.

11   Lubricate the new seals with brake fluid.

12   Assemble the brake cylinder components, making sure the boots are properly seated.

INSTALLATION

13   Place the wheel cylinder in position (see illustration).

14   Connect the brake lines loosely.

15   Install the wheel cylinder bolts and tighten them securely.

16   Tighten the brake line fittings.

17   Install the brake shoes (see Section 5).

18   Bleed the brakes (see Section 9).

---

7   Master cylinder - removal, overhaul and installation

▶ **Refer to illustrations 7.2, 7.6, 7.9a, 7.9b, 7.11, 7.12, 7.16, 7.18, 7.19a, 7.19b, 7.19c, 7.19d, 7.19e, 7.19f and 7.20**

✴✴ CAUTION:

On models equipped with an Anti-lock Brake System (ABS), have some plugs ready to cap the metal lines that connect the master cylinder to the hydraulic modulator. Failure to do so will allow air into the ABS modulator and can allow dirt and moisture to enter the ABS system.

➡**Note: Before deciding to overhaul the master cylinder, check on the availability and cost of a new or factory-rebuilt unit and the availability of a rebuild kit.**

REMOVAL

1   Detach the cable from the negative battery terminal.

✴✴ CAUTION:

If the vehicle is equipped with a Delco Loc II or Theftlock audio system, make sure you have the correct activation code before disconnecting the battery. See the information at the front of this manual for the radio re-activation procedure.

2   Unplug the fluid level sensor switch connector (see illustration).

3   Place rags under the line fittings and prepare caps or plastic bags to cover the ends of the lines once they're disconnected.

7.2 Unplug the fluid level sensor connector (arrow) and unscrew the brake line fittings

7.6 Remove the master cylinder mounting nuts (arrows)

7.9a Pry the plastic reservoir from the cylinder body

✳✳ CAUTION:

Brake fluid will damage paint. Cover all painted parts and be careful not to spill fluid during this procedure.

4   Loosen the fittings at the ends of the brake lines where they enter the master cylinder. To prevent rounding off the flats on the fittings, use a flare-nut wrench, which wraps around the hex.

5   Pull the brake lines away from the master cylinder and plug the ends to prevent contamination.

6   Remove the two mounting nuts (see illustration) and detach the master cylinder from the vehicle.

7   Remove the reservoir cover and reservoir diaphragm, then discard any remaining fluid in the reservoir.

8   Mount the master cylinder in a vise. Be sure to line the vise jaws with blocks of wood to prevent damage to the cylinder body.

9   On models through 1999, pull straight up on the reservoir assembly and separate it from the master cylinder body (see illustration). It may be necessary to gently pry it out of the grommets. Remove and discard the two grommets (see illustration). On 2000 and later models, the reservoir is attached to the master cylinder body with two pins. Tap the retaining pins out with a hammer and punch.

10   Remove the proportioner valves and the O-rings. Set each proportioner valve assembly aside.

11   Remove the primary piston lock ring by depressing the piston and prying the ring out with a screwdriver (see illustration).

12   Remove the primary piston assembly from the bore (see illustration).

13   Remove the secondary piston assembly from the bore. It may be necessary to remove the master cylinder from the vise and invert it, carefully tapping it against a block of wood to expel the piston.

OVERHAUL

14   Clean the master cylinder body, the primary and secondary piston assemblies, the proportioner valve assemblies and the reservoir with brake system cleaner.

✳✳ WARNING:

DO NOT, under any circumstances, use petroleum-based solvents to clean brake parts.

15   Inspect the master cylinder piston bore for corrosion and score marks. If any corrosion or damage in the bore is evident, replace the master cylinder body - don't use abrasives to try to clean it up.

16   Remove the old seals from the secondary piston assembly and install the new seals with the cup lips facing out (see illustration).

7.9b Remove the reservoir grommets

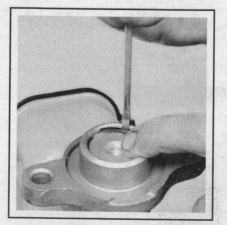

7.11 Press down on the piston and remove the primary piston lock ring

7.12 Remove the primary piston assembly

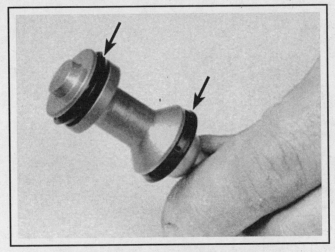

7.16 The secondary piston seals must be installed with the lips facing out as shown

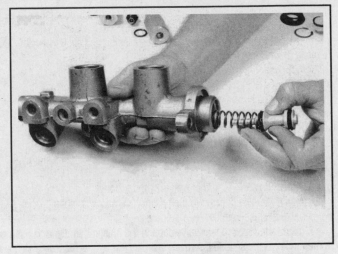

7.18 Install the secondary piston assembly

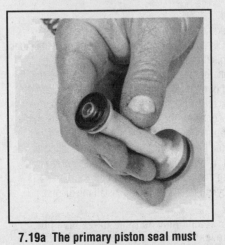

7.19a The primary piston seal must be installed with the lip facing away from the piston

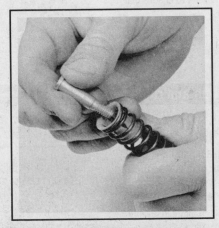

7.19b Install the seal guard over the seal

7.19c Place the primary piston spring in position

17 Attach the spring retainer to the secondary piston assembly.

18 Lubricate the cylinder bore with clean brake fluid and install the spring and secondary piston assembly (see illustration).

19 Disassemble the primary piston assembly, noting the locations of the parts, then lubricate the new seals with clean brake fluid and install them on the piston (see illustrations).

7.19d Insert the spring retainer into the spring

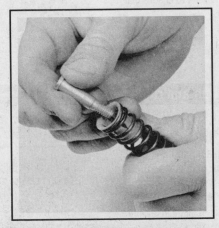

7.19e Insert the spring retaining bolt through the retainer and spring and thread it into the piston

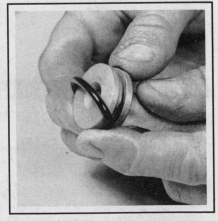

7.19f Install the O-ring on the piston

20 Install the primary piston assembly in the cylinder bore (see illustration), depress it and install the lock ring.

21 Inspect the proportioner valves for corrosion and score marks. Replace them if necessary.

22 Lubricate the new O-rings and proportioner valve seals with the silicone grease supplied with the rebuild kit (if no silicone grease is available, use clean brake fluid). Also lubricate the stem of the proportioner valve pistons.

23 Install the new seals on the proportioner valve pistons with the seal lips facing toward the cap assembly.

24 Install the proportioner valve pistons and seals in the master cylinder body.

25 Install the springs in the master cylinder body.

26 Install the new O-rings in their respective grooves in the proportioner valve cap assemblies.

27 Install the proportioner valve caps in the master cylinder and tighten them to the specified torque.

28 Inspect the reservoir for cracks and distortion. If any damage is evident, replace it.

29 Lubricate the new reservoir O-rings with clean brake fluid and press them into their respective grooves in the master cylinder body. Make sure they're properly seated.

30 Lubricate the reservoir fittings with clean brake fluid and install the reservoir on the master cylinder body by pressing it straight down. On 2000 and later models, tap the reservoir retaining pins in place with a hammer and punch.

31 Inspect the reservoir diaphragm and cover for cracks and deformation. Replace any damaged parts with new ones and attach the diaphragm to the cover.

➡Note: Whenever the master cylinder is removed, the complete hydraulic system must be bled. The time required to bleed the system can be reduced if the master cylinder is filled with fluid and bench bled (refer to Steps 32 through 35) before it's installed on the vehicle.

32 Insert threaded plugs of the correct size into the brake line outlet holes and fill the reservoirs with brake fluid. The master cylinder should be supported so brake fluid won't spill during the bench bleeding procedure.

33 Loosen one plug at a time and push the piston assembly into the bore to force air from the master cylinder. To prevent air from being drawn back in, the appropriate plug must be replaced before allowing the piston to return to its original position.

34 Stroke the piston three or four times for each outlet to ensure that

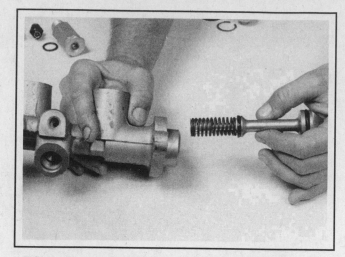

7.20 Insert the primary piston assembly into the body

all air has been expelled.

35 Since high pressure isn't involved in the bench bleeding procedure, there is an alternative to the removal and replacement of the plugs with each stroke of the piston assembly. Before pushing in on the piston assembly, remove one of the plugs completely. Before releasing the piston, however, instead of replacing the plug, simply put your finger tightly over the hole to keep air from being drawn back into the master cylinder. Wait several seconds for the brake fluid to be drawn from the reservoir into the piston bore, then repeat the procedure. When you push down on the piston it'll force your finger off the hole, allowing the air inside to be expelled. When only brake fluid is being ejected from the hole, replace the plug and go on to the other port.

36 Refill the master cylinder reservoirs and install the diaphragm and cover assembly.

➡Note: The reservoirs should only be filled to the top of the reservoir divider to prevent overflowing when the cover is installed.

INSTALLATION

37 Carefully install the master cylinder by reversing the removal steps, then bleed the brakes at each wheel (see Section 9).

---

8   Brake hoses and lines - inspection and replacement

◆ **Refer to illustrations 8.2 and 8.11**

1   About every six months, raise the vehicle and support it securely on jackstands, then check the flexible hoses that connect the steel brake lines to the front and rear brake assemblies. Look for cracks, chafing of the outer cover, leaks, blisters and other damage. The hoses are important and vulnerable parts of the brake system and the inspection should be thorough. A light and mirror will be helpful to see into restricted areas. If a hose exhibits any of the above conditions, replace it with a new one.

FRONT BRAKE HOSE

2   Using a back-up wrench, disconnect the brake line from the hose fitting, being careful not to bend the frame bracket or brake line (see illustration).

3   Use pliers to remove the U-clip from the female fitting at the bracket, then remove the hose from the bracket.

4   At the caliper end of the hose, remove the bolt from the fitting block, then remove the hose and the copper gaskets on either side of the fitting block.

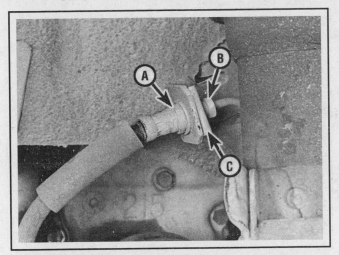

8.2 Using a back-up wrench on the flexible hose side of the fitting (A), loosen the tube nut (B) with a flare nut wrench and remove the U-clip (C) from the hose fitting

5   When installing the hose, always use new copper gaskets on either side of the fitting block and lubricate all bolt threads with clean brake fluid before installation.

6   With the fitting flange engaged with the caliper locating ledge, attach the hose to the caliper.

7   Without twisting the hose, install the female fitting in the hose bracket. It'll fit the bracket in only one position.

8   Install the U-clip retaining the female fitting to the frame bracket.

9   Using a back-up wrench, attach the brake line to the hose fitting.

10 When the brake hose installation is complete, there shouldn't be any kinks in the hose. Make sure the hose doesn't contact any part of the suspension. Check it by turning the wheels to the extreme left and right positions. If the hose makes contact, remove the hose and correct the installation as necessary.

REAR BRAKE HOSE

11 Using a back-up wrench, if necessary, disconnect the hose at both ends, being careful not to bend the bracket or steel lines (see illustration).

12 Remove the two U-clips with pliers and separate the female fittings from the brackets.

13 Unbolt the hose retaining clip and remove the hose.

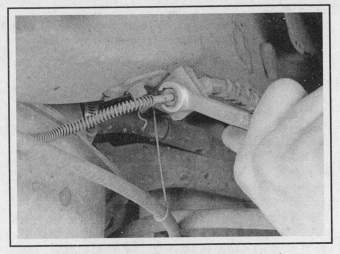

8.11 To loosen the rear brake hose/line fitting, simply loosen the nut and pull the clip off the bracket

14 Without twisting the hose, install the female ends in the frame brackets. It'll fit the bracket in only one position.

15 Install the U-clips retaining the female end to the bracket.

16 Using a back-up wrench, attach the steel line fittings to the female fittings. Again, be careful not to bend the bracket or steel line.

17 Make sure the hose installation didn't loosen the frame bracket. Tighten the bracket if necessary.

18 Fill the master cylinder reservoir and bleed the system (refer to Section 9).

METAL BRAKE LINES

19 When replacing brake lines, be sure to buy the correct replacement parts. Don't use copper or any other tubing for brake lines.

20 Auto parts stores and brake supply houses carry various lengths of prefabricated brake line. These sections can be bent with a tubing bender.

21 When installing the new line, make sure it's securely supported in the brackets with plenty of clearance between moving or hot components.

22 After installation, check the master cylinder fluid level and add fluid as necessary. Bleed the brake system as outlined in the next Section and test the brakes carefully before driving the vehicle in traffic.

9   Brake system bleeding

▶ Refer to illustration 9.8

⁂ WARNING:

Wear eye protection when bleeding the brake system. If you get fluid in your eyes, rinse them immediately with water and seek medical attention.

➡Note: Bleeding the brakes is necessary to remove air that manages to find its way into the system when its been opened during removal and installation of a hose, line, caliper or master cylinder.

1   It'll probably be necessary to bleed the system at all four brakes if air has entered the system due to low fluid level, or if the brake lines have been disconnected at the master cylinder.

2   If a brake line was disconnected at only one wheel, then only that caliper or wheel cylinder must be bled.

3   If a brake line is disconnected at a fitting located between the master cylinder and any of the brakes, that part of the system served by the disconnected line must be bled.

4   Remove any residual vacuum from the power brake booster by applying the brake several times with the engine off.

5   Remove the master cylinder reservoir cover and fill the reservoir with brake fluid. Reinstall the cover.

→Note: Check the fluid level often during the bleeding procedure and add fluid as necessary to prevent the level from falling low enough to allow air bubbles into the master cylinder.

6   Have an assistant on hand, as well as a supply of new brake fluid, an empty, clear plastic container, a length of plastic, rubber or vinyl tubing to fit over the bleeder valve and a wrench to open and close the bleeder valve.

7   Beginning at the right rear wheel, loosen the bleeder valve slightly, then tighten it to a point where it's snug but can still be loosened quickly and easily.

8   Place one end of the tubing over the bleeder valve and submerge the other end in brake fluid in the container (see illustration).

9   Have your assistant pump the brakes slowly a few times to get pressure in the system, then hold the pedal down firmly.

10   While the pedal is held down, open the bleeder valve just enough to allow fluid to flow out of the valve. Watch for air bubbles to exit the submerged end of the tube. When the fluid slows after a couple of seconds, close the valve and have your assistant release the pedal.

11   Repeat Steps 9 and 10 until no more air is seen leaving the tube, then tighten the bleeder valve and proceed to the left rear wheel, the right front wheel and the left front wheel, in that order, and perform the same procedure. Be sure to check the fluid in the master cylinder reservoir frequently.

12   Never use old brake fluid. It contains moisture which can boil, rendering the brakes useless.

13   Refill the master cylinder with fluid at the end of the operation.

14   Check the operation of the brakes. The pedal should feel firm when depressed. If necessary, repeat the entire procedure.

9.8 When bleeding the brakes, a hose is connected to the bleeder valve at the caliper (or wheel cylinder) and then submerged in brake fluid - air will be seen as bubbles in the container or in the tube (all air must be expelled before continuing to the next wheel)

⋇⋇ WARNING:

Don't operate the vehicle if you're in doubt about the effectiveness of the brake system.

---

10 Parking brake - adjustment

→Note: On later models with rear disc brakes, the parking brake cables operate the rear calipers directly. There is no adjustment at the caliper - the parking brake pedal in the vehicle automatically adjusts the parking brake system.

VEHICLES WITH ANCHOR PLATE (DUO-SERVO) TYPE REAR BRAKES

♦ Refer to illustration 10.5

1   Adjust the brakes (see Step 10 in Section 5).

2   Apply the parking brake pedal exactly three ratchet clicks.

3   Raise the vehicle and support it securely on jackstands.

4   Before adjusting the parking brake, make sure the equalizer nut groove is lubricated with multi-purpose grease.

5   Tighten the adjusting nut (see illustration) until the right rear wheel can barely be turned backwards with two hands, but locks when turned forward.

6   Release the parking brake pedal and check to make sure the rear wheels turn freely in both directions.

7   Lower the vehicle.

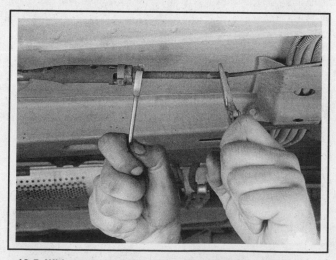

10.5 With a pair of locking pliers clamped to the end of the threaded rod to hold it still, turn the adjusting nut until the right rear wheel can just barely be turned backwards but not forward

VEHICLES WITH LEADING/TRAILING TYPE REAR BRAKES

▶ Refer to illustration 10.14

8   Adjust the brakes (see Step 10 in Section 5).

9   Apply and release the parking brake six times to ten ratchet clicks.

10   Check the parking brake pedal assembly for full release by turning the ignition to On and noting whether the Brake warning light is off. If it's on even though the brake appears to be released, operate the pedal release lever and pull down on the front parking brake cable to remove slack from the assembly. Check both rear wheels to make sure they still turn freely.

11   Apply the parking brake to four clicks.

12   Raise the vehicle and place it securely on jackstands.

13   Remove the access hole plug.

14   Adjust the parking brake cable until you can insert a 1/8-inch drill bit - but not a 1/4-inch bit - through the access hole into the space between the shoe web and the parking brake (see illustration).

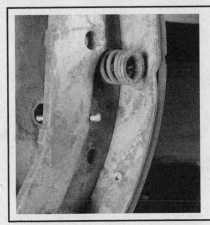

10.14 Turn the parking brake cable adjuster until you can insert a 1/8-inch drill bit through the access hole in the backing plate and between the parking brake lever and the trailing brake shoe (drum removed for clarity)

➡ Note: The drill bit must be perpendicular (at a right angle) to the backing plate.

15   Release the parking brake and verify that both wheels rotate freely.

16   Replace the access hole plug.

17   Lower the vehicle.

11   Parking brake cables - removal and installation

1   Detach the cable from the negative battery terminal.

✳ CAUTION:

If the vehicle is equipped with a Delco Loc II or Theftlock audio system, make sure you have the correct activation code before disconnecting the battery. See the information at the front of this manual for the radio re-activation procedure.

2   If you're going to remove or replace a rear cable, loosen the wheel lug nuts.

3   Raise the vehicle and place it securely on jackstands.

FRONT CABLE

▶ Refer to illustrations 11.4, 11.5 and 11.6

4   Locate the equalizer assembly along the left side of the underbody (see illustration).

5   Loosen the equalizer assembly (see illustration) and detach the front cable.

6   Remove the nut from the front cable (see illustration).

7   Working inside the vehicle, detach the cable housing and cable from the parking brake lever assembly.

8   Installation is the reverse of removal.

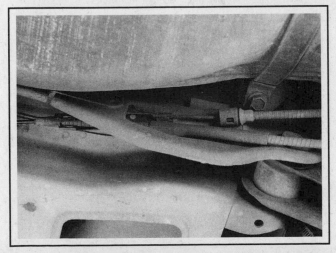

11.4   The parking brake cable equalizer is located on the left side of the rear suspension crossmember

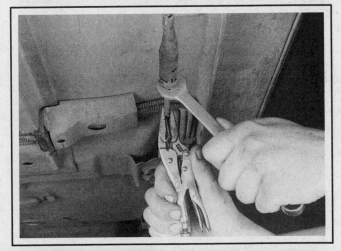

11.5   To disconnect the front or intermediate cable from the equalizer assembly, simply loosen the threaded adjuster until the cables are slack

11.6 Before the front cable will slide through the floor, you'll have to unscrew this nut from the underbody

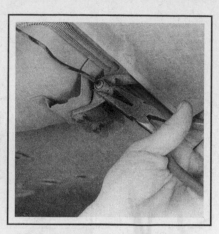

11.10 To detach the intermediate cable from this mounting bracket, pinch the tabs on the cable housing with a pair of pliers

11.11a To detach the intermediate cable from the underbody, detach this guide . . .

INTERMEDIATE CABLE

▶ **Refer to illustrations 11.10, 11.11a, 11.11b, 11.12a and 11.12b**

9 Disconnect the intermediate cable from the forward equalizer assembly (see illustration 11.5).

11.11b . . . and detach this cable clip

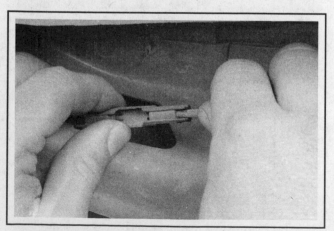

11.12a Disconnect the intermediate and right rear cables . . .

10 Detach the intermediate cable housing from the bracket (see illustration).

11 Detach the intermediate cable guide and clip from the underbody (see illustrations).

12 Disconnect the intermediate and right rear cables (see illustration), then disconnect the intermediate cable from the left rear cable (see illustration).

13 Installation is the reverse of removal.

LEFT REAR CABLE

▶ **Refer to illustrations 11.18 and 11.19**

➥**Note: On later models with rear disc brakes, the parking brake cable housing attaches to a bracket on the caliper. To remove or install the cable, lift the cable eye from the slot on the caliper, then squeeze the tabs at the bracket until the cable housing can be removed from the bracket.**

14 Back off the equalizer nut until cable tension is eliminated (see illustration 11.5).

15 Remove the left rear wheel.

16 Remove the left brake drum (see Section 5).

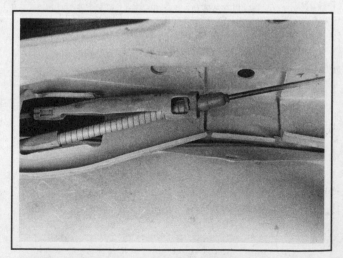

11.12b . . . then disconnect the intermediate cable from the left rear cable

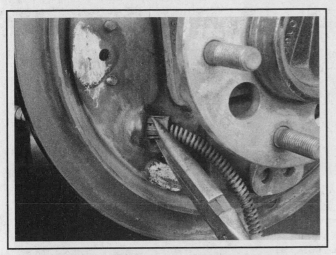

11.18 To detach the cable from the rear drum brake assembly, pinch the tabs on the housing with a pair of pliers and slide it through the hole in the brake backing plate

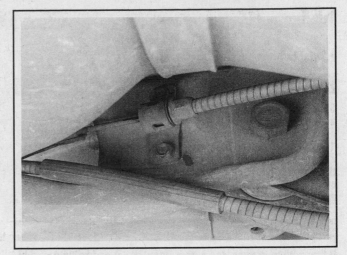

11.19 To remove the left rear cable from the underbody, detach this cable clip

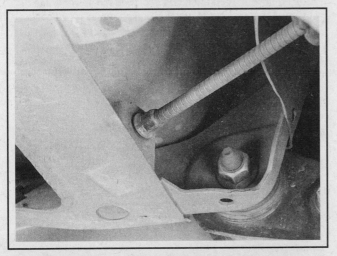

11.27 To remove the right rear cable from the underbody, pinch the tabs on this housing (not visible in this photo, they're on the other side of the housing bracket) with a pair of pliers and slide the housing and cable out of the bracket

17 Remove the left rear brake shoe and the parking brake lever as an assembly and detach the cable from the lever (see Section 5).

18 Detach the cable fitting from the backing plate (see illustration).

19 Detach the cable bracket (see illustration).

20 Disconnect the intermediate and right rear cables, then discon-nect the left rear cable from the intermediate cable (see illustrations 11.12a and 11.12b).

21 Installation is the reverse of removal.

RIGHT REAR CABLE

▶ **Refer to illustration 11.27**

22 Back off the equalizer nut until cable tension is eliminated (see illustration 11.5).

23 Remove the right rear wheel.

24 Remove the right brake drum (see Section 5).

25 Remove the right front brake shoe and the parking brake lever as an assembly and detach the cable from the lever (see Section 5).

26 Detach the cable fitting from the backing plate (see illustration 11.18).

27 Detach the right rear cable housing from the underbody (see illustration).

28 Disconnect the intermediate and right rear cables, then discon-nect the left rear cable from the intermediate cable (see illustrations 11.12a and 11.12b).

29 Installation is the reverse of removal.

ALL CABLES

30 On drum brake models, be sure to adjust the parking brake cable after installation (see Section 10).

12 Power brake booster - inspection, removal and installation

♦ **Refer to illustration 12.6**

1   The power brake booster unit requires no special maintenance apart from periodic inspection of the vacuum hose and the case. Early models have an in-line filter which should be inspected periodically and replaced if clogged or damaged.

2   Dismantling of the power unit requires special tools and is not ordinarily done by the home mechanic. If a problem develops, install a new or factory rebuilt unit.

3   On later models, remove the fuel-injection sight shield and also disconnect the battery. Remove the nuts attaching the master cylinder to the booster (see Section 7) and carefully pull the master cylinder forward until it clears the mounting studs. Be careful to avoid bending or kinking the brake lines.

4   Disconnect the vacuum hose where it attaches to the power brake booster. On 2001 and later models, disconnect the accelerator cable and tie it aside, disconnect the EGR valve electrical connector and posi-

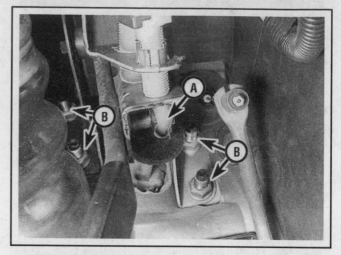

12.6 Remove the retaining clip and slide the power brake pushrod (A) off the brake pedal pin, then remove the booster-to-firewall nuts (B)

tion the wiring harness out of the way, then unbolt the cruise control module from the firewall and set it aside.

5   From the passenger compartment, disconnect the power brake pushrod from the top of the brake pedal.

6   Also from this location, remove the nuts attaching the booster to the firewall (see illustration).

7   Carefully lift the booster unit away from the firewall and out of the engine compartment.

8   To install the booster, place it in position and tighten the mounting nuts. Reconnect the brake pedal.

9   Install the master cylinder and vacuum hose.

10 Carefully test the operation of the brakes before driving the vehicle in traffic.

13 Brake light switch - removal, installation and adjustment

REMOVAL

♦ **Refer to illustration 13.3**

1   The brake light switch is located on a bracket at the top of the brake pedal. The switch activates the brake lights at the rear of the vehicle when the pedal is depressed.

2   Remove the under dash cover and disconnect the wiring to the courtesy light in the panel.

3   Locate the switch at the top of the brake pedal (see illustration). If the vehicle is equipped with cruise control, there will be another switch very similar in appearance. The brake light switch is the one towards the end of the bracket.

4   Disconnect the negative battery cable from the battery.

13.3 The brake light switch (arrow) is located to the right of the steering column at the end of the mounting bracket

❄ CAUTION:

If the vehicle is equipped with a Delco Loc II or Theftlock audio system, make sure you have the correct activation code before disconnecting the battery. See the information at the front of this manual for the radio re-activation procedure.

5   Detach the wiring connectors at the brake light switch.

6   Depress the brake pedal and pull the switch out of the clip. The switch appears to be threaded, but it's designed to be pushed into and out of the clip, not turned.

INSTALLATION AND ADJUSTMENT

7   With the brake pedal depressed, push the new switch into the

clip. Note that audible clicks will be heard as this is done.

8   Pull the brake pedal all the way to the rear, against the pedal stop until the clicking sounds can no longer be heard. This action will automatically move the switch the proper amount and no further adjustment will be required.

❄ CAUTION:

Don't apply excessive force during this adjustment procedure, as power booster damage may result.

9   Connect the wiring at the switch and the battery. Make sure the brake lights are functioning properly.

14   Anti-lock Brake System (ABS) and Traction Control System (TCS) - general information

This system is available as an option. It is designed to reduce lost traction during heavy braking or on slippery surfaces. The system is similar to the non-ABS system except for the Electronic Brake Control Module (EBCM) and related wiring, speed sensors and the Brake Pressure Modulator Valve (BPMV).

Anti-lock braking occurs only when a wheel is about to lock up (lose traction). Input signals from the wheel speed sensors to the computer are used to determine when a wheel is about to lose traction during braking. Hydraulic pressure will be reduced for the wheel about to lose traction.

Some models are also equipped with a Traction Control System

(TCS). This system uses the ABS speed sensors to monitor the speed of the drive wheels. If the computer senses one of the drive wheels spinning significantly faster than the other, the computer, through the ABS system, will apply brake pressure to the more rapidly spinning wheel. This allow torque to be transferred to the wheel with the most traction.

Due to the special tools required and the extremely involved diagnostic procedures, all diagnosis and service to either of these systems must be performed by a dealer service department or other qualified repair shop.

Specifications

General

Brake fluid type	See Chapter 1

Disc brakes

Brake pad lining minimum thickness	See Chapter 1
Disc thickness*	
1985 through 1991	
Standard	1.043 in
Minimum thickness after resurfacing	0.972 in
Discard thickness*	0.957 in
1992 and later	
Front	
Standard	1.276 in
Minimum thickness after resurfacing	1.224 in
Discard thickness*	1.209 in
Rear	
Standard	0.433 in
Minimum thickness after resurfacing	0.423 in
Discard thickness*	0.374 in

Disc brakes (continued)

Disc runout
    1985 through 1993   0.004 in
    1994 and later   0.002 in
Disc thickness variation limit   0.0005 in
Caliper-to-bracket stop clearance   0.005 to 0.012 in

Refer to marks cast into the disc (they supersede information printed here)

Drum brakes

Brake shoe lining minimum thickness   See Chapter 1
Drum diameter
    Standard   8.860 in
    Service limit
        1985 through 1993   8.880 in
        1994 and later   8.920 in
    Discard diameter*   8.909 in
Drum out-of-round limit   0.006 in

Refer to marks cast into the drum (they supersede information printed here)

Torque Specifications   Ft-lbs

Caliper mounting bolts
    Front
        1996 and earlier models   38
        1997 and later models   63
    Rear   20
Caliper bracket bolts   137
Brake hose-to-caliper bolt   33
Master cylinder-to-booster nuts   20
Proportioner valve caps   20
Booster-to-pedal bracket nuts
    1996 and earlier models   15
    1997 through 2000   15
    2001 and later   17
Wheel lug nuts   See Chapter 1

Section

Reference to other Chapters

10

SUSPENSION AND STEERING SYSTEMS

1.1 Front suspension components

1	*Control arm*	2	*Stabilizer bar*	3	*Outer tie-rod*	4	*Balljoint*

1.2a Rear suspension components

1	*Stabilizer bar*	2	*Control arm*	3	*Suspension adjustment links*

1 General information

▶ **Refer to illustrations 1.1, 1.2a and 1.2b**

✷✷ WARNING:

Whenever any of the suspension or steering fasteners are loosened or removed, they must be inspected and, if necessary, replaced with new ones of the same part number or of original equipment quality and design. Torque specifications must be followed for proper reassembly and component retention. Never attempt to heat or straighten any suspension or steering components. Instead, replace any bent or damaged part with a new one.

✷✷ CAUTION:

If the vehicle is equipped with a Delco Loc II or Theftlock audio system, make sure you have the correct activation code before disconnecting the battery. See the information at the front of this manual for the radio re-activation procedure.

The front suspension is a combination strut and spring design. The steering knuckles are located by lower control arms which are mounted to longitudinally positioned, removable frame members. The lower end of the steering knuckle pivots on a balljoint riveted to the control arm. The balljoint is fastened to the steering knuckle with a castellated nut. The control arms are connected by a stabilizer bar, which reduces body lean during cornering (see illustration).

The rear suspension is fully independent with each suspension knuckle supported by a lower control arm, coil spring and strut. A stabilizer bar minimizes body roll. Each control arm is equipped with a suspension adjustment link to provide for toe adjustment and to minimize alignment variation with suspension movement. The rear control arm is attached to the suspension knuckle through a balljoint to reduce friction.

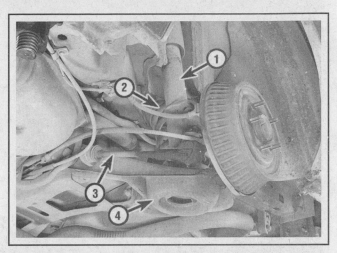

1.2b Details of the rear suspension assembly

1	Strut	3	Toe adjustment link
2	Coil spring	4	Control arm

On 2001 and later models, the rear suspension components attach to a suspension support assembly or subframe, which is attached to the body.

The power rack-and-pinion steering system is located behind the engine/transaxle assembly on the firewall and actuates the tie-rods, which transmit steering inputs to the steering knuckles. The steering column is connected to the steering gear through an insulated coupler. The steering column is designed to collapse in the event of an accident.

➡**Note: These vehicles have a combination of standard and metric fasteners on the various suspension and steering components, so it would be a good idea to have both types of tools available when beginning work.**

2 Front stabilizer bar and bushings - removal and installation

▶ **Refer to illustrations 2.2 and 2.3**

REMOVAL

1 Loosen the lug nuts on both front wheels, raise the front of the vehicle and support it securely on jackstands. Apply the parking brake and block the rear wheels to keep the vehicle from rolling off the jackstands. Remove the front wheels.

2 Remove the stabilizer bar-to-control arm bolts. Note how the link bushings, spacers and washers are arranged (see illustration).

3 Remove the stabilizer bar bushing clamp bolts (see illustration).

4 Detach the tie-rods (only the left tie-rod needs to be disconnected on 2000 and later models) from the steering knuckles (see Section 16).

5 Detach the exhaust pipe from the vehicle (see Chapter 4).

6 Turn the right strut to the right.

2.2 The stabilizer bar link has washers, rubber bushings and spacers to connect the stabilizer bar to the control arm

7   Slide the bar over the right steering knuckle (the left steering knuckle on 2000 and later models), then pull down until the bar clears the frame.

8   Inspect the bushings for wear and damage and replace them if necessary.

INSTALLATION

9   Guide the bar through the wheel well, over the suspension supports and into position.

10   Loosely install the bushings and clamps.

11   Center the bar in the vehicle and install the stabilizer bar-to-control arm bolts, spacers, bushings and washers. Tighten all of the fasteners to the torque figures listed in this Chapter's Specifications at this time.

12   Install the wheels and lower the vehicle. Tighten the lug nuts to the torque specified.

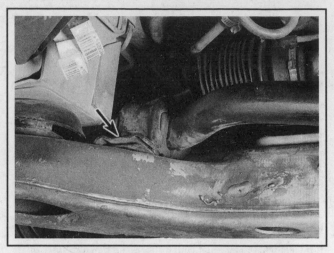

2.3 Remove the two bolts from each stabilizer bar bushing clamps (arrow)

3   Balljoint - check and replacement

◆ **Refer to illustrations 3.3a, 3.3b and 3.11**

CHECK

1   Raise the front of the vehicle and support it securely on jackstands. Apply the parking brake and block the rear wheels to keep the vehicle from rolling off the jackstands.

2   Visually inspect the rubber seal for damage, deterioration and leaking grease. If any of these conditions are noticed, the balljoint should be replaced.

3   Place a large prybar under the balljoint and attempt to push the balljoint up. Next, position the prybar between the steering knuckle and control arm and pry down (see illustrations). If any movement is seen or felt during either of these checks, a worn out balljoint is indicated.

4   Have an assistant grasp the tire at the top and bottom and move the top of the tire in-and-out. Touch the balljoint stud castellated nut. If any looseness is felt, suspect a worn out balljoint stud or a widened

hole in the steering knuckle boss. If the latter problem exists, the steering knuckle should be replaced as well as the balljoint.

5   Separate the control arm from the steering knuckle (see Section 4). Using your fingers (don't use pliers), try to twist the stud in the socket. If the stud turns, replace the balljoint.

REPLACEMENT

6   Loosen the wheel lug nuts, raise the front of the vehicle and support it securely on jackstands. Apply the parking brake and block the rear wheels to keep the vehicle from rolling off the jackstands. Remove the wheel.

7   Separate the control arm from the steering knuckle (see Section 4). Temporarily insert the balljoint stud back into the steering knuckle (loosely). This will ease balljoint removal after Step 9 has been performed, as well as hold the assembly stationary while drilling out the rivets.

3.3a Check for movement between the balljoint and steering knuckle (arrow) when prying up

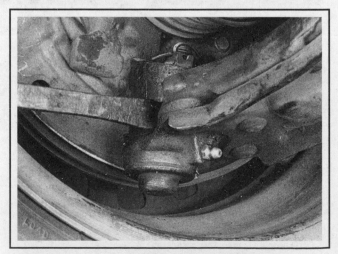

3.3b With the prybar positioned between the steering knuckle boss and the balljoint, pry down and check for play in the balljoint - if there's any play, replace the balljoint

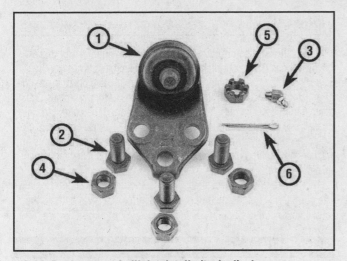

3.11 Replacement balljoint details (typical) - be sure to tighten the bolts to the torque specified on the instruction sheet

1	*Replacement balljoint*	4	*Nut*
2	*Bolt*	5	*Castle nut*
3	*Grease fitting*	6	*Cotter pin*

8   Using a 1/8-inch drill bit, drill a pilot hole into the center of each balljoint-to-control arm rivet. Be careful not to damage the CV joint boot in the process.

9   Using a 1/2-inch drill bit, drill the head off each rivet. Work slowly and carefully to avoid deforming the holes in the control arm.

10 Loosen (but don't remove) the stabilizer bar-to-control arm nut. Pull the control arm and balljoint down to remove the balljoint stud from the steering knuckle, then dislodge the balljoint from the control arm.

11 Position the new balljoint on the control arm and install the bolts (supplied in the balljoint kit) from the top of the control arm (see illustration). Tighten the bolts to the torque specified in the new balljoint instruction sheet.

12 Insert the balljoint into the steering knuckle, install the castellated nut, tighten it to the torque listed in this Chapter's Specifications and install a new cotter pin. It may be necessary to tighten the nut some to align the cotter pin hole with an opening in the nut, which is acceptable. Never loosen the castellated nut to allow cotter pin insertion.

13 Tighten the stabilizer bar-to-control arm nut to the torque listed in this Chapter's Specifications.

14 Install the wheel, lower the vehicle and tighten the lug nuts to the torque listed in the Chapter 1 Specifications.

4   Control arm - removal and installation

▶ **Refer to illustrations 4.3, 4.4 and 4.5**

REMOVAL

1   Loosen the wheel lug nuts, raise the front of the vehicle and support it securely on jackstands. Apply the parking brake and block the rear wheels to keep the vehicle from rolling off the jackstands. Remove the wheel.

2   If only one control arm is being removed, disconnect only that end of the stabilizer bar. If both control arms are being removed, disconnect both ends (see Section 2 if necessary).

3   Remove the balljoint stud-to-steering knuckle castellated nut and cotter pin (see illustration).

4   Using a large prybar positioned between the control arm and steering knuckle, "pop" the balljoint out of the knuckle (see illustration).

✳ CAUTION:

When removing the balljoint from the knuckle, be careful not to overextend the inner CV joint or it may be damaged. If the balljoint stud won't come out, try striking the steering knuckle boss with a hammer. If that doesn't work, use a "picklefork" type balljoint separator, but note that the use of this tool will almost surely damage the balljoint boot.

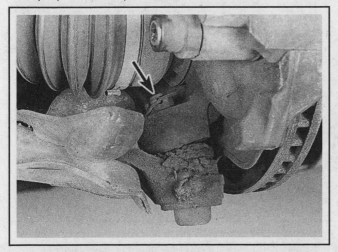

4.3 Remove the cotter pin and castellated nut (arrow) from the balljoint stud

4.4 Pry the balljoint out of the steering knuckle - if it's stubborn and won't come out, strike the steering knuckle boss on both sides (arrow) simultaneously with two hammers, then try again

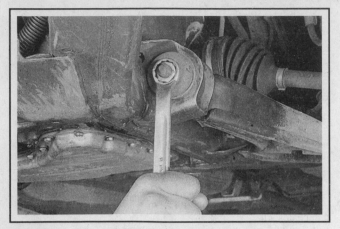

4.5 Remove the control arm pivot bolts

5   Remove the two control arm pivot bolts and detach the control arm (see illustration).

6   The control arm bushings are replaceable, but special tools and expertise are necessary to do the job. Carefully inspect the bushings for hardening, excessive wear and cracks. If they appear to be worn or deteriorated, take the control arm to a dealer service department or repair shop.

INSTALLATION

7   Position the control arm in the suspension support and install the pivot bolts. Do not tighten them completely at this time.

8   Insert the balljoint stud into the steering knuckle boss, install the castellated nut and tighten it to the torque listed in this Chapter's Specifications. If necessary, tighten the nut a little more if the cotter pin hole doesn't line up with an opening on the nut. Install a new cotter pin.

9   Install the stabilizer bar-to-control arm bolt, spacer, bushings and washers and tighten the nut to the torque listed in this Chapter's Specifications.

10   Using a floor jack, raise the outer end of the control arm to simulate normal ride height, then tighten the control arm pivot bolts to the torque listed in this Chapter's Specifications.

> ❋❋ **CAUTION:**
>
> **If the bolts aren't tightened with the control arm raised to simulate normal ride height, control arm bushing damage may occur.**

11   Install the wheel and lower the vehicle. Tighten the lug nuts to the torque listed in the Chapter 1 Specifications.

5   Front strut and spring assembly - removal, inspection and installation

REMOVAL

▶ **Refer to illustrations 5.2, 5.4a, 5.4b, 5.6 and 5.7**

1   Loosen the wheel lug nuts, raise the front of the vehicle and support it securely on jackstands. Apply the parking brake and block the rear wheels to keep the vehicle from rolling off the jackstands. Remove the wheel.

2   Remove the brake line bracket from the strut (see illustration).

3   If the vehicle is equipped with anti-lock brakes, disconnect the front sensors. On later models, remove the ABS harness and bracket from the strut first. If the vehicle is equipped with the computer command ride (CCR), disconnect the CCR electrical connector from the front strut.

4   Using white paint or a scribe, mark the strut-to-steering knuckle relationship and make a line around the strut-to-steering knuckle nuts (see illustration). Also mark the relationship of the upper strut mounting studs to the body (see illustration).

5   Separate the tie-rod end from the steering arm as described in Section 16.

6   Remove the strut-to-knuckle nuts (see illustration) and knock the bolts out with a soft-face hammer.

5.2 Detach the front brake line bracket (arrow) from the strut

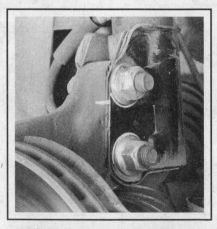

5.4a Mark the strut-to-steering knuckle relationship and draw a line around the nuts with paint or a scribe

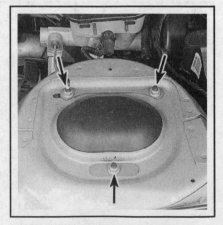

5.4b Before you remove the three strut-to-shock tower nuts (arrows), be sure to mark their relationship to the body

5.6 Remove the strut-to-knuckle nuts and bolts - the bolts are splined and must be driven out with a brass, lead or plastic hammer

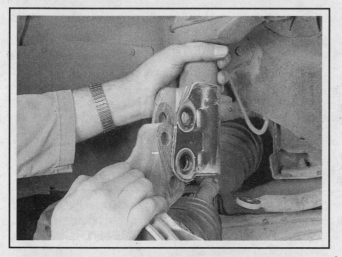

5.7 Push in on the strut while pulling out on the top of the brake rotor to separate the knuckle and strut

7   Separate the strut from the steering knuckle (see illustration). Be careful not to overextend the inner CV joint or stretch the brake hose.

8   Have an assistant support the strut assembly. Remove the three strut-to-shock tower nuts. Remove the assembly out through the fenderwell.

INSPECTION

9   Check the strut body for leaking fluid, dents, cracks and other obvious damage which would warrant repair or replacement.

10   Check the coil spring for chips and cracks in the spring coating (this will cause premature spring failure due to corrosion). Inspect the spring seat for hardening, cracks and general deterioration.

11   If wear or damage is evident, proceed to Section 6 for the strut disassembly procedure.

INSTALLATION

12   Install the strut. Once the three studs protrude from the shock tower, install the nuts so the strut won't fall back through. This may require an assistant, since the strut is quite heavy and awkward. Be sure to align the marks you made on disassembly.

13   Slide the steering knuckle into the strut flange and insert the two bolts. They should be positioned with the flats situated horizontally. Install the nuts, align the marks and tighten the nuts to the torque listed in this Chapter's Specifications.

14   Install the tie-rod end to the steering knuckle and tighten the castellated nut to the specified torque. Install a new cotter pin. If the cotter pin won't pass through, tighten the nut a little more, but just enough to align the hole in the stud with a castellation on the nut (don't loosen the nut).

15   Install the wheel, lower the vehicle and tighten the lug nuts to the torque listed in the Chapter 1 Specifications.

16   Tighten the three upper mounting nuts to the specified torque.

17   Have the front end alignment checked and, if necessary, adjusted.

6   Strut cartridge (1996 and earlier models) - replacement

▶ **Refer to illustrations 6.4, 6.5a, 6.5b, 6.6, 6.7, 6.15a and 6.15b**

➡**Note:** Strut cartridges are available for most 1996 and earlier models. However, rebuilt strut assemblies (some complete with springs) are available on an exchange basis which eliminates much time and work. Whichever route you choose to take, check on the cost and availability of parts before disassembling anything.

1   If the struts exhibit the telltale signs of wear (leaking fluid, loss of dampening capability) the strut cartridges can be replaced.

❊❊❊ **WARNING:**

Disassembling a strut is a dangerous job. Be very careful and follow the instructions closely or serious injury may result. Use only a high-quality spring compressor and carefully follow the manufacturer's instructions furnished with the tool. After removing the coil spring from the strut assembly, set it aside in a safe, isolated area.

2   Remove the strut and spring assembly following the procedure described in Section 5. Mount the strut assembly in a vise.

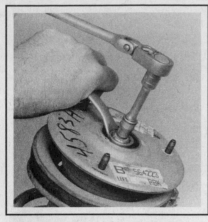

6.4 After the spring has been compressed, remove the damper shaft nut

6.5a Remove the bearing cap . . .

6.5b . . . and the upper spring seat and insulator from the damper shaft

6.6 Remove the compressed spring assembly - be EXTREMELY CAREFUL when handling the spring!

6.7 Using a tubing cutter, cut the end cap off the strut body at the groove (arrow)

6.15a Install the upper spring seat with the flat (arrow) facing the steering knuckle flange

Cushion the vise jaws with rags or blocks of wood.

3   Following the tool manufacturer's instructions, install the spring compressor (which can be obtained at most auto parts stores or equipment yards on a daily rental basis) on the spring and compress it sufficiently to relieve all pressure from the spring seat. This can be verified by wiggling the spring seat.

4   Loosen the damper shaft nut while using a socket wrench on the shaft hex to prevent it from turning (see illustration).

5   Lift the bearing cap, upper spring seat and upper insulator off the damper shaft (see illustrations). Inspect the bearing in the spring seat for smooth operation and replace it if necessary.

6   Carefully remove the compressed spring assembly (see illustration) and set it in a safe place.

❋❋ WARNING:

Don't position your head near the end of the spring!

7   Locate the groove cut in the strut reservoir tube, 3/4-inch from the top of the tube (see illustration). Using a tubing cutter, cut around the groove until the reservoir tube is severed. Lift out the piston rod assembly with the cylinder and end cap. Discard these items.

8   Remove the strut reservoir tube from the vise and pour the damper fluid into an approved oil container.

9   Place the strut back in the vise and lightly file around the inner edge of the opening to eliminate any burrs that may have resulted from the cutting operation. Be careful not to damage the internal threads in the strut body.

6.15b The bearing cap must also be positioned with the flat (arrow) facing the knuckle flange

10 Thread the cartridge retaining nut into the reservoir tube, as straight as possible, to establish a clean path in the existing threads. Remove the nut.

11 Insert the replacement strut cartridge into the reservoir tube and turn it until you feel the pads on the bottom of the cartridge seat in the depressions at the bottom of the reservoir tube.

12 Slide the nut over the cartridge and thread it into the tube, tightening it to the torque specified in the kit instructions.

13 Stroke the damper shaft up-and-down a few times to verify proper operation.

14 Extend the damper shaft all the way and hold it in place with a clothes pin at the bottom of the rod.

15 Assemble the strut beginning with the lower spring insulator and spring, then the upper spring insulator, spring seat and bearing cap. Position the spring seat and bearing cap with the flats facing the steering knuckle flange (see illustrations).

16 Install the damper shaft nut and tighten it securely. Remove the clothes pin from the damper shaft.

17 Install the strut and spring assembly as outlined in Section 5.

7   Front hub and wheel bearing assembly - removal and installation

▶ **Refer to illustrations 7.6, 7.7, 7.8, 7.9 and 7.10**

➡**Note: The front hub and wheel bearing assembly is sealed-for-life and must be replaced as a unit.**

1 Remove the hubcap or wheel cover and loosen the driveaxle/hub nut. Loosen the wheel lug nuts, raise the front of the vehicle and support it securely on jackstands placed under the frame. Apply the parking brake and block the rear wheels to keep the vehicle from rolling off the jackstands. Remove the wheel.

2 Disconnect the stabilizer bar from the control arm (see Section 2 if necessary). On models with ABS brakes, disconnect the wheel speed sensor and release the ABS harness clip at the brake dust shield.

3 Remove the balljoint-to-steering knuckle nut and separate the control arm from the knuckle (see Section 4).

4 Remove the caliper from the steering knuckle and hang it out of the way with a piece of wire (see Chapter 9).

5 Pull the rotor off the hub and remove the driveaxle (see Chapter 8 if necessary).

6 Using a no. 55 Torx bit, remove the three hub retaining bolts through the opening in the flange (see illustration).

➡**Note: On 2001 and later models, the bolts are removed from behind the knuckle, not the front.**

7 Wiggle the hub and bearing assembly back-and-forth and pull it out of the steering knuckle, along with the rotor shield (see illustration).

8 If the hub and bearing assembly is being replaced with a new

7.6 A no. 55 Torx bit is required to remove the hub bolts - DO NOT use an Allen wrench or the bolts will be damaged

one, it's a good idea to replace the dust seal in the back of the steering knuckle. Pry it out of the knuckle with a screwdriver (see illustration).

9 Drive the new dust seal into the knuckle with a large socket or a seal driver and a hammer (see illustration). Try not to cock the seal in the bore.

7.7 Pull the hub and bearing assembly and the rotor shield out of the steering knuckle

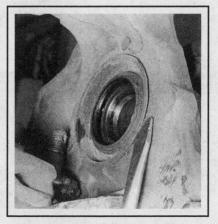

7.8 Pry the seal out of the knuckle with a screwdriver

7.9 Using a large socket, drive the new seal into place

10 Install a new O-ring around the rear of the bearing and push it up against the bearing flange (see illustration).

11 Clean the mating surfaces on the steering knuckle, bearing flange and knuckle bore. Lubricate the outside diameter of the bearing and the seal lips with high-temperature grease and insert the hub and bearing into the steering knuckle. Position the rotor shield and install the three bolts, tightening them to the torque listed in this Chapter's Specifications.

12 Install the driveaxle (see Chapter 8).

13 Attach the control arm to the steering knuckle (see Section 4).

14 Reconnect the stabilizer bar to the control arm (see Section 2).

15 Install the brake rotor and caliper (see Chapter 9).

16 Install the driveaxle/hub nut and tighten it securely. Prevent the axle from turning by inserting a screwdriver through the caliper and into a rotor cooling vane.

17 Install the wheel, lower the vehicle and tighten the lug nuts to the torque listed in the Chapter 1 Specifications.

18 Tighten the driveaxle/hub nut to the torque listed in the Chapter 8 Specifications.

7.10 Install a new O-ring (arrow) on the wheel bearing assembly

8 Steering knuckle and hub - removal and installation

REMOVAL

1 Remove the hubcap or wheel cover and loosen the driveaxle/hub nut. Loosen the wheel lug nuts, raise the front of the vehicle and support it securely on jackstands. Apply the parking brake and block the rear wheels to keep the vehicle from rolling off the jackstands. Remove the wheel.

2 Remove the driveaxle/hub nut. Insert a screwdriver through the caliper and into a rotor cooling vane to prevent the driveaxle from turning.

3 Remove the caliper and suspend it out of the way with a piece of wire. Lift the rotor off the hub.

4 Mark the position of the two strut-to-knuckle nuts and remove them (see illustration 5.4a). Don't drive out the bolts at this time.

5 Separate the control arm balljoint from the steering knuckle (see Section 4 if necessary).

6 Attach a puller to the hub flange and push the driveaxle out of the hub (see Chapter 8). Hang the driveaxle with a piece of wire to prevent damage to the inner CV joint.

7 Support the knuckle and drive out the two strut-to-knuckle bolts

with a soft-face hammer. Remove the steering knuckle assembly from the strut.

INSTALLATION

8 Position the knuckle in the strut and insert the two splined bolts, with the flats on the bolt heads in the horizontal position. Tap the bolts into place and install the nuts, but don't tighten them at this time.

9 Install the driveaxle in the hub.

10 Connect the control arm to the steering knuckle and tighten the castellated nut to the torque listed in this Chapter's Specifications. Install a new cotter pin.

11 Align the strut-to-knuckle nuts with the previously applied marks and tighten them to the specified torque.

12 Install the brake rotor and caliper.

13 Tighten the driveaxle/hub nut securely.

14 Install the wheel, lower the vehicle and tighten the lug nuts to the torque listed in the Chapter 1 Specifications.

15 Tighten the driveaxle/hub nut to the torque specified in Chapter 8.

16 Have the front end alignment checked and, if necessary, adjusted.

9 Rear stabilizer bar - removal and installation

▶ **Refer to illustrations 9.5, 9.6a and 9.6b**

1 Loosen the rear wheel lug nuts.

2 Raise the rear of the vehicle and place it securely on jackstands.

3 Block the front wheels.

4 Remove the rear wheels. On 2000 and later models, remove the electronic level control sensor from the rear control arm and position it aside.

5 Detach the stabilizer bar link bolt assemblies from the knuckle brackets on 1999 and earlier models (see illustration) or from the rear control arm on 2000 and later models.

6 On 1999 and earlier models, if you're replacing the stabilizer bar and/or the lower bushing from the hanger clamp only, remove the pinch bolt from each hanger clamp (see illustration), bend the end of the clamp open and remove the stabilizer bar and bushings. If you plan to reinstall the same bar or replace both bushings, simply remove the upper bolt from the hanger clamp (see illustration) and leave the hanger attached to the stabilizer bar.

7 On 2000 and later models, simply remove the clamp bolt and detach the stabilizer bar from the vehicle.

8 Installation is the reverse of removal. Tighten the fasteners to the torque values listed in this Chapter's Specifications.

9.5 To detach the rear stabilizer bar link bolt assembly from the knuckle, simply remove the nut on the upper end (arrow) and tap the link bolt out - don't lose any of the bushings or washers

9.6a If you're replacing the rear stabilizer bar and/or lower bushing, remove this bolt on each hanger clamp, bend the clamp open and remove the bar

9.6b If you're going to reinstall the same stabilizer bar or replace both bushings, simply remove the upper hanger clamp bolt and leave the clamp attached to the bar during removal

10 Rear strut/shock absorber - removal and installation

▶ **Refer to illustrations 10.2, 10.9 and 10.10**

1 Remove the rear speaker assembly (see Chapter 12).
2 Remove the trunk side cover (see illustration).
3 Loosen the rear wheel lug nuts.
4 Raise the rear of the vehicle and support it securely on jackstands placed under the frame.
5 Block the front wheels.
6 Remove the rear wheels.
7 If the vehicle is equipped with electronic level control (ELC), remove the ELC air tube from the strut/shock absorber air tube fitting. If the vehicle is equipped with the computer command ride (CCR), dis-

connect the CCR electrical connector from the strut.
8 Support the control arm at the balljoint with a floor jack (see illustration 11.9).
9 Remove the strut/shock absorber upper mounting nuts from inside the trunk (see illustration). On some models this may require removing a plastic cover to access the upper retaining nuts.
10 Remove the nuts, bolts and washers from the knuckle or rear control arm (see illustration).
11 Remove the strut/shock absorber.
12 Installation is the reverse of removal. Tighten the fasteners to the torque values listed in this Chapter's Specifications.

10.2 Remove the trunk side cover

10.9 Remove the strut upper mounting nuts (arrows)

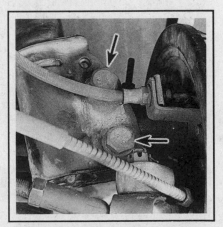

10.10 On 1999 and earlier models, remove the nuts and pull out the bolts (arrows) from the knuckle - don't lose the washers

11 Rear springs and insulators - removal and installation

♦ **Refer to illustration 11.9**

1 Loosen the rear wheel lug nuts.

2 Raise the rear of the vehicle and support it securely on jackstands placed under the frame. Make sure the jackstands aren't under the control arms.

3 Block the front wheels.

4 Remove the rear wheels.

5 Detach the parking brake cable retaining clip from the control arm if necessary.

6 If the vehicle is equipped with electronic level control (ELC), detach the ELC height sensor link from the control arm.

7 Detach the rear stabilizer bar (see Section 9).

8 Detach the suspension adjustment link (see Section 13).

9 Place a floor jack under the spring pocket of the control arm (see illustration). Raise the jack just enough to remove tension from the control arm.

10 Place a chain around the spring and through the control arm as a safety measure. Bolt the chain together, but make sure there is enough slack in it to allow the spring to fully extend.

11 On 1999 and earlier models, remove the strut-to-knuckle bolts and detach the strut from the knuckle (see Section 10). On 2000 and later models, remove the bolts securing the shock to the control arm.

12 Slowly lower the jack to relieve tension on the control arm. When all compression is removed from the spring, remove the safety chain,

11.9 Support the outer end of the control arm with a floor jack or a jackstand - place a chain through the spring and the control arm to prevent the chain from flying off when the control arm is lowered

spring and insulators.

13 Installation is the reverse of removal. Tighten the fasteners to the torque values listed in this Chapter's Specifications.

12 Rear hub and wheel bearing assembly - removal and installation

♦ **Refer to illustration 12.6**

➡ **Note: The rear hub and bearing assembly is sealed for life and must be replaced as a unit.**

1 Loosen the rear wheel lug nuts.

2 Raise the rear of the vehicle and support it securely on jackstands placed under the frame.

3 Block the front wheels.

4 Remove the rear wheel.

5 Remove the brake drum or disc (see Chapter 9).

6 To remove the hub and bearing assembly, you'll need to rotate the stud flange to align one of its holes with each of the four mounting flange Torx bolts (see illustration). When the last bolt is removed, support the backing plate assembly, remove the hub and bearing assembly and reinstall two of the bolts finger tight. This will prevent the brake line from being strained under the weight of the backing plate.

7 Installation is the reverse of removal. Be sure tighten the bolts to the torque listed in this Chapter's Specifications.

12.6 To remove the rear hub and wheel bearing assembly, rotate the stud flange until one of its holes is aligned with each of the four flange mounting Torx bolts (the brake shoe assembly is shown removed for clarity)

13 Rear suspension adjustment link - removal and installation

♦ **Refer to illustrations 13.5 and 13.7**

1 Loosen the rear wheel lug nuts.

2 Raise the vehicle and place it securely on jackstands.

3 Block the front wheels.

4 Remove the wheel.

5 Remove the cotter pin and nut from the outer end of the adjustment link (see illustration).

6 Separate the suspension adjustment link from the knuckle on 1999 and earlier models or from the control arm on 2000 and later models with a puller (see illustration 16.2b).

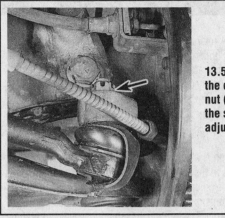

13.5 Remove the cotter pin and nut (arrow) from the suspension adjustment link stud

13.7 To detach the inner end of the suspension adjustment link from the control arm, remove this nut (arrow) - don't lose the washer and spacer

7   To detach the inner end of the link on 1999 and earlier models, remove the retaining nut, washer and spacer (see illustration). To detach the the inner end of the link on 2000 models, remove the nut, the adjustment cam washer and the adjustment cam bolt.

8   Remove the suspension adjustment link from the control arm on 1999 and earlier models or from the chassis on 2000 and later models.

9   Installation is the reverse of removal. Tighten the fasteners to the torque values listed in this Chapter's Specifications, and use a new cotter pin.

14 Rear control arm and suspension support subframe - removal and installation

CONTROL ARMS

▶ **Refer to illustration 14.9**

➡**Note: Replacement of the bushings in the rear control arms requires a number of special tools. Balljoint replacement on 1999 and earlier models is difficult if special tools are not available. If the bushing or balljoint must be replaced, take the control arm to a dealer service department.**

1   Loosen the rear wheel lug nuts.
2   Raise the rear of the vehicle and support it securely on jackstands placed under the frame.
3   Block the front wheels.
4   Remove the electronic level control (ELC) height sensor link, if equipped, from the control arm.

5   Remove the parking brake cable retaining clip from the control arm if equipped.
6   Remove the suspension adjustment link from the control arm (see Section 13).
7   Remove the coil spring (see Section 11).
8   To separate the knuckle from the balljoint stud on 1999 and earlier models, refer to Section 4 (the procedure for splitting the rear knuckle and ballstud is identical to the procedure for splitting a front knuckle and stud).
9   Remove the control arm pivot bolts (see illustration).

➡**Note: On 2001 and later models, you must lower the rear suspension subframe to remove the control arms.**

10 Remove the control arm from the vehicle.
11 Installation is the reverse of removal. Tighten the fasteners to the torque values listed in this Chapter's Specifications.

REAR SUSPENSION SUBFRAME

12 Raise and support the vehicle on sturdy jackstands and remove the rear wheels.
13 Disconnect the ride-height link (if equipped with ride control) and remove the exhaust system (see Chapter 4).
14 Refer to Chapter 9 and disconnect the rear parking brake cables, disconnect the ABS electrical connectors and remove the brake discs.
15 With a wide jack (like a transmission jack) supporting the rear suspension subframe, remove the coil springs. Unbolt the rear struts from the control arms and lower the control arms with a jack as far as they will go. Use a long prybar to pry the bottom end of the coil springs from the control arm and remove the rubber isolators.
16 Remove the subframe-to-body bolts and lower the assembly.
17 Installation is the reverse of the removal procedure.

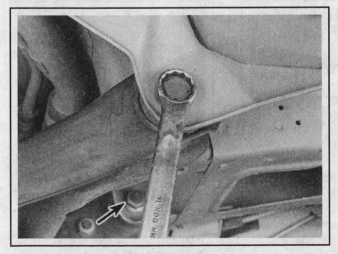

14.9 Remove the rear control arm pivot bolts (arrow)

15 Steering system - general information

⁎⁎⁎ WARNING:

Whenever any of the steering fasteners are removed, they must be inspected and, if necessary, replaced with new ones of the same part number or of original equipment quality and design. Torque specifications must be followed for proper reassembly and component retention. Never attempt to heat or straighten any suspension or steering components. Instead, replace any bent or damaged part with a new one.

All vehicles covered by this manual have power rack-and-pinion steering systems. The components making up the system are the steering wheel, steering column, rack-and-pinion assembly, tie-rods and tie-rod ends. The power steering system has a belt-driven pump to provide hydraulic pressure.

In the power steering system, the motion of turning the steering wheel is transferred through the column to the pinion shaft in the rack-and-pinion assembly. Teeth on the pinion shaft are meshed with teeth on the rack, so when the shaft is turned, the rack is moved left or right in the housing. A rotary control valve in the rack-and-pinion unit directs hydraulic fluid under pressure from the power steering pump to either side of the integral rack piston, which is connected to the rack, thereby reducing manual steering force. Depending on which side of the piston this hydraulic pressure is applied to, the rack will be forced either left or right, which moves the tie-rods, etc. If the power steering system loses hydraulic pressure it will still function manually, though with increased effort.

The steering column is a collapsible, energy-absorbing type, designed to compress in the event of a front end collision to minimize injury to the driver. The column also houses the ignition switch lock, key warning buzzer, turn signal controls, headlight dimmer control and windshield wiper controls. The ignition and steering wheel can both be locked while the vehicle is parked.

Due to the column's collapsible design, it's important that only the specified screws, bolts and nuts be used as designated and that they're tightened to the specified torque. Other precautions particular to this design are noted in appropriate Sections.

16 Tie-rod ends - removal and installation

⬧ **Refer to illustrations 16.2a, 16.2b and 16.3**

REMOVAL

1   Loosen the wheel lug nuts, raise the front of the vehicle and support it securely on jackstands. Apply the parking brake and block the rear wheels to keep the vehicle from rolling off the jackstands. Remove the wheel.

2   Remove the cotter pin and castellated nut from the tie-rod, then disconnect the tie-rod from the steering knuckle arm with a puller (see illustrations).

3   Mark the relationship of the tie-rod end to the threaded adjuster (see illustration). This will ensure the toe-in setting is restored when reassembled.

4   Unscrew the tie-rod end from the tie-rod.

INSTALLATION

5   Thread the tie-rod end onto the tie-rod to the marked position and connect the tie-rod end to the steering arm. Install the castellated nut and tighten it to the torque listed in this Chapter's Specifications. Install a new cotter pin.

6   Install the wheel. Lower the vehicle and tighten the lug nuts to the torque listed in the Chapter 1 Specifications.

7   Have the front end steering geometry checked and, if necessary, adjusted.

16.2a Remove the cotter pin and castellated nut from the tie-rod end stud . . .

16.2b . . . then separate the tie-rod end from the steering knuckle arm with a two-jaw puller - DO NOT pound on the stud!

16.3 Using white paint, mark the relationship of the tie-rod end and the threaded adjuster

17 Steering gear - removal and installation

▸ Refer to illustrations 17.5, 17.11, 17.12a and 17.12b

※ WARNING:

On models equipped with an airbag, make sure the steering shaft is not turned while the steering gear is removed or the airbag system could become damaged. To prevent the shaft from turning, place the ignition key in the LOCK position or thread the seat belt through the steering wheel and clip it into place.

1 Detach the cable from the negative battery terminal.

※ CAUTION:

If the vehicle is equipped with a Delco Loc II or Theftlock audio system, make sure you have the correct activation code before disconnecting the battery. See the information at the front of this manual for the radio re-activation procedure.

2 Loosen the wheel lug nuts.
3 Raise the vehicle and place it securely on jackstands.
4 Remove the wheels. Lock the steering column in the straight-ahead position.
5 Remove the pinch bolt at the steering gear (see illustration). On 2001 and later models, remove the steering gear heat shield.
6 Unplug the compressor cutout switch wire.
7 Detach the power steering line fittings from the steering gear.
8 Detach the tie-rod ends from the steering knuckle (see Section 16).
9 Support the engine/transaxle assembly with a floor jack. Put a block of wood between the jack and the pan to prevent damage.

➡Note: Steps 9 through 11 do not apply to 2001 and later models. On these models, the steering gear may be unbolted from the subframe and drawn out through the left wheelwell, after disconnecting the left stabilizer bar link.

10 Loosen - but do not remove - the two front subframe mounting bolts (see Section 8, Chapter 7).
11 Remove the four rear subframe-to-body bolts (see illustration).
12 Remove the two mounting bolts from the left side of the steering gear assembly and the four bolts from the two clamps on the right (see illustrations).
13 Lower the subframe about three inches.
14 Slide the steering gear assembly out the left wheel well to remove it.
15 Installation is the reverse of removal. Tighten the fasteners to the torque values listed in this Chapter and in Chapter 7.

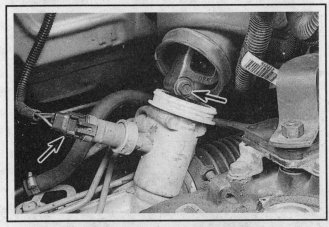

17.5 Remove the pinch bolt from the end of the steering column and unplug the compressor cutout switch wire (arrows)

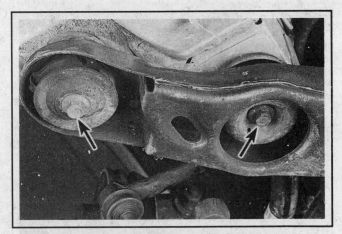

17.11 Remove the two subframe-to-body mounting bolts (arrows) at each rear corner of the subframe

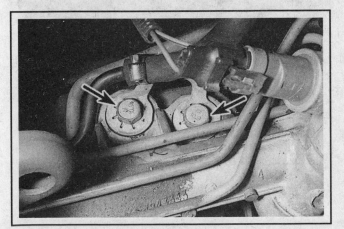

17.12a Remove the two mounting bolts (arrows) from the left end of the rack-and-pinion assembly

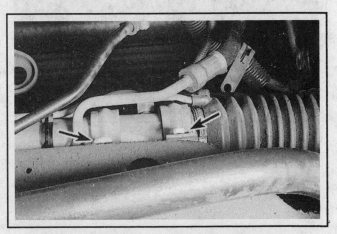

17.12b Remove the two bolts from each of the clamps (arrows) on the right end of the rack-and-pinion assembly

18 Steering gear boots - replacement

1 Remove the steering gear from the vehicle (see Section 17).
2 Detach the tie-rod ends from the steering gear (see Section 16).
3 Cut off both boot clamps and discard them.
4 Slide the old boots off (they can be pulled off the ends of the vent cross tube).
5 Before installing the new boots, wrap the threads and serrations on the ends of the steering rods with a layer of tape so the small ends of the new boots aren't damaged.

6 Slide the new boots on.
7 Loosely install the clamps.
8 Push the boots onto both ends of the vent tube.
9 Tighten the clamps.
10 Install the tie-rods (see Section 16).
11 Install the steering gear assembly.

19 Power steering pump - removal and installation

▸ **Refer to illustrations 19.3 and 19.4**

REMOVAL

1 Disconnect the cable from the negative battery terminal.

✳ CAUTION:

If the vehicle is equipped with a Delco Loc II or Theftlock audio system, make sure you have the correct activation code before disconnecting the battery. See the information at the front of this manual for the radio re-activation procedure.

2 Remove the pump drivebelt (see Chapter 1). On 2001 and later models, remove the coolant recovery tank (see Chapter 3).

3 Remove the pump mounting bolts and detach the pump from the engine (see illustration).
4 Detach the return line from the pump (see illustration), and drain the fluid in the line.
5 Disconnect the pressure line from the pump (see illustration 19.4).

INSTALLATION

6 Installation is the reverse of removal.
➡**Note: It's a good idea to attach (but not tighten) the hoses to the pump before installing it in its bracket.**

7 Fill the reservoir with the recommended fluid and bleed the system, following the procedure described in the next Section.

19.3 Remove the power steering pump mounting bolts (one is shown with a socket on it and the other one is indicated by an arrow)

19.4 Detach the power steering pump return line (1), drain the contents into a container, then disconnect the pressure line (2)

20 Power steering system - bleeding

1 Following any operation in which the power steering fluid lines have been disconnected, the power steering system must be bled to remove air and obtain proper steering performance.

2 With the front wheels turned all the way to the left, check the power steering fluid level and, if low, add fluid until it reaches the Cold mark on the dipstick.

3 Start the engine and allow it to run at fast idle. Recheck the fluid level and add more if necessary to reach the Cold mark on the dipstick.

4 Bled the system by turning the wheels from side-to-side, without hitting the stops. This will work the air out of the system. Don't allow

the reservoir to run out of fluid.

5 When the air is worked out of the system, return the wheels to the straight ahead position and leave the engine running for several minutes before shutting it off. Recheck the fluid level.

6 Road test the vehicle to be sure the steering system is functioning normally with no noise.

7 Recheck the fluid level to be sure it's up to the Hot mark on the dipstick while the engine is at normal operating temperature. Add fluid if necessary.

21 Steering wheel - removal and installation

▶ **Refer to illustrations 21.3a, 21.3b, 21.4a, 21.4b, 21.5a, 21.5b, 21.6, 21.7, 21.8 and 21.9**

❋❋ WARNING:

Some models covered by this manual are equipped with airbags. Always turn the steering wheel to the straight-ahead position, place the ignition switch in the Lock position and disable the airbag system before working in the vicinity of the impact sensors, steering column or instrument panel to avoid the possibility of accidental deployment of the airbag(s), which could cause personal injury (see Chapter 12 for the airbag disarming procedure).

❋❋ CAUTION:

If the vehicle is equipped with a Delco Loc II or Theftlock audio system, make sure you have the correct activation code before disconnecting the battery. See the information at the front of this manual for the radio re-activation procedure.

1 Park the vehicle with the wheels pointing straight ahead. Disconnect the cable from the negative terminal of the battery. On airbag-equipped models, also disconnect the positive terminal, then wait two minutes before proceeding.

2 On models without an airbag, remove the two screws from the backside of the steering wheel and detach the horn pad from the steering wheel.

3 On 1996 and earlier airbag-equipped models, use a no. 30 Torx bit to remove the screws that secure the airbag module to the steering wheel (see illustration). On 1997 and later models, insert a flat-bladed screwdriver through each of the openings in the back of the steering wheel, one at a time, and twist it to disengage the retaining spring from the module retaining pins (see illustration).

4 Lift the airbag module carefully away from the steering wheel and

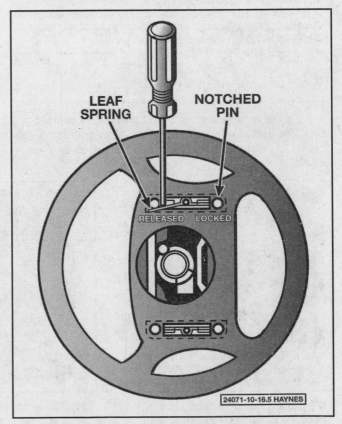

21.3b To detach the airbag module from the steering wheel on 1997 and later models, insert a screwdriver into each of the four holes in the backside of the steering wheel and pry each leaf spring aside to release it from its notched pin

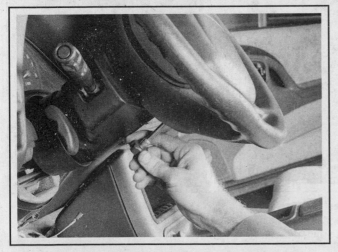

21.3a Use a Torx bit to remove the airbag module screws from behind the steering wheel (1996 and earlier models)

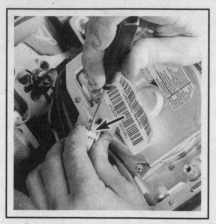

21.4a Use a small screwdriver to release the plastic locking clip (arrow) from the airbag module connector . . .

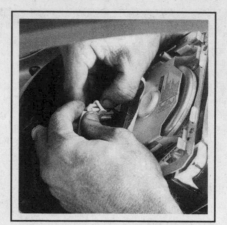

21.4b . . . then squeeze the tab and separate the airbag connector

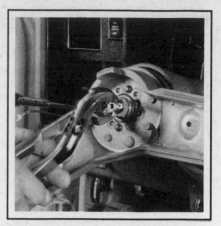

21.5a Remove the steering wheel nut retainer, if equipped

disconnect the yellow airbag electrical connector. This is a two-part disconnection, as there is a plastic clip that must be removed before the connector can be disconnected (see illustrations). Remove the module.

⁑ WARNING:

When carrying the airbag module, keep the driver's side of it away from your body, and when you set it down (in an isolated area), have the driver's side facing up.

5   Remove the steering wheel nut retainer (if equipped) and the steering wheel nut (see illustrations).
6   Mark the relationship of the steering wheel to the shaft (see illustration).
7   Install a steering wheel puller (available at most auto parts stores) and turn the center bolt until the wheel is free (see illustration).
8   Remove the puller and disconnect the horn and ground connector (see illustration). On later models with steering-wheel mounted switches, disconnect the electrical connector. Remove the steering wheel.

21.5b Remove the steering wheel nut with a deep socket

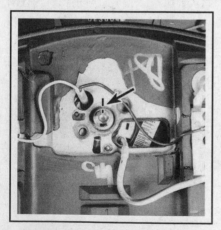

21.6 Mark the relationship of the steering wheel to thesteering shaft

21.7 A steering wheel puller threads into two holes in the steering wheel - tightening the center bolt removes the wheel

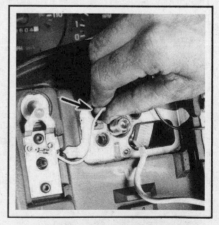

21.8 Disconnect the horn and ground wire connector

On airbag-equipped models, don't allow the steering shaft to turn with the steering wheel removed. If the shaft turns, the air-bag SIR coil assembly (a ribbon-like mechanism which allows electrical current to flow to the airbag module regardless of steering wheel angle) will become uncentered and may break when the vehicle is returned to service.

9   Installation is the reverse of removal. On airbag-equipped models, before the steering wheel is installed, make sure the SIR coil is centered (see illustration). If it isn't, see Chapter 12, Section 24 for the centering procedure. Connect the airbag connector to the back of the airbag module just as it was before steering wheel removal, with the plastic locking device in place. Be sure to tighten the steering wheel nut and airbag screws (if equipped) to the torque listed in this Chapter's Specifications.

10   If equipped with an airbag, refer to Chapter 12 for the airbag enabling procedure.

21.9 When properly aligned, the airbag coil will be centered with the marks aligned (in the circle here) and the tab fitted between the projections on the top of the steering column (arrow)

22   Wheel studs - replacement

▶ **Refer to illustrations 22.3 and 22.4**

➡**Note: This procedure applies to both the front and rear wheel studs.**

1   Remove the hub and wheel bearing assembly (see Section 7 or 12).

2   Install a lug nut part way onto the stud being replaced.

3   Push the stud out of the hub flange with a press tool (see illustration).

4   Insert the new stud into the hub flange from the back side and install four flat washers and a lug nut on the stud (see illustration).

5   Tighten the lug nut until the stud is seated in the flange.

6   Reinstall the hub and wheel bearing assembly.

22.3 Use a press tool to push the stud out of the flange

1	Hub flange	3 Press tool
2	Lug nut on stud	

22.4 Install four washers and a lug nut on the stud, then tighten the nut to draw the stud into place

1 Hub flange	2 Spacer/washers

23 Wheels and tires - general information

◆ **Refer to illustration 23.1**

All vehicles covered by this manual are equipped with metric-size fiberglass or steel belted radial tires (see illustration). The use of other size or type tires may affect the ride and handling of the vehicle. Don't mix different types of tires, such as radials and bias belted, on the same vehicle, since handling may be seriously affected. Tires should be replaced in pairs on the same axle, but if only one tire is being replaced, be sure it's the same size, structure and tread design as the other.

Because tire pressure affects handling and wear, the tire pressures should be checked at least once a month or before any extended trips (see Chapter 1).

Wheels must be replaced if they're bent, dented, leak air, have elongated bolt holes, are heavily rusted, out of vertical symmetry or if the lug nuts won't stay tight. Wheel repairs by welding or peening aren't recommended.

Tire and wheel balance is important to the overall handling, braking and performance of the vehicle. Unbalanced wheels can adversely affect handling and ride characteristics as well as tire life. Whenever a tire is installed on a wheel, the tire and wheel should be balanced by a shop with the proper equipment.

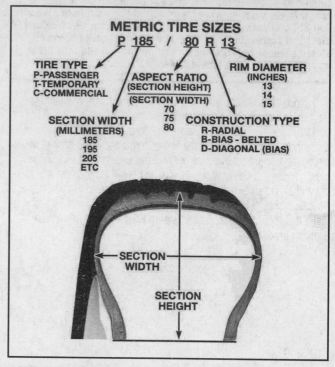

23.1 Metric tire size code

24 Wheel alignment - general information

◆ **Refer to illustration 24.1**

Wheel alignment refers to the adjustments made to the wheels so they're in proper angular relationship to the suspension and the ground. Wheels that are out of proper alignment not only affect steering control, but also increase tire wear. Camber, caster and toe-in are adjustable on the front end (see illustration). On the rear, toe-in and camber are the only adjustments possible.

Getting the proper wheel alignment is a very exacting process, one in which complicated and expensive machines are necessary to perform the job properly. Because of this, you should have a technician with the proper equipment perform these tasks. We will, however, attempt to give you a basic idea of what's involved with front end alignment so you can better understand the process and deal intelligently with the shop that does the work.

Toe-in is the turning in of the wheels. The purpose of a toe specification is to ensure parallel rolling of the wheels. In a vehicle with zero toe-in, the distance between the front edges of the wheels will be the same as the distance between the rear edges of the wheels. The actual amount of toe-in is normally only a fraction of an inch. On the front wheels, toe-in adjustment is controlled by the tie-rod end position on the inner tie-rod; on the rear wheels, toe-in is controlled by the tie-rod end position on the suspension adjustment link. Incorrect toe-in will cause the tires to wear improperly by making them scrub against the road surface.

Camber is the tilting of the wheels from the vertical when viewed from the front or rear of the vehicle. When the wheels tilt out at the top, the camber is said to be positive (+). When the wheels tilt in at the top the camber is negative (-). The amount of tilt is measured in degrees

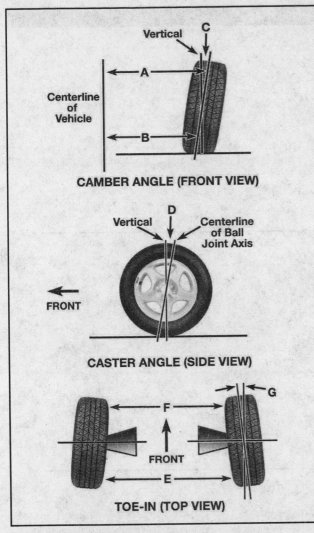

CAMBER ANGLE (FRONT VIEW)

CASTER ANGLE (SIDE VIEW)

TOE-IN (TOP VIEW)

from the vertical and this measurement is called the camber angle. This angle affects the amount of tire tread which contacts the road and compensates for changes in the suspension geometry when the vehicle is cornering or traveling over an undulating surface. Camber is adjusted by loosening the strut-to-knuckle bolts/nuts and altering the relationship between the strut and knuckle.

Caster is the tilting of the top of the front steering axis from the vertical. A tilt toward the rear is positive caster and a tilt toward the front is negative caster. Caster is adjustable on the front end by elongating the holes for the strut upper mounting studs and moving the top of the strut towards the front or rear of the vehicle, as necessary. Caster is not adjustable on the rear of these vehicles.

24.1 Front end alignment details

A minus B = C (degrees camber)
D = caster (expressed in degrees)
E minus F = toe-in (measured in inches)
G = toe-in (expressed in degrees)

Specifications

Torque specifications   Ft-lbs (unless otherwise indicated)

Front suspension

Balljoint mounting nuts	50
Balljoint-to-steering knuckle nut	88 in-lbs (plus an additional 2/3-turn)
Brake line bracket bolt	156 in-lbs
Control arm	
Front mounting nut	140
Rear mounting nut (2000 and earlier models)	90
Rear mounting bolt (2001 and later models)	117
Hub retaining bolts	70
Stabilizer bar bushing bolt/nut	156 in-lbs

Torque specifications · Ft-lbs (unless otherwise indicated)

Stabilizer bar mounting bracket bolts	37
Steering knuckle-to-strut assembly bolts	
1985 through 1990	144
1991	180
1992 and later	140
Strut assembly upper mounting nuts	
Models through 2000	18
2001 and later	35
Tie-rod end nut	
Models through 2000	35
2001 and later	35 to 52

Rear suspension

Adjustment link-to-knuckle nut	33
Adjustment link-to-chassis bolts	
2000 and later	55
Adjustment link-to-control arm nut	
1999 and earlier	63
2000 and later	38
Balljoint stud nuts	22
Control arm pivot nuts	85
Control arm pivot bolts	
1999 and earlier	125
2000 and later	78
Hub and bearing assembly bolts	52

Rear suspension

Stabilizer bar bushing assembly nut	37
Stabilizer bar bushing pinch bolt	37
Stabilizer bar clamp bolt	
2000 and later	24
Stabilizer bar link nut	156 in-lbs
Stabilizer bar mounting bracket bolt	156 in-lbs
Shock-to-lower control arm	
2000 and later	18
Upper shock mounting nut	15
Strut-to-knuckle nuts (1999 and earlier)	144
Strut tower mounting nuts	18
Suspension support (2001 and later)	
Bolts to bracket	63
Assembly-to-body bolts, front	141
Assembly-to-body bolts, rear	191

Steering

Airbag module screws	27 in-lbs
Steering gear shaft pinch bolt	35
Outer tie-rod jam nut	30
Steering wheel nut	30
Tie-rod end castellated nut	35
Steering gear mounting bolts	68
Wheel lug nuts	See Chapter 1

Section

11

BODY

1   General information

These models are two or four-door sedans. The vehicle is a "uni-body" type, which means the body is designed to provide vehicle rigid-ity so a separate frame isn't necessary.

Body maintenance is an important part of the retention of the vehi-cle's market value. It's far less costly to handle small problems before they grow into larger ones.

Major body components which are particularly vulnerable in acci-dents are removable. These include the hood, front fenders, grille, doors, trunk lid and tail light assembly. It's often cheaper and less time consuming to replace an entire panel than it is to attempt a restoration of the old one. However, this must be decided on a case-by-case basis.

2   Maintenance - body

1   The condition of the body is very important, because the value of the vehicle is dependent on it. It's much more difficult to repair a neglected or damaged body than it is to repair mechanical components. The hidden areas of the body, such as the fender wells and the engine compartment, are equally important, although they obviously don't require as frequent attention as the rest of the body.

2   Once a year, or every 12,000 miles, it's a good idea to have the underside of the body steam cleaned. All traces of dirt and oil will be removed and the underside can then be inspected carefully for rust, damaged brake lines, frayed electrical wiring, damaged cables and other problems. The front suspension components should be greased after completion of this job.

3   At the same time, clean the engine and the engine compartment with a water soluble degreaser.

4   The fenderwells should be given particular attention, as under-coating can peel away and stones and dirt thrown up by the tires can cause the paint to chip and flake, allowing rust to set in. If rust is found, clean down to the bare metal and apply an anti-rust paint.

5   The body should be washed as needed. Wet the vehicle thor-oughly to soften the dirt, then wash it down with a soft sponge and plenty of clean soapy water. If the surplus dirt isn't washed off very carefully, it will in time wear down the paint.

6   Spots of tar or asphalt coating thrown up from the road should be removed with a cloth soaked in solvent.

7   Once every six months, wax the body thoroughly. If a chrome cleaner is used to remove rust from any of the vehicle's plated parts, remember that the cleaner also removes part of the chrome, so use it sparingly.

3   Maintenance - upholstery and carpets

1   Every three months remove the carpets or mats and clean the interior of the vehicle (more frequently if necessary). Vacuum the upholstery and carpets to remove loose dirt and dust.

2   If the upholstery is soiled, apply upholstery cleaner with a damp sponge and wipe it off with a clean, dry cloth.

4   Vinyl trim - maintenance

Vinyl trim should not be cleaned with detergents, caustic soaps or petroleum-based cleaners. Plain soap and water or a mild vinyl cleaner is best for stains. Test a small area for color fastness. Bubbles under the vinyl can be eliminated by piercing them with a pin and then work-ing the air out.

5   Body repair - minor damage

♦ **See photo sequence**

REPAIR OF MINOR SCRATCHES

1   If the scratch is superficial and does not penetrate to the metal of the body, repair is very simple. Lightly rub the scratched area with a fine rubbing compound to remove loose paint and built up wax. Rinse the area with clean water.

2   Apply touch-up paint to the scratch, using a small brush. Con-tinue to apply thin layers of paint until the surface of the paint in the scratch is level with the surrounding paint. Allow the new paint at least two weeks to harden, then blend it into the surrounding paint by rub-bing with a very fine rubbing compound. Finally, apply a coat of wax to the scratch area.

3   If the scratch has penetrated the paint and exposed the metal of the body, causing the metal to rust, a different repair technique is

required. Remove all loose rust from the bottom of the scratch with a pocket knife, then apply rust inhibiting paint to prevent the formation of rust in the future. Using a rubber or nylon applicator, coat the scratched area with glaze-type filler. If required, the filler can be mixed with thinner to provide a very thin paste, which is ideal for filling narrow scratches. Before the glaze filler in the scratch hardens, wrap a piece of smooth cotton cloth around the tip of a finger. Dip the cloth in thinner and then quickly wipe it along the surface of the scratch. This will ensure that the surface of the filler is slightly hollow. The scratch can now be painted over as described earlier in this section.

REPAIR OF DENTS

4   When repairing dents, the first job is to pull the dent out until the affected area is as close as possible to its original shape. There is no point in trying to restore the original shape completely as the metal in the damaged area will have stretched on impact and cannot be restored to its original contours. It is better to bring the level of the dent up to a point which is about 1/8-inch below the level of the surrounding metal. In cases where the dent is very shallow, it is not worth trying to pull it out at all.

5   If the back side of the dent is accessible, it can be hammered out gently from behind using a soft-face hammer. While doing this, hold a block of wood firmly against the opposite side of the metal to absorb the hammer blows and prevent the metal from being stretched.

6   If the dent is in a section of the body which has double layers, or some other factor makes it inaccessible from behind, a different technique is required. Drill several small holes through the metal inside the damaged area, particularly in the deeper sections. Screw long, self tapping screws into the holes just enough for them to get a good grip in the metal. Now the dent can be pulled out by pulling on the protruding heads of the screws with locking pliers.

7   The next stage of repair is the removal of paint from the damaged area and from an inch or so of the surrounding metal. This is easily done with a wire brush or sanding disk in a drill motor, although it can be done just as effectively by hand with sandpaper. To complete the preparation for filling, score the surface of the bare metal with a screwdriver or the tang of a file or drill small holes in the affected area. This will provide a good grip for the filler material. To complete the repair, see the Section on filling and painting.

REPAIR OF RUST HOLES OR GASHES

8   Remove all paint from the affected area and from an inch or so of the surrounding metal using a sanding disk or wire brush mounted in a drill motor. If these are not available, a few sheets of sandpaper will do the job just as effectively.

9   With the paint removed, you will be able to determine the severity of the corrosion and decide whether to replace the whole panel, if possible, or repair the affected area. New body panels are not as expensive as most people think and it is often quicker to install a new panel than to repair large areas of rust.

10 Remove all trim pieces from the affected area except those which will act as a guide to the original shape of the damaged body, such as headlight shells, etc. Using metal snips or a hacksaw blade, remove all loose metal and any other metal that is badly affected by rust. Hammer the edges of the hole inward to create a slight depression for the filler material.

11 Wire brush the affected area to remove the powdery rust from the surface of the metal. If the back of the rusted area is accessible, treat it with rust-inhibiting paint.

12 Before filling is done, block the hole in some way. This can be done with sheet metal riveted or screwed into place, or by stuffing the hole with wire mesh.

13 Once the hole is blocked off, the affected area can be filled and painted. See the following sub-section on filling and painting.

FILLING AND PAINTING

14 Many types of body fillers are available, but generally speaking, body repair kits which contain filler paste and a tube of resin hardener are best for this type of repair work. A wide, flexible plastic or nylon applicator will be necessary for imparting a smooth and contoured finish to the surface of the filler material. Mix up a small amount of filler on a clean piece of wood or cardboard (use the hardener sparingly). Follow the manufacturer's instructions on the package, otherwise the filler will set incorrectly.

15 Using the applicator, apply the filler paste to the prepared area. Draw the applicator across the surface of the filler to achieve the desired contour and to level the filler surface. As soon as a contour that approximates the original one is achieved, stop working the paste. If you continue, the paste will begin to stick to the applicator. Continue to add thin layers of paste at 20-minute intervals until the level of the filler is just above the surrounding metal.

16 Once the filler has hardened, the excess can be removed with a body file. From then on, progressively finer grades of sandpaper should be used, starting with a 180-grit paper and finishing with 600-grit wet-or-dry paper. Always wrap the sandpaper around a flat rubber or wooden block, otherwise the surface of the filler will not be completely flat. During the sanding of the filler surface, the wet-or-dry paper should be periodically rinsed in water. This will ensure that a very smooth finish is produced in the final stage.

17 At this point, the repair area should be surrounded by a ring of bare metal, which in turn should be encircled by the finely feathered edge of good paint. Rinse the repair area with clean water until all of the dust produced by the sanding operation is gone.

18 Spray the entire area with a light coat of primer. This will reveal any imperfections in the surface of the filler. Repair the imperfections with fresh filler paste or glaze filler and once more smooth the surface with sandpaper. Repeat this spray-and-repair procedure until you are satisfied that the surface of the filler and the feathered edge of the paint are perfect. Rinse the area with clean water and allow it to dry completely.

19 The repair area is now ready for painting. Spray painting must be carried out in a warm, dry, windless and dust free atmosphere. These conditions can be created if you have access to a large indoor work area, but if you are forced to work in the open, you will have to pick the day very carefully. If you are working indoors, dousing the floor in the work area with water will help settle the dust which would otherwise be in the air. If the repair area is confined to one body panel, mask off the surrounding panels. This will help minimize the effects of a slight mismatch in paint color. Trim pieces such as chrome strips, door handles, etc., will also need to be masked off or removed. Use masking tape and several thicknesses of newspaper for the masking operations.

20 Before spraying, shake the paint can thoroughly, then spray a test area until the spray painting technique is mastered. Cover the repair area with a thick coat of primer. The thickness should be built up using several thin layers of primer rather than one thick one. Using 600-grit wet-or-dry sandpaper, rub down the surface of the primer until it is very smooth. While doing this, the work area should be thoroughly rinsed with water and the wet-or-dry sandpaper periodically rinsed as well. Allow the primer to dry before spraying additional coats.

These photos illustrate a method of repairing simple dents. They are intended to supplement Body repair - minor damage in this Chapter and should not be used as the sole instructions for body repair on these vehicles.

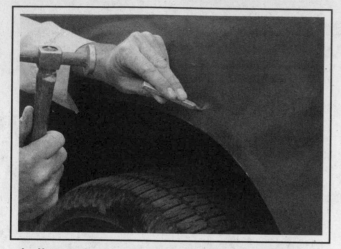

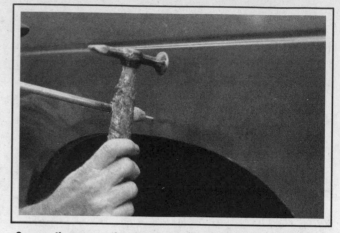

1   If you can't access the backside of the body panel to hammer out the dent, pull it out with a slide-hammer-type dent puller. In the deepest portion of the dent or along the crease line, drill or punch hole(s) at least one inch apart . . .

2   . . . then screw the slide-hammer into the hole and operate it. Tap with a hammer near the edge of the dent to help 'pop' the metal back to its original shape. When you're finished, the dent area should be close to its original contour and about 1/8-inch below the surface of the surrounding metal

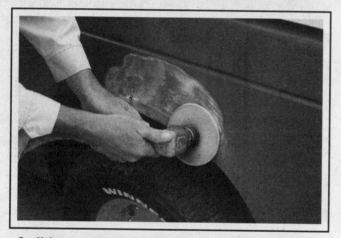

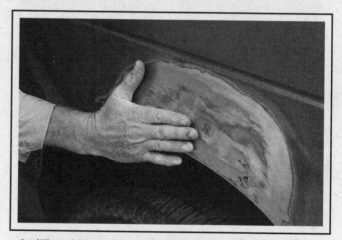

3   Using coarse-grit sandpaper, remove the paint down to the bare metal. Hand sanding works fine, but the disc sander shown here makes the job faster. Use finer (about 320-grit) sandpaper to feather-edge the paint at least one inch around the dent area

4   When the paint is removed, touch will probably be more helpful than sight for telling if the metal is straight. Hammer down the high spots or raise the low spots as necessary. Clean the repair area with wax/silicone remover

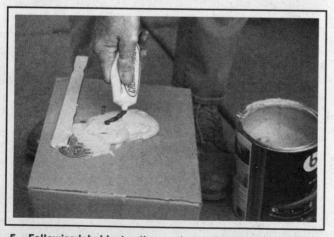

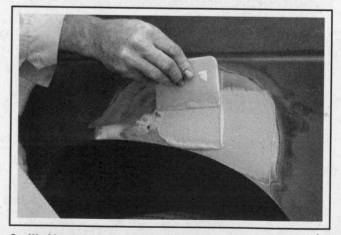

5   Following label instructions, mix up a batch of plastic filler and hardener. The ratio of filler to hardener is critical, and, if you mix it incorrectly, it will either not cure properly or cure too quickly (you won't have time to file and sand it into shape)

6   Working quickly so the filler doesn't harden, use a plastic applicator to press the body filler firmly into the metal, assuring it bonds completely. Work the filler until it matches the original contour and is slightly above the surrounding metal

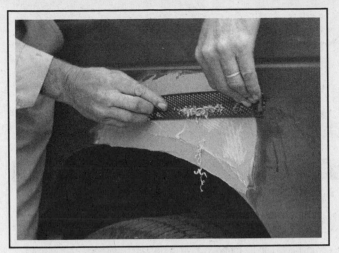

7   Let the filler harden until you can just dent it with your fingernail. Use a body file or Surform tool (shown here) to rough-shape the filler

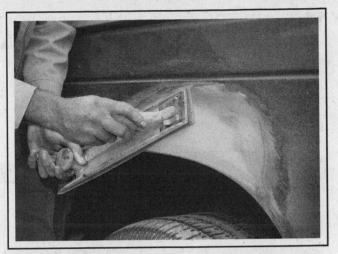

8   Use coarse-grit sandpaper and a sanding board or block to work the filler down until it's smooth and even. Work down to finer grits of sandpaper - always using a board or block - ending up with 360 or 400 grit

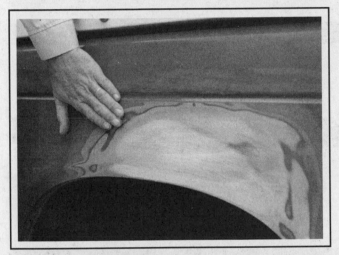

9   You shouldn't be able to feel any ridge at the transition from the filler to the bare metal or from the bare metal to the old paint. As soon as the repair is flat and uniform, remove the dust and mask off the adjacent panels or trim pieces

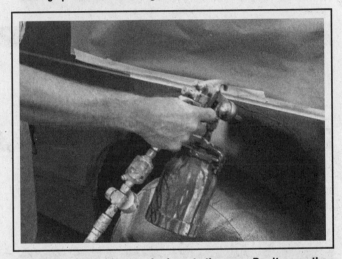

10   Apply several layers of primer to the area. Don't spray the primer on too heavy, so it sags or runs, and make sure each coat is dry before you spray on the next one. A professional-type spray gun is being used here, but aerosol spray primer is available inexpensively from auto parts stores

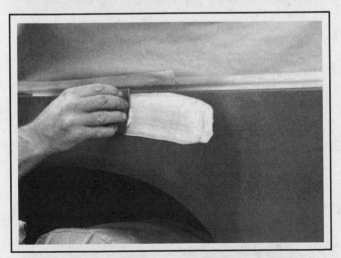

11   The primer will help reveal imperfections or scratches. Fill these with glazing compound. Follow the label instructions and sand it with 360 or 400-grit sandpaper until it's smooth. Repeat the glazing, sanding and respraying until the primer reveals a perfectly smooth surface

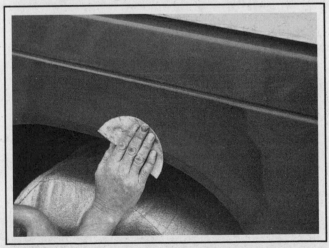

12   Finish sand the primer with very fine sandpaper (400 or 600-grit) to remove the primer overspray. Clean the area with water and allow it to dry. Use a tack rag to remove any dust, then apply the finish coat. Don't attempt to rub out or wax the repair area until the paint has dried completely (at least two weeks)

21 Spray on the top coat, again building up the thickness by using several thin layers of paint. Begin spraying in the center of the repair area and then, using a circular motion, work out until the whole repair area and about two inches of the surrounding original paint is covered.

Remove all masking material 10 to 15 minutes after spraying on the final coat of paint. Allow the new paint at least two weeks to harden, then use a very fine rubbing compound to blend the edges of the new paint into the existing paint. Finally, apply a coat of wax.

6 Body repair - major damage

1 Major damage must be repaired by an auto body/frame repair shop with the necessary welding and hydraulic straightening equipment.

2 If the damage has been serious, it is vital that the structure be checked for proper alignment or the vehicle's handling characteristics may be adversely affected. Other problems, such as excessive tire wear and wear in the driveline and steering may occur.

3 Due to the fact that all of the major body components (hood, fenders, etc.) are separate and replaceable units, any seriously damaged components should be replaced rather than repaired. Sometimes these components can be found in a wrecking yard that specializes in used vehicle components, often at considerable savings over the cost of new parts.

7 Maintenance - hinges and locks

Every 3000 miles or three months, the door, hood and trunk lid hinges should be lubricated with a few drops of oil. The door striker

plates should also be given a thin coat of white lithium-base grease to reduce wear and ensure free movement.

8 Windshield and fixed glass - replacement

1 Replacement of the windshield and fixed glass requires the use of special fast-setting adhesive/caulk materials. These operations should be left to a dealer or a shop specializing in glass work.

2 Windshield-mounted rear view mirror support removal is also best left to experts, as the bond to the glass also requires special tools and adhesives.

9 Hood - removal and installation

◆ **Refer to illustrations 9.1 and 9.2**

1 Use rags or pads to protect the windshield from the rear of the hood (see illustration).

2 Scribe or draw alignment marks around the hinge bolts (see illustration).

3 On models so equipped, detach the assist strut and unplug the hood lamp connector.

4 Remove the bolts and, with the help of an assistant, detach the hood from the vehicle.

5 Installation is the reverse of removal.

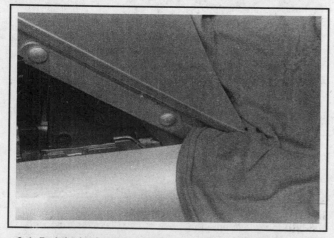

9.1 Pad the back corners of the hood with rags so the windshield won't be damaged it the hood accidentally swings to the rear

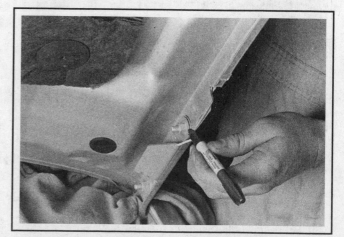

9.2 Use scribe or felt-tip pen to mark the hood bolt positions

10 Hood latch cable - replacement

▶ Refer to illustrations 10.1, 10.3 and 10.4

1   In the passenger compartment, remove the latch handle retaining screws and detach the handle (see illustration).

2   Remove the radiator upper panel (see Chapter 3).

3   In the engine compartment, remove the cable retainer from the bracket (see illustration).

4   Spread the clip with a screwdriver and detach the end of the cable from the latch (see illustration).

5   Connect a piece of string or thin wire of suitable length to the end of the wire to the cable and pull the cable through into the passenger compartment. On 2001 and later models, remove the left fenderwell liner and front fascia end cap to access the clips along the cable's path.

6   Connect the string or wire to the new cable and pull it back into the engine compartment. Connect the new cable to the latch release in the driver's kick panel.

7   Connect the cable and install the latch screws and trim panel.

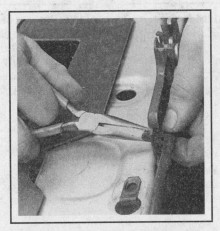

10.1 Remove the two screws with a Phillips screwdriver

10.3 The hood latch cable retainer can be detached from the bracket by squeezing the locking tab with a pair of needle-nose pliers

10.4 Pry the clip back and lift the cable up to detach it (arrow)

11 Front fender liner - removal and installation

▶ Refer to illustrations 11.2, 11.4 and 11.6

1   Raise the vehicle, support it securely on jackstands and remove the front wheel.

2   The fender liner is held in place with special plastic retainers or screws. To remove the plastic retainers, use wire cutters or a similar tool to pry the heads of the retainers out of the retainer bodies to release them (see illustration). Pry the heads out, do not cut them off to remove them.

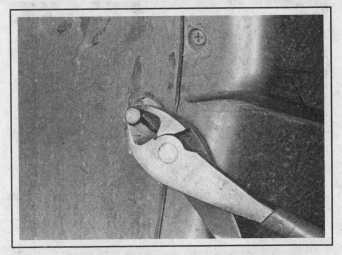

11.2 Pry the head out of the retainer body - do not cut he head off

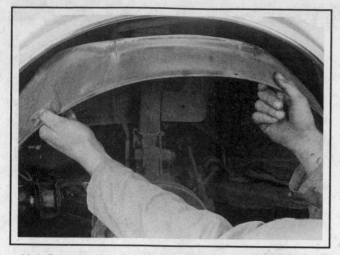

11.4 Grasp the fender liner and pull it out of the fender

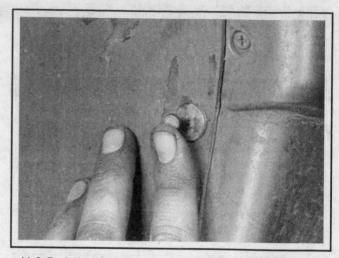

11.6 Push the head of the retainer into the body until it locks in place

3   Once all of the retainers are released, remove them from the fender liner.
4   Detach the liner and remove it from the vehicle (see illustration).

5   To install, place the liner in position and align the retainer holes.
6   Install the screws or retainers and push the heads in to securely lock them in place (see illustration).

12 Front fender - removal and installation

▶ Refer to illustration 12.4

✳✳✳ WARNING:

Some models covered by this manual are equipped with an airbag system. Always disable the airbag system before working in the vicinity of the impact sensors, steering column or instrument panel to avoid the possibility of accidental deployment of the airbag(s), which could cause personal injury. See Chapter 12 for the airbag disarming procedure.

1   Raise the front of the vehicle, support it securely on jackstands and remove the front wheel.
2   Remove the front fender liner (see Section 11).
3   On some models, it may be necessary to remove the grille and headlight assembly (see Section 13). On 2001 and later models, remove the rocker panel trim for access to two lower fender mounting bolts.
4   Remove the retaining bolts and detach the fender from the vehicle (see illustration).
5   To install, place the fender in place and install the retaining bolts. Tighten the bolts securely.
6   The remainder of installation is the reverse of removal.

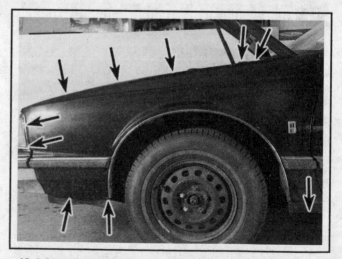

12.4 Location of the front fender mounting bolts (arrows)

13 Radiator grille - removal and installation

▶ **Refer to illustration 13.3**

1   Open the hood.
2   On some models, it will be necessary to remove the headlight trim panel.
3   Remove the grille retaining screws (see illustration). On 2001 and later models, the grille is retained by five nuts accessed on the radiator side of the grille.
4   Rotate the top of the grille out and lift it from the vehicle.
5   Installation is the reverse of removal.

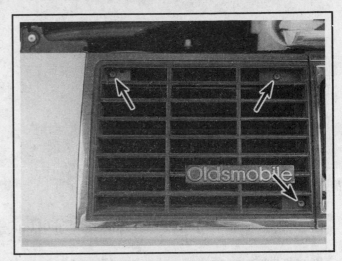

13.3 The radiator grille is held in place by Torx head screws (arrows)

14 Door trim panel - removal and installation

▶ **Refer to illustrations 14.1, 14.4 and 14.7**

1   Remove door glass regulator handle (if equipped) (see Section 17), and the visible door panel screws from the door trim panel (see illustration). On 2001 and later models, remove the triangular trim panel that covers the small "tweeter" speaker, then release the pin at the top, lift the panel straight up and disconnect the electrical connector.
2   On models with a passive seat belt system, remove the lower arm rest trim panel and the seat belt escutcheon and retainer.
3   Remove the reflector(s) from the rear end of the trim panel.
4   Remove the electrical switch panel and unplug the connectors (see illustration).
5   Remove the screws from the arm rest.
6   Remove the door handle cover assembly by removing the screw(s).

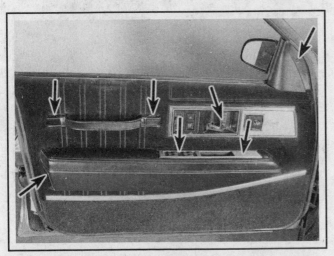

14.1 Typical door panel screw locations (some screws are concealed under panels, courtesy lights and straps)

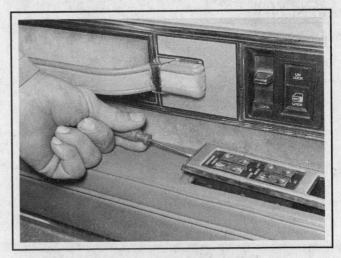

14.4 To remove the switch panel, pry it up gently

7   Pry around the outer circumference of the door panel with a trim panel tool, prybar or screwdriver to disengage the clips (see illustration).

8   Grasp the door trim panel securely and lift up to disengage it from the door upper edge.

9   Carefully peel the water shield from the door for access to the inner door components. Take care not to tear the water shield as it must be reinstalled.

10 Installation is the reverse of removal.

14.7 Pry around the door panel with a trim panel tool or a screwdriver (pry only in the area of the plastic clips, or the panel could be damaged)

15 Door lock assembly - removal and installation

▶ **Refer to illustration 15.2**

1   With the window glass in the full up position, remove the door trim panel and water shield (see Section 14).

2   Remove the lock assembly-to-door bolts (see illustration).

3   Working through the large access hole, position the lock as necessary to disengage the rods. The rods are attached by plastic or pressed metal clips. On 2001 and later models, remove one bolt at the front door-glass run channel and move the channel enough to remove the lock assembly.

4   To install, place the assembly in the door and connect the lock rods.

5   Guide the lock into position and install the bolts, tightening them securely.

6   Install the water shield and door trim panel.

15.2 Remove the lock retaining bolts (arrows) from the end of the door, then detach the actuating rods and pull the lock through the access hole

16 Door window glass - removal and installation

1   With the window glass in the full up position, remove the door trim panel and water shield (see Section 14).

EARLY MODELS

Front door

▶ **Refer to illustration 16.3**

2   Remove the trim panel retainer and sealer strip.

3   Remove the rubber stop bumper, screws and front run channel (see illustration).

4   Lower the glass halfway, slide the window regulator guide block off the sash channel and tilt the glass outboard of the door frame to remove it.

5   To install, insert the glass into the door and engage the regulator guide block to the sash channel. Engage the rear guide clip on the glass to the rear run channel weatherstrip.

6   Raise the glass to the half way up position and install the front run channel bolts finger tight. Engage the front clip on the glass in the front run channel and tighten the bolts securely. Install the rubber stop bumper

7   The remainder of installation is the reverse of removal.

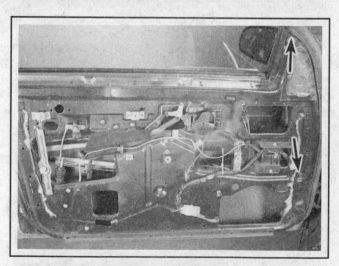

16.3 Locations of the front door glass run channel bolts (arrows)

Rear door

8 Remove the sash screws while supporting the glass.

9 Disengage the glass and lower it to the bottom of the door.

10 Remove the vent glass assembly.

11 Lift the glass from the door.

12 To install, insert the glass into the door and connect the channel, making sure to engage the glass guide securely to the channel.

13 The remainder of installation is the reverse of removal.

LATE MODELS

14 Remove the inner weatherstrip. Secure the window glass in the up position with strong adhesive tape fastened to the glass and wrapped over the door frame.

15 Loosen the regulator to window clamp bolts three to four turns and lower the window regulator.

16 Remove the tape then lower the glass in the door. Remove the window by tilting it forward and out of the door.

17 Installation is the reverse of the removal

---

17 Door glass regulator - removal and installation

▶ **Refer to illustrations 17.2a and 17.2b**

1 On power window equipped models, disconnect the negative cable at the battery.

✳✳ CAUTION:

If the vehicle is equipped with a Delco Loc II or Theftlock audio system, make sure you have the correct activation code before disconnecting the battery. See the information at the front of this manual for the radio re-activation procedure.

2 On manual window glass regulator models, remove the handle by pressing the bearing plate and door trim panel in and, with a piece of hooked wire, pulling off the spring clip (see illustration). A special tool is available for this purpose (see illustration) but its use is not essential. With the clip removed, take off the handle and the bearing plate.

3 With the window glass in the full up position, remove the door trim panel and water shield (see Section 14).

4 Secure the window glass in the up position with strong adhesive tape fastened to the glass and wrapped over the door frame.

EARLY MODELS

Removal

5 Punch out the center pins of the rivets that secure the window regulator and drill the heads of the rivets off with a 1/4-inch drill bit.

✳✳ CAUTION:

Be careful not to enlarge the holes in the door sheetmetal.

6 On power window equipped models, unplug the electrical connector.

7 Remove the retaining rivets or bolts and move the regulator until it is disengaged from the sash channel. Lift the regulator from the door.

8 The window motor on models with power windows can be replaced if necessary. This involves drilling out the rivets securing the motor to the regulator frame. The new motor should come with screws and nuts for attaching the motor to the regulator.

✳✳ WARNING:

The regulator arms are under extreme pressure and can cause serious injury if the motor is removed without locking the sector gear to the regulator frame. This can be done by inserting a bolt through one of the holes in the regulator frame and sector gear, then fastening it in place with a nut.

Installation

9 Place the regulator in position in the door and engage it in the sash channel.

10 Secure the regulator to the door using 3/16-inch rivets and a rivet tool.

11 Install the bolts and tighten them securely.

12 Plug in the electrical connector (if equipped).

13 Install the water shield, door trim panel and window regulator handle. Connect the negative battery cable.

LATE MODELS

14 With the glass taped in the Up position in the door, loosen each

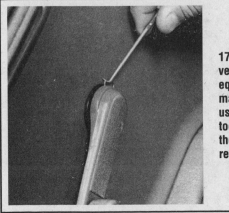

17.2a If your vehicle is equipped with manual windows, use a hooked tool to remove the window crank retaining clip . . .

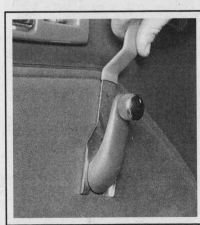

17.2b . . . or use a special removal tool like this one, available at most auto parts stores

glass-to-regulator bolt three to four turns, then lower the regulator mechanism to the 3/4 down position.

15 Disconnect the electrical connection from the window motor.

16 Remove the window regulator bolts and remove the regulator from the door.

17 Installation is the reverse of the removal

18 Door handles - removal and installation

1 With the window glass in the full up position, remove the door trim panel and water shield (see Section 14).

OUTSIDE HANDLE

▶ **Refer to illustration 18.2**

2 Unclip the remote rod from the handle with a small screwdriver, remove the nuts and lift the handle off (see illustration).

3 Installation is the reverse of removal.

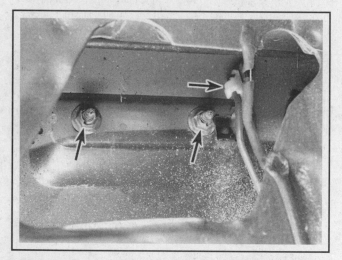

18.2 Detach the rod from the handle, then remove the nuts (arrows)

INSIDE HANDLE

▶ **Refer to illustration 18.5**

4 Disconnect the rod from the handle. On 2001 and later models, remove the door's electronic module and set it aside.

5 Punch out the center pins of the rivets that secure the window regulator and drill the rivets out with a 3/16-inch drill bit (see illustration). On later models, the handle is secured by bolts, not rivets.

6 Lift the handle from the door.

7 To install, place the handle in position and secure it to the door, using 3/16-inch rivets and a rivet tool.

8 The remainder of installation is the reverse of removal.

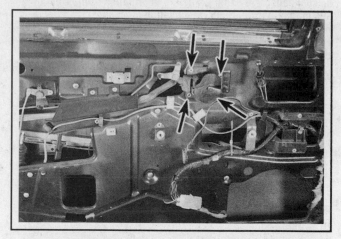

18.5 The inside door handle is retained by rivets (arrows)

19 Door lock cylinder - removal and installation

▶ **Refer to illustration 19.2**

1 With the window glass in the full up position, remove the door trim panel and water shield (see Section 14).

2 Disconnect the rod from the lock cylinder (see illustration).

3 Use a screwdriver to pry the retainer off and withdraw the lock cylinder from the door.

4 Installation is the reverse of removal.

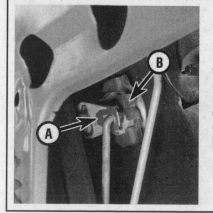

19.2 After removing the clip (A) and detaching the lock rod, the retaining clip (B) can be pried off to remove the lock cylinder

20 Door lock striker - removal and installation

♦ **Refer to illustration 20.2**

1   Mark the position of the striker bolt on the door pillar.
2   It will be necessary to use a large Torx bit to fit the star-shaped recess in the striker bolt head. Unscrew the bolt and remove it (see illustration).
3   To install, screw the lock striker bolt into the tapped cage plate in the door pillar and tighten it finger tight at the marked position. Close the door until the lock almost contacts the striker; the striker should be positioned in the center of the lock opening. If not, move the striker up or down, as necessary.

➡**Note: When the door closes, the striker should not cause the door to lift or drop.**

The striker should not be used to compensate for poor door alignment. If the rear of the door is too high or too low, adjust the door at the hinges then readjust the striker). Be sure to tighten the striker bolt securely.

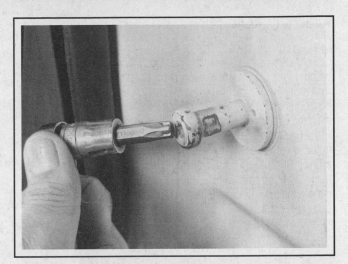

20.2 Use a Torx bit to loosen the striker bolt

21 Door - removal and installation

1   Remove the door trim panel and water shield.
2   Unplug any wiring connectors.
3   Open the door all the way and support it on a floor jack or blocks covered with cloth or pads to prevent damage to the paint.
4   Scribe around the hinges to ensure correct realignment

during installation.
5   Remove the bolts and nuts retaining the hinges to the door and with the help of an assistant lift the door away.
6   Install the door by reversing the removal procedure. After obtaining the proper adjustment, tighten the nuts and bolts securely.

22 Trunk lid - removal and installation

♦ **Refer to illustration 22.2**

1   Open the trunk lid and unplug any electrical connectors and disconnect the solenoid (if equipped).
2   Scribe or mark around the heads of the retaining bolts to mark their locations for ease of reinstallation (see illustration).
3   With an assistant supporting the trunk lid, remove the bolts. Lift the trunk lid from the vehicle.
4   Installation is the reverse of removal.

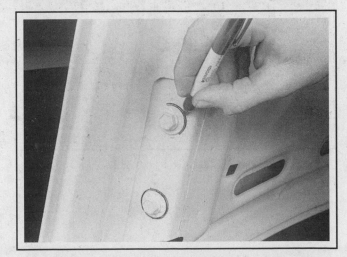

22.2 Mark the position of the trunk lid bolts with a scribe or a felt-tip pen before removing them

23 Trunk lock cylinder - removal and installation

▶ **Refer to illustration 23.2**

1   Open the trunk lid.
2   On early models, pry the retaining clip off and withdraw the lock cylinder from the vehicle. Some retaining clips will be secured with a rivet which must be drilled out with a 5/32-inch drill bit (see illustration). To install, place the lock cylinder in place and secure it with the retaining clip. Install a new rivet.
3   On later models it will be necessary to remove the trim panel from the outside of the trunk lid, then detach the cable from the lock cylinder (if equipped) and remove the retaining nuts. To remove the trunk outer trim panel, which also holds the taillights/backup lights, remove the trunk inner trim panel to access the nuts securing the outer panel to the trunklid.

23.2 Drill out the rivet, then pry the retaining clip out with a screwdriver

24 Trunk latch and striker - removal and installation

▶ **Refer to illustrations 24.1 and 24.2**

1   Remove the electric release solenoid (if equipped) and unbolt and remove the latch and (if equipped) the ajar switch (see illustration).
2   Remove the retaining nut and lift off the striker (see illustration).
3   Installation is the reverse of removal.

24.2 The trunk latch striker is retained by one nut

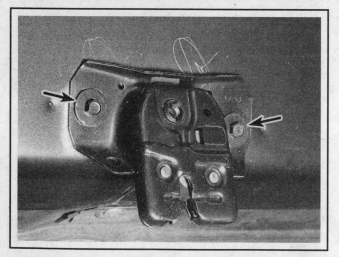

24.1 The trunk latch is secured by two bolts

25 Rear lens assembly - removal and installation

▶ **Refer to illustrations 25.2a and 25.2b**

1　Open the trunk lid.
2　Unscrew the plastic wing nuts, pull the lens assembly out and

lean it back (see illustrations). Disconnect the bulb holders (Chapter 12) and lift the assembly from the vehicle.
3　Installation is the reverse of removal.

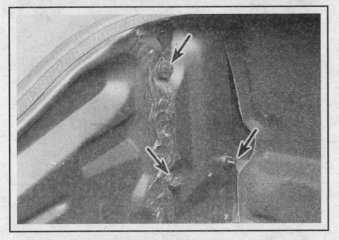

25.2a　Unscrew the plastic retaining wing nuts and . . .

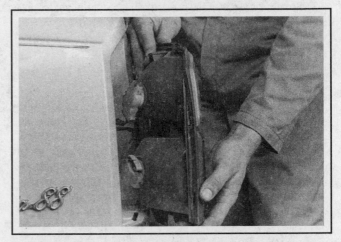

25.2b　. . . remove the lens assembly from the body of the vehicle

26 Console - removal and installation

✸✸ WARNING:

Some models covered by this manual are equipped with an airbag system. Always disable the airbag system before working in the vicinity of the impact sensors, steering column or instrument panel to avoid the possibility of accidental deployment of the airbag(s), which could cause personal injury. See Chapter 12 for the airbag disarming procedure.

1　Disconnect the negative cable at the battery.

✸✸ CAUTION:

If the vehicle is equipped with a Delco Loc II or Theftlock audio system, make sure you have the correct activation code before disconnecting the battery. See the information at the front of this manual for the radio re-activation procedure.

2　Remove the retaining clip from the shift handle and pull the handle or knob off the shift lever (see Chapter 7).
3　Remove the insert compartments.
4　Remove the console trim plates.
5　Remove the mounting bolts and pull the console up. On 2001 and later models, move the seats forward to access the rear bolts (on each lower side of the console), then move the seat rearward to access the lower front console bolts.
6　Lift the console up for access and unplug the electrical connectors.
7　Some or all of the retaining clips will probably come out during removal, so be sure to reinstall them prior to console installation.
8　Installation is the reverse of removal.

27 Seats - removal and installation

FRONT SEAT

1　Move the seat all the way forward.
2　Remove the seat track covers and pull the carpet away from the adjuster and retaining nuts.
3　Remove the seat adjuster-to-floor panel retaining nuts.
4　Move the seat all the way to the rear.
5　Remove the front seat retaining nuts. On power seats, unplug the

electrical connector. Lift the seat from the vehicle.
6　Installation is the reverse of removal.

REAR SEAT

7　Remove the seat cushion retaining bolts, detach the seat cushion and remove it from the vehicle.
8　Installation is the reverse of removal.

28 Outside mirror - removal and installation

♦ **Refer to illustration 28.3**

1   Remove the door trim panel (see Section 14).
2   On power mirrors, unplug the electrical connector.
3   Remove the two nuts and one bolt and lift off the mirror assembly (see illustration).
4   Installation is the reverse of removal.

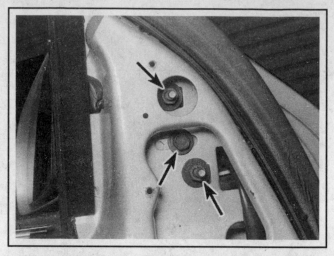

28.3 Remove the mirror retaining nuts and bolt (arrows)

Section

Reference to other Chapters

12

CHASSIS ELECTRICAL SYSTEM

1   General information

The electrical system is a 12-volt, negative ground type. Power for the lights and all electrical accessories is supplied by a lead/acid-type battery which is charged by the alternator.

This Chapter covers repair and service procedures for the various electrical components not associated with the engine. Information on the battery, alternator, ignition system and starter motor can be found in Chapter 5.

It should be noted that when portions of the electrical system are serviced, the negative battery cable should be disconnected from the battery to prevent electrical shorts and/or fires.

> **✳✳ CAUTION:**
>
> **If the vehicle is equipped with a Delco Loc II or Theftlock audio system, make sure you have the correct activation code before disconnecting the battery. See the information at the front of this manual for the radio re-activation procedure.**

2   Electrical troubleshooting - general information

A typical electrical circuit consists of an electrical component, any switches, relays, motors, fuses, fusible links or circuit breakers related to that component and the wiring and connectors that link the component to both the battery and the chassis. To help you pinpoint an electrical circuit problem, wiring diagrams are included at the end of this book.

Before tackling any troublesome electrical circuit, first study the appropriate wiring diagrams to get a complete understanding of what makes up that individual circuit. Trouble spots, for instance, can often be narrowed down by noting if other components related to the circuit are operating properly. If several components or circuits fail at one time, chances are the problem is in a fuse or ground connection, because several circuits are often routed through the same fuse and ground connections.

Electrical problems usually stem from simple causes, such as loose or corroded connections, a blown fuse, a melted fusible link or a bad relay. Visually inspect the condition of all fuses, wires and connections in a problem circuit before troubleshooting it.

If testing instruments are going to be utilized, use the diagrams to plan ahead of time where you will make the necessary connections in order to accurately pinpoint the trouble spot.

The basic tools needed for electrical troubleshooting include a circuit tester or voltmeter (a 12-volt bulb with a set of test leads can also be used), a continuity tester, which includes a bulb, battery and set of test leads, and a jumper wire, preferably with a circuit breaker incorporated, which can be used to bypass electrical components. Before attempting to locate a problem with test instruments, use the wiring diagram(s) to decide where to make the connections.

VOLTAGE CHECKS

Voltage checks should be performed if a circuit is not functioning properly. Connect one lead of a circuit tester to either the negative battery terminal or a known good ground. Connect the other lead to a connector in the circuit being tested, preferably nearest to the battery or fuse. If the bulb of the tester lights, voltage is present, which means that the part of the circuit between the connector and the battery is problem

free. Continue checking the rest of the circuit in the same fashion. When you reach a point at which no voltage is present, the problem lies between that point and the last test point with voltage. Most of the time the problem can be traced to a loose connection.

➡**Note: Keep in mind that some circuits receive voltage only when the ignition key is in the Accessory or Run position.**

FINDING A SHORT

One method of finding shorts in a circuit is to remove the fuse and connect a test light or voltmeter in its place to the fuse terminals. There should be no voltage present in the circuit. Move the wiring harness from side-to-side while watching the test light.

If the bulb goes on, there is a short to ground somewhere in that area, probably where the insulation has rubbed through. The same test can be performed on each component in the circuit, even a switch.

GROUND CHECK

Perform a ground test to check whether a component is properly grounded. Disconnect the battery and connect one lead of a self-powered test light, known as a continuity tester, to a known good ground. Connect the other lead to the wire or ground connection being tested. If the bulb goes on, the ground is good. If the bulb does not go on, the ground is not good.

CONTINUITY CHECK

A continuity check is done to determine if there are any breaks in a circuit - if it is passing electricity properly. With the circuit off (no power in the circuit), a self-powered continuity tester can be used to check the circuit. Connect the test leads to both ends of the circuit (or to the "power" end and a good ground), and if the test light comes on the circuit is passing current properly. If the light doesn't come on, there is a break somewhere in the circuit. The same procedure can be used to test a switch, by connecting the continuity tester to the switch terminals. When the switch is turned On, the test light should come on.

FINDING AN OPEN CIRCUIT

When diagnosing for possible open circuits, it is often difficult to locate them by sight because oxidation or terminal misalignment are hidden by the connectors. Merely wiggling a connector on a sensor or in the wiring harness may correct the open circuit condition. Remember this when an open circuit is indicated when troubleshooting a circuit. Intermittent problems may also be caused by oxidized or loose connections.

Electrical troubleshooting is simple if you keep in mind that all electrical circuits are basically electricity running from the battery, through the wires, switches, relays, fuses and fusible links to each electrical component (light bulb, motor, etc.) and to ground, from which it is passed back to the battery. Any electrical problem is an interruption in the flow of electricity to and from the battery.

For more information about electrical troubleshooting, refer to the *Haynes Automotive Electrical Manual.*

3 Fuses - general information

▶ **Refer to illustrations 3.1 and 3.3**

The electrical circuits of the vehicle are protected by a combination of fuses, circuit breakers and fusible links. The interior fuse block is located under the instrument panel on the left side of the dashboard (see illustration). Most models also have one or more fuse/relay blocks in the engine compartment, Later models may also have fuse/relay blocks under the right side of the instrument panel and in the trunk compartment behind the rear seat. On 2003 and later models, the rear fuse panel is located under the rear seat cushion, on the left side of the vehicle.

Each of the fuses is designed to protect a specific circuit, as identified on the fuse cover. Spare fuses and a special removal tool are included in the fuse box cover.

Miniaturized fuses are employed in the fuse blocks. These compact fuses, with blade terminal design, allow fingertip removal and replacement. If an electrical component fails, always check the fuse first. The best way to check the fuses is with a test light. Check for power at the exposed terminal tips of each fuse. If power is present at one side of the fuse but not the other, the fuse is blown. A blown fuse can also be identified by visually inspecting it (see illustration).

Fuses are replaced by simply pulling out the old one and pushing in the new one.

Be sure to replace blown fuses with the correct type. Fuses of different ratings are physically interchangeable, but only fuses of the proper rating should be used. Replacing a fuse with one of a higher or lower value than specified is not recommended. Each electrical circuit needs a specific amount of protection. The amperage value of each fuse is molded into the fuse body.

If the replacement fuse immediately fails, don't replace it again until the cause of the problem is isolated and corrected. In most cases, this will be a short circuit in the wiring caused by a broken or deteriorated wire.

3.1 Here's a typical fuse box located in the left side of the dashboard behind a small protective panel

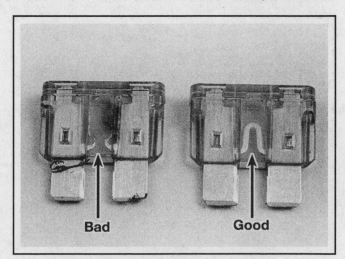

3.3 When a fuse blows, the element between the terminals melts - the fuse on the left is blown, the fuse on the right is good

4 Fusible links - general information

Some circuits are protected by fusible links. The links are used in circuits which are not ordinarily fused, such as the ignition circuit.

Although the fusible links appear to be a heavier gauge than the wire they are protecting, the appearance is due to the thick insulation. All fusible links are several wire gauges smaller than the wire they are designed to protect.

Fusible links cannot be repaired, but a new link of the same size wire can be put in its place. The procedure is as follows:

a) Disconnect the negative cable from the battery.

❊ CAUTION:

If the vehicle is equipped with a Delco Loc II or Theftlock audio system, make sure you have the correct activation code before disconnecting the battery. See the information at the front of this manual for the radio re-activation procedure.

b) Disconnect the fusible link from the wiring harness.
c) Cut the damaged fusible link out of the wiring just behind the connector.
d) Strip the insulation back approximately 1/2-inch.
e) Position the connector on the new fusible link and crimp it into place.
f) Use rosin core solder at each end of the new link to obtain a good solder joint.
g) Use plenty of electrical tape around the soldered joint. No wires should be exposed.
h) Connect the battery ground cable. Test the circuit for proper operation.

5 Circuit breakers - general information

Circuit breakers protect components such as power windows, power door locks and headlights. Some circuit breakers are located in the fuse box.

On some models the circuit breaker resets itself automatically, so an electrical overload in a circuit breaker protected system will cause the circuit to fail momentarily, then come back on. If the circuit doesn't come back on, check it immediately. Once the condition is corrected, the circuit breaker will resume its normal function. Some circuit breakers must be reset manually.

6 Relays - general information and testing

▶ **Refer to illustrations 6.1a, 6.1b, 6.1c and 6.4**

GENERAL INFORMATION

1   Several electrical accessories in the vehicle, such as the fuel injection system, horns, starter, and fog lamps use relays to transmit the electrical signal to the component. Relays use a low-current cir-cuit (the control circuit) to open and close a high-current circuit (the power circuit). If the relay is defective, that component will not operate properly. The various relays are mounted in the instrument panel, on the firewall and possibly several other locations throughout the vehicle, depending on the year and model of vehicle (see illustrations). On later models, most relays are mounted in the underhood fuse/relay panel or the rear fuse/relay panel. If a faulty relay is suspected, it can be removed and tested using the procedure below. Defective relays must be replaced as a unit.

TESTING

2   It's best to refer to the wiring diagram for the circuit to determine the proper hook-ups for the relay you're testing. However, if you're not able to determine the correct hook-up from the wiring diagrams, you may be able to determine the test hook-ups from the information that follows.

3   On most relays, two of the terminals are the relay's control circuit (they connect to the relay coil which, when energized, closes the large contacts to complete the circuit). The other terminals are the power circuit (they are connected together within the relay when the control-circuit coil is energized).

4   Some relays are marked as an aid to help you determine which terminals are the control circuit and which are the power cir-cuit (see illustration). If the relay isn't marked, check for continuity between the terminals of the relay; the terminals with continuity between them are the ones for the control circuit (this assumes that the control

6.1a These relays (arrows) on the firewall are typical of the relays you'll find throughout this vehicle

6.1b A typical horn relay, located near the steering column

6.1c A typical array of relays mounted on the heater module programmer

6.4 Most relays are marked on the outside to easily identify the control circuit and power circuits

circuit coil is operational). The other terminals are the power circuit.

5   Connect a fused jumper wire between one of the two control circuit terminals and the positive battery terminal. Connect another jumper wire between the other control circuit terminal and ground. When the connections are made, the relay should click. On some relays, polarity may be critical, so, if the relay doesn't click, try swapping the jumper wires on the control circuit terminals.

6   With the jumper wires connected, check for continuity between the power circuit terminals as indicated by the markings on the relay.

7   If the relay fails any of the above tests, replace it.

7   Turn signal/hazard flashers - check and replacement

♦ **Refer to illustration 7.5**

1   Small canister-shaped flasher units are incorporated into the electrical circuits for the directional signals and hazard warning lights.

2   When the units are functioning properly, an audible click can be heard with the circuit in operation. If the turn signals fail on one side only and the flasher unit cannot be heard, a faulty bulb is indicated.

3   If the turn signal fails on both sides, the problem may be due to a blown fuse, faulty flasher or a broken or loose connection. If the fuse has blown, check the wiring for a short before installing a new fuse.

4   The hazard warning lights are checked in the same way.

5   The hazard warning flasher is located in the convenience center adjacent to the fuse block under the dash on early models. On later models (up to 1996) it's mounted on the right side of the steering column and retained by a spring clip (see illustration). On 1997 through 1999 models the turn signal and hazard flashers are combined into a single module, located behind the left side of the instrument panel, adjacent to the lower steering column carrier. On 2000 and later models, the hazard/turn signal flasher is an integral part of the hazard switch and if defective the entire hazard switch assembly must be replaced.

6   The turn signal flasher on early models is located behind the instrument panel, clipped to the left side of the steering column brace. On later models it's mounted on the left side under-dash panel, retained

7.5 A typical hazard flasher unit, mounted on the steering column brace

by a spring clip, or near the fuse block.

7   When replacing either of the flasher units, be sure to buy a replacement of the same capacity. Compare the new flasher to the old one before installing it.

8 Multi-function/turn signal lever (1996 and earlier models only) - replacement

▶ Refer to illustrations 8.3 and 8.5

1 Detach the cable from the negative terminal of the battery.

2 The sound insulators under the instrument panel must be removed to gain access to the multi-function/turn signal lever pigtail connector.

3 Locate the multi-function lever electrical connector (see illustration) and unplug it.

4 Attach a suitable length of wire to the pigtail to pull the pigtail back through on installation.

5 Pull the lever straight out through the steering column (see illustration).

6 Pull the pigtail lead up through the steering column. Detach the wire from the pigtail lead and attach it to the lead of the new lever.

7 Carefully thread the new connector and lead back through the steering column.

8 The remainder of installation is the reverse of removal.

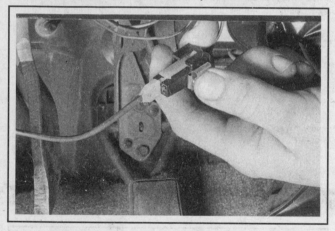

8.3 Before you pull off the multi-function switch lever, be sure to unplug the pigtail connector under the dash

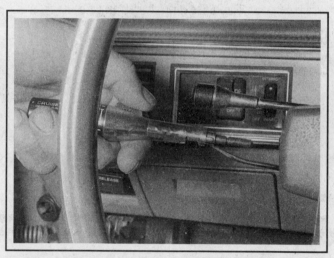

8.5 To detach the multi-function switch lever, simply grasp it firmly and pull it straight off

9 Multi-function/turn signal switch assembly - replacement

1 Detach the cable from the negative terminal of the battery.

2 Remove the steering wheel (see Chapter 10).

1996 AND EARLIER MODELS

▶ Refer to illustrations 9.3a, 9.3b, 9.4a, 9.4b, 9.5, 9.6, 9.7, 9.8 and 9.10

3 Depress the shaft lock plate and cover, then remove the retaining clip (see illustration). Remove the plate and cover (see illustration).

4 Remove the canceling cam and the spring (see illustrations).

5 Remove the hazard flasher button (see illustration).

6 Remove the switch lever screw (see illustration) and remove the lever.

7 Remove the three turn signal switch screws (see illustration).

8 Remove the sound insulator and unplug the multi-function switch connector (see illustration).

9 Pull the turn signal switch wire harness up through the steering column until there is sufficient slack to remove the turn signal switch.

10 Pull the turn signal switch out (see illustration).

11 Installation is the reverse of removal.

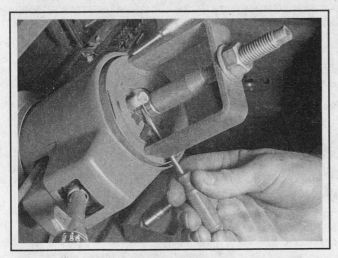

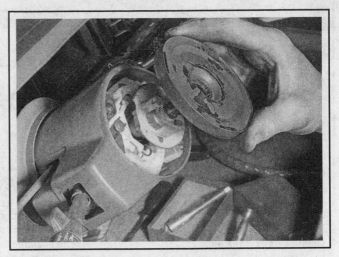

9.3a Depress the shaft lock plate and cover (you may need a special tool like the one shown), then remove the retaining clip from the steering shaft with a small screwdriver . . .

9.3b . . . and remove the shaft lock and cover

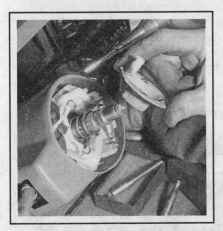

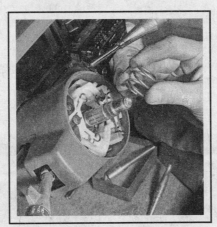

9.4a Remove the canceling cam assembly . . .

9.4b . . . and the spring

9.5 To detach the hazard flasher button from the steering column, simply remove the retaining screw in the middle of the button

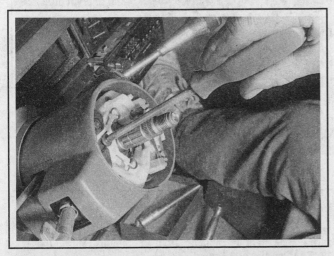

9.6 Remove the multi-function/turn signal lever screw

9.7 Remove the multi-function/turn signal switch assembly mounting screws (arrows)

9.8 Unplug the multi-function switch connector under the dash

9.10 Pull the switch lead through the steering column and remove the switch assembly

9.14 Remove the turn signal retaining screws (arrow points to the screw facing back; there's another screw at the top of the switch)

1997 AND LATER MODELS

▶ **Refer to illustration 9.14**

➡ **Note: Some 1997 and later vehicles are equipped with the earlier style steering columns. On these models refer to the 1996 and earlier procedures.**

12 Remove the airbag and steering wheel (see Chapter 10). Remove the snap-ring and slide the airbag coil off the steering shaft. Depress the lock plate, remove the snap-ring and remove the lock plate.

13 Remove the steering column covers.

14 Remove the Torx screws, unplug the electrical connector and detach the switch from the column (see illustration). Disconnect the turn signal connectors from the larger bulkhead connector.

15 Installation is the reverse of the removal procedure. When installing the airbag coil, make sure it is centered (see Chapter 10, Section 21).

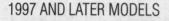

10 Ignition switch/key lock cylinder - replacement

❊❊ **WARNING:**

Some models covered by this manual are equipped with an airbag system. Always disable the airbag system before working in the vicinity of the impact sensors, steering column or instrument panel to avoid the possibility of accidental deployment of the airbag(s), which could cause personal injury. See Section 24 for the airbag disarming procedure.

1 Disconnect the cable from the negative terminal of the battery.

❊❊ **CAUTION:**

If the vehicle is equipped with a Delco Loc II or Theftlock audio system, make sure you have the correct activation code before disconnecting the battery. See the information at the front of this manual for the radio re-activation procedure.

1996 AND EARLIER MODELS

▶ **Refer to illustrations 10.2, 10.3 and 10.4**

2 Remove the multi-function/turn signal switch assembly (see Section 9). Remove the key warning buzzer switch (see illustration). The easiest way to get the buzzer switch out is to use a paper clip to pry it out.

➡ **Note: Don't lose the small retainer clip that holds the buzzer switch in place. This clip must be installed in exactly the same position it is in the accompanying illustration.**

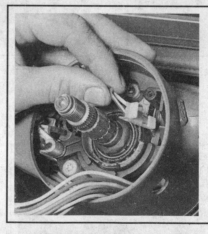

10.2 Remove the key warning buzzer switch (if you can't get it out with your fingers, use a paper clip to pry it out)

3 Remove the lock retaining screw (see illustration).

4 Turn the ignition switch to the Run position and pull it out (see illustration).

5 Installation is the reverse of removal.

1997 AND LATER MODELS

▶ **Refer to illustration 10.8**

➡ **Note: Some 1997 and later vehicles are equipped with the earlier style steering columns. On these models refer to the 1996 and earlier procedures.**

6 Remove the steering wheel (see Chapter 10).

10.3 Remove the lock retaining screw

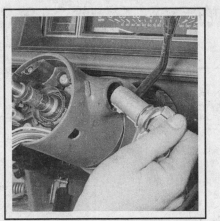

10.4 Turn the ignition switch to the Run position and pull the lock out

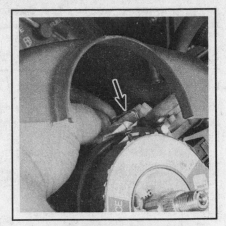

10.8 Use a 3/16-inch Allen wrench (arrow) to press the lock cylinder release pin

7   Remove the steering column covers.

8   With the ignition switch in the Start position, insert a 3/16-inch Allen wrench into the hole in the top of the casting, press the release pin, rotate the key to the Run position and pull the lock cylinder out (see illustration).

➡Note: On 2001 and later models, lift the upper steering column trim cover enough to use the short (right-angle) end of an Allen wrench to push down the pin. On 2001 and later Bonneville models, remove the radio and access the pin through the radio opening in the instrument panel.

9   Installation is the reverse of the removal procedure.

11   Headlight switch - replacement

➤ **Refer to illustrations 11.3a and 11.3b**

1   Detach the cable from the negative terminal of the battery.

❊❊ CAUTION:

If the vehicle is equipped with a Delco Loc II or Theftlock audio system, make sure you have the correct activation code before disconnecting the battery. See the information at the front of this manual for the radio re-activation procedure.

2   On models with switches in the instrument cluster, detach the instrument cluster bezel from the instrument panel, then remove the headlight switch from the bezel by prying out the tabs.

3   On early models, remove the dash trim plate. Unscrew the headlight switch assembly and pull it out from the dash (see illustrations). On 2001 and later models, pry the switch bezel (left of the instrument cluster) out and release the clips to remove the switch.

4   Installation is the reverse of removal.

11.3a After you've removed the dash trim plate, remove the headlight switch mounting screws (arrows) . . .

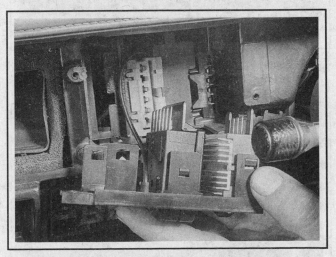

11.3b . . . and simply pull the switch assembly straight out of the dash (the switch assemblies on the models covered by this manual are modular plug-in types - there are no wires or connectors)

12 Headlight - removal and installation

SEALED BEAM

▶ **Refer to illustrations 12.1, 12.2 and 12.3**

1   Remove the headlight bezel retaining screws (see illustration). On some models you'll need to unplug the connector for the turn signal bulb.

2   Remove the headlight mounting screws (see illustration) and tilt the headlight assembly forward.

3   Unplug the electrical connector (see illustration) and remove the headlight.

4   Installation is the reverse of removal.

COMPOSITE HEADLIGHT

▶ **Refer to illustrations 12.6, 12.7 and 12.8**

✵✵ WARNING:

Halogen gas filled bulbs are under pressure and may shatter if the surface is scratched or the bulb is dropped. Wear eye protection and handle the bulbs carefully, grasping only the base whenever possible. Do not touch the surface of the bulb with your fingers because the oil from your skin could cause it to overheat and fail prematurely. If you do touch the bulb surface, clean it with rubbing alcohol.

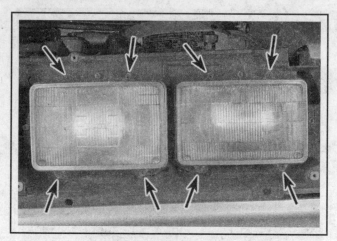

12.2 Remove the headlight retaining screws (arrows) (Oldsmobile shown, other models similar)

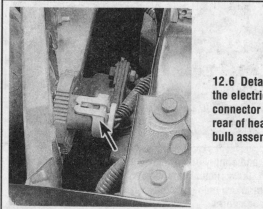

12.6 Detach the electrical connector from the rear of headlight bulb assembly

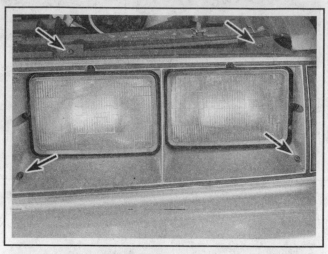

12.1 Remove the headlight bezel retaining screws (arrows) (Oldsmobile shown, other models similar)

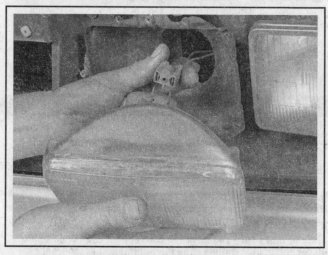

12.3 Unplug the electrical connector and remove the headlight (Oldsmobile shown, other models similar)

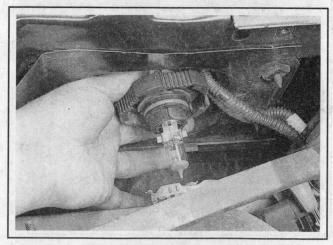

12.7 Rotate the headlight bulb retaining ring counterclockwise and pull the bulb assembly out of the housing - when installing the new bulb, don't touch the surface, but clean it with rubbing alcohol if you do

5   Open the hood.

6   Disconnect the electrical connector from the bulb assembly (see illustration).

7   Rotate the headlight bulb retaining ring counterclockwise as viewed from the rear (see illustration).

8   Withdraw the bulb assembly and retaining ring from the headlight housing.

→**Note: On some later models it may be necessary to remove the headlight housing screws and pull the housing outward to access the headlight bulbs. On 2000 and later models, the bulbs can be accessed easily on the driver's side, but the headlight housing must be removed for access on the passenger side (see illustration).**

9   Without touching the glass with your bare fingers, insert the new bulb assembly into the headlight housing, install and tighten the retaining ring.

10 Plug in the electrical connector. Test headlight operation, then close the hood.

12.8 On some later models the headlight bulbs (arrows) can only be accessed after the headlight housing is removed

13 Headlights - adjustment

▶ **Refer to illustration 13.1a, 13.1b and 13.3**

→**Note 1: The headlights must be aimed correctly. If adjusted incorrectly they could blind the driver of an oncoming vehicle and cause a serious accident or seriously reduce your ability to see the road. The headlights should be checked for proper aim every 12 months and any time a new headlight is installed or front end body work is performed. It should be emphasized that the following procedure is only an interim step which will provide temporary adjustment until the headlights can be adjusted by a properly equipped shop.**

→**Note 2: Some later models have a bubble-type level on the back of the headlight housing. On these models, simply park on a level surface and adjust the horizontal aiming screw until the bubble is at the "0" mark. The vertical aiming should then be performed with the following "screen" method.**

1   Headlights have two adjusting screws. Models with sealed-beam headlights have one on the top controlling up-and-down movement and one on the side controlling left-and-right movement (see illustration). Models with composite headlights (the kind with replaceable halogen bulbs) also have two adjusting screws - one on the outer end of the

13.1a Vertical and horizontal headlight adjustment screws (arrows) (sealed beam type)

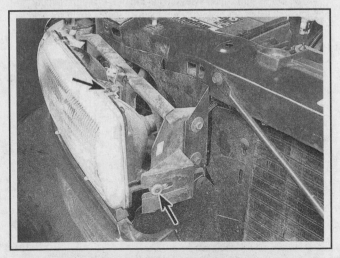

13.1b The headlight vertical adjustment screw is located at the top of the headlight (arrow) and the horizontal screw(s) is on the side of the headlight (arrows) - a Torx-head tool will be required for making headlight adjustments (composite headlight type)

headlight housing (controlling horizontal aim) and one in the top center of the housing controlling vertical aim) (see illustration).

2   There are several methods of adjusting the headlights. The simplest method requires a blank wall 25-feet in front of the vehicle and a level floor.

3   Position masking tape vertically on the wall in reference to the vehicle centerline and the centerlines of both headlights (see illustration).

4   Position a horizontal tape line in reference to the centerline of all the headlights.

➡Note: It may be easier to position the tape on the wall with the vehicle parked only a few inches away.

5   Adjustment should be made with the vehicle sitting level, the gas tank half-full and no unusually heavy load in the vehicle.

6   Starting with the low beam adjustment, position the high intensity zone so it's two inches below the horizontal line and two inches to the right of the headlight vertical line. Adjustment is made by turning the top adjusting screw clockwise to raise the beam and counterclockwise to lower the beam. The adjusting screw on the side should be used in the same manner to move the beam left or right.

7   With the high beams on, the high intensity zone should be vertically centered with the exact center just below the horizontal line.

➡Note: It may not be possible to position the headlight aim exactly for both high and low beams. If a compromise must be made, keep in mind that the low beams are the most used and have the greatest effect on driver safety.

8   Have the headlights adjusted by a dealer service department or service station at the earliest opportunity.

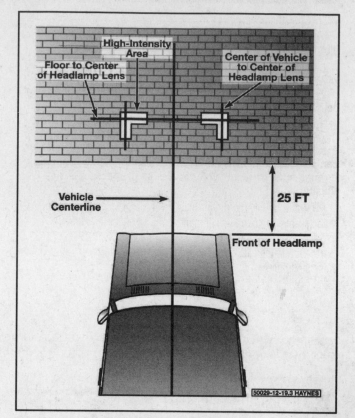

13.3 Headlight aiming details

14 Bulb replacement

▶ **Refer to illustrations 14.2a, 14.2b, 14.3a, 14.3b and 14.3c**

1   The lenses of many lights are held in place by screws, which makes it a simple procedure to gain access to the bulbs.

2   On some lights, the lenses are held in place by tabs. Simply pop them off with your fingers or pry them off with a small screwdriver (see illustrations). On 2001 and later models, remove the headlight housing to access the park/turn signal bulbs.

3   Several types of bulbs are used (see illustrations). Some are removed by pushing in and turning counterclockwise; others can simply be pulled straight out of the socket.

4   To gain access to the instrument panel lights, the instrument cluster will have to be removed first (see Section 17).

14.2a To remove a typical dome light bulb, pry off the lens with a small screwdriver . . .

14.2b . . . then pull the bulb straight out

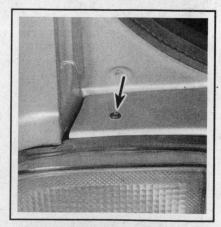

14.3a To remove the holder for the rear brake/turn signal bulb from the tail light lens assembly, remove the lens retaining screw (arrow)

14.3b Turn the holder counterclockwise and pull it out, then push down on the bulb, turn it counterclockwise and remove it from the holder (Oldsmobile shown, others similar)

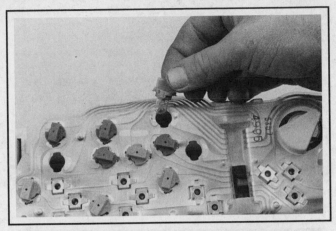

14.3c To replace an instrument cluster bulb, remove the instrument cluster (see Section 17), then twist the holder counterclockwise, pull it out of the cluster and pull the bulb straight out of the holder

15 Radio and speakers - removal and installation

✳✳ WARNING:

Some models covered by this manual are equipped with an airbag system. Always disable the airbag system before working in the vicinity of the impact sensors, steering column or instrument panel to avoid the possibility of accidental deployment of the airbag(s), which could cause personal injury. See Section 24 for the airbag disarming procedure.

1   Detach the cable from the negative terminal of the battery prior to performing any of the following procedures.

✳✳ CAUTION:

If the vehicle is equipped with a Delco Loc II or Theftlock audio system, make sure you have the correct activation code before disconnecting the battery. See the information at the front of this manual for the radio re-activation procedure.

RADIO

◗ **Refer to illustrations 15.4 and 15.5**

2   Carefully pry off the trim plate (see illustrations).
3   On earlier models, remove the knobs and heater control switches. On later models, use a flat-bladed tool to release the two clips on each side of the radio, then pull the radio out.
4   Remove the radio mounting screws (see illustration).
5   Pull the radio out of the dash and unplug the antenna, speaker, power and ground connectors from the radio (see illustration).
6   Installation is the reverse of removal.

SPEAKERS

Dashboard speakers

7   Pry off the speaker cover.

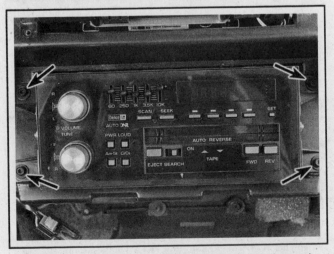

15.4 Remove the radio assembly mounting screws (arrows)

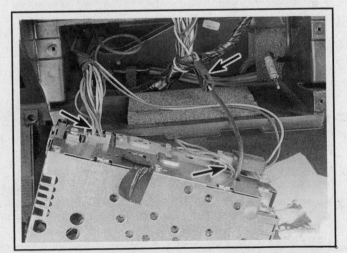

15.5 Pull the radio out from the dash and unplug the antenna and all electrical connectors

8 Remove the speaker mounting screws.

9 Lift the speaker out of its enclosure, unplug the electrical connector and remove the speaker.

10 Installation is the reverse of removal.

Rear speakers

11 Open the trunk lid and locate the speakers - they're affixed to the underside of the package tray.

12 Remove the speaker enclosure retaining clips or screws and lower the enclosure to the trunk floor.

13 Remove the speaker mounting screws and remove the speaker from the enclosure.

14 Unplug the speaker electrical connector.

15 Installation is the reverse of removal.

16 Radio antenna - removal and installation

▶ **Refer to illustration 16.5**

⁘ **WARNING:**

Some models covered by this manual are equipped with an airbag system. Always disable the airbag system before working in the vicinity of the impact sensors, steering column or instrument panel to avoid the possibility of accidental deployment of the airbag(s), which could cause personal injury. See Section 24 for the airbag disarming procedure.

➡ Note: On 2001 and later models, the antenna and its electronic module are mounted in the roof. Since access to these components involves removing all or part of the interior headliner, it is recommended that cable and/or module replacement be done by a dealership or qualified shop.

1 Detach the cable from the negative battery terminal.

⁘ **CAUTION:**

If the vehicle is equipped with a Delco Loc II or Theftlock audio system, make sure you have the correct activation code before disconnecting the battery. See the information at the front of this manual for the radio re-activation procedure.

2 If the antenna is mounted in the fender, remove the screws and clips and pull out the inner fender liner. If the antenna is rear mounted, remove the right side trunk liner.

3 Remove the sound insulator from underneath the right side of the dash (see Section 7).

4 Unplug the antenna leads from the relay and radio (see Section 15). If the antenna is rear mounted, unplug it from the body harness.

5 Remove the antenna upper nut (see illustration).

6 Remove the antenna bracket screws.

7 Pull the antenna assembly up slightly to separate it, but don't remove it.

8 Wrap the lead with wire so it can be threaded back to it's proper location.

9 Remove the antenna assembly and the old grommet or bezel.

10 Install a new grommet or bezel.

11 Feed the new lead through the grommet or bezel and carefully thread it through with the wire.

12 The remainder of installation is the reverse of removal.

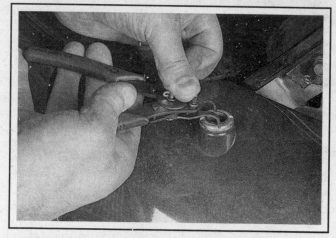

16.5 The antenna base retaining nut can be removed using a pair of snap-ring pliers or similar tool

17 Instrument cluster - removal and installation

▶ **Refer to illustrations 17.3, 17.4 and 17.6**

⁘ **WARNING:**

Some models covered by this manual are equipped with an airbag system. Always disable the airbag system before working in the vicinity of the impact sensors, steering column or instrument panel to avoid the possibility of accidental deployment of the airbag(s), which could cause personal injury. See Section 24 for the airbag disarming procedure.

➡ Note 1: Vehicles covered by this manual are equipped with either a standard gauge or digital instrument cluster. They may also have a trip calculator or driver information display center next to the cluster. Installation of all instrument clusters is basically the same.

➡ Note 2: On 2001 and later models, the end caps and upper portion of the instrument panel must be removed to access the cluster bezel and cluster, and this procedure is not recommended for the home mechanic.

1 Detach the cable from the negative battery terminal.

17.3 Detach the gear indicator wire

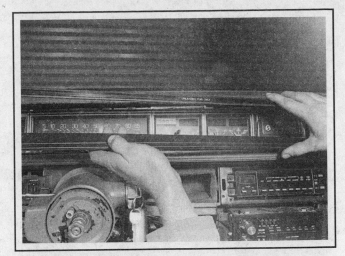

17.4 Remove the cluster trim plate

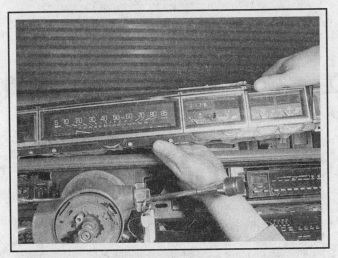

17.6 Remove the instrument cluster (on newer units, the connectors are the plug-in type; on older models, you'll have to unplug the leads before you can detach the cluster from the dash)

✳✳ CAUTION:

If the vehicle is equipped with a Delco Loc II or Theftlock audio system, make sure you have the correct activation code before disconnecting the battery. See the information at the front of this manual for the radio re-activation procedure.

2   Remove all screws from the trim plate around the cluster.

3   If the gear indicator window is integral with the instrument cluster, remove the sound insulator under the dash and detach the gear indicator wire (see illustration).

4   Move the gear selector lever to "1" (manual Low) and carefully remove the cluster trim plate (see illustration).

5   Remove the mounting screws from the cluster.

6   Pull the cluster out of the dash (see illustration).

➡ **Note: If the cluster is on an older model, it may not have modular plug-in boards on back like newer units - so you'll have to unplug the connectors on back.**

7   Installation is the reverse of removal. Don't forget to attach the gear indicator wire.

18 Rear window defogger - check and repair

▶ **Refer to illustrations 18.4 and 18.9**

1   This option consists of a rear window with a number of horizontal elements baked into the glass surface during the glass forming operation.

2   Small breaks in the element can be successfully repaired without removing the rear window.

3   To test the grids for proper operation, start the engine and turn on the system.

4   Ground one lead of a test light and carefully touch the other lead to each element line (see illustration).

5   The brilliance of the test light should increase as the lead is moved across the element. If the test light glows brightly at both ends of the lines, check for a loose ground wire. All of the lines should be checked in at least two places.

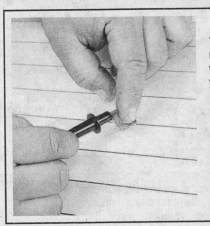

18.4 When measuring the voltage at the rear window defogger grid, wrap a piece of aluminum foil around the negative probe of the voltmeter and press the foil against the wire with your finger

6   To repair a break in a line, it is recommended that a repair kit specifically for this purpose be purchased from your local auto parts store. Included in the repair kit will be a decal, a container of silver plastic and hardener, a mixing stick and instructions.

7   To repair a break, first turn off the system and allow it to de-energize for a few minutes.

8   Lightly buff the element area with fine steel wool, then clean it thoroughly with rubbing alcohol.

9   Use the decal supplied in the repair kit or apply strips of electrician's tape above and below the area to be repaired (see illustration). The space between the pieces of tape should be the same width as the existing lines. This can be checked from outside the vehicle. Press the tape tightly against the glass to prevent seepage.

10   Mix the hardener and silver plastic thoroughly.

11   Using the wood spatula, apply the silver plastic mixture between the pieces of tape, overlapping the undamaged area slightly on either end.

12   Carefully remove the decal or tape and apply a constant stream of hot air directly to the repaired area. A heat gun set at 500 to 700 degrees Fahrenheit is recommended. Hold the gun one inch from the glass for two minutes.

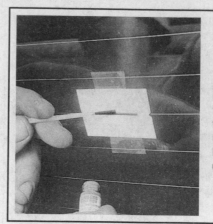

18.9 To use a defogger repair kit, apply masking tape to the inside of the window at the damaged area, then brush on the special conductive coating

13   If the new element appears off color, tincture of iodine can be used to clean the repair and bring it back to the proper color. This mixture should not remain on the repair for more than 30 seconds.

14   Although the defogger is now fully operational, the repaired area should not be disturbed for at least 24 hours.

19   Windshield wiper motor/washer pump - removal and installation

▶ Refer to illustrations 19.2, 19.5, 19.6, 19.7a, 19.7b, 19.8, 19.9, 19.13a and 19.13b

➡ Note: The windshield washer pump on 1987 and later models is located in the windshield washer reservoir. To replace the pump on 1987 and later models, first drain the reservoir then remove the reservoir, replace the pump and re-install the reservoir.

1   Detach the cable from the negative terminal of the battery.

✳✳ CAUTION:

If the vehicle is equipped with a Delco Loc II or Theftlock audio system, make sure you have the correct activation code before disconnecting the battery. See the information at the front of this manual for the radio re-activation procedure.

2   Unplug the electrical connectors from the motor and the pump (see illustration).

3   Detach the windshield washer fluid hoses from the washer pump.

4   Disconnect the connector for the windshield washer fluid line at the wiper blade.

5   Pry off the windshield wiper arm cap over the nut and remove the nut (see illustration).

6   Using a small battery post puller or similar tool, remove the left windshield wiper blade assembly. If you don't have a battery post puller, pry the blade off with a screwdriver (see illustration).

7   Some newer models don't use the same wiper arm installation. On these units the wiper is attached to the wiper motor shaft by a spring-loaded rotating clip (see illustration). To release the wiper arm, simply push in on the clip tang that protrudes from the underside of the arm assembly and lift straight up (see illustration).

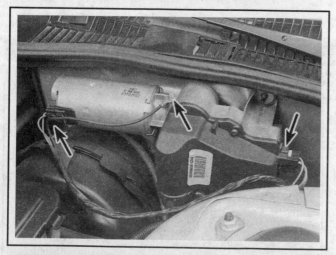

19.2 Unplug the electrical connectors from the windshield wiper motor and pump (arrows) and detach the windshield washer fluid hoses from the pump

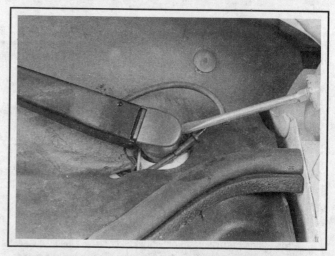

19.5 Pry the windshield wiper arm retaining nut cap loose with a screwdriver

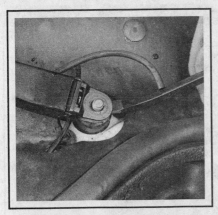

19.6 If you don't have a battery cable puller or some other type of small puller, you can pry the windshield wiper arm off with a screwdriver, but be careful

19.7a This type of windshield arm uses a special spring-loaded rotating clip to lock it onto the shaft

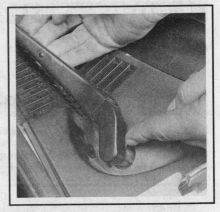

19.7b To release the clip from the wiper arm shaft, push in on the tang that protrudes from underneath the wiper arm and lift up at the same time

8   Remove the vent grille (see illustration). Be careful when removing the vent grille fasteners - they break easily. On 2001 and later models, remove the bolts/nuts and the windshield reinforcement plate that is over the wiper assembly.

9   Remove the nut that fastens the transmission linkage arm to the motor shaft (see illustration) and detach the linkage from the motor.

10   Remove the motor mounting bolts and remove the motor (see illustration 19.9).

11   If you're replacing the motor, be sure to switch the washer pump to the new motor (see Step 13).

12   Installation is the reverse of removal.

13   If you're only replacing the windshield washer pump, remove the small locking clip (see illustration) and pull the pump from the bottom of the motor (see illustration).

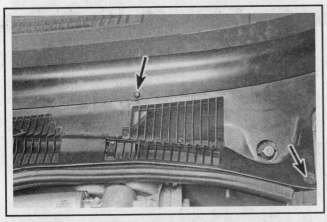

19.8 To get at the windshield wiper motor transmission (linkage), remove the rubber molding along the upper edge of the firewall, pry loose the fasteners (arrows) and remove the vent grille (left half of grille, right half similar)

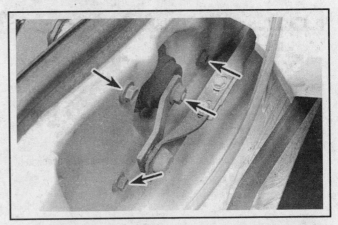

19.9 Remove the nut that fastens the transmission linkage arm to the motor shaft (center) - to remove the windshield wiper motor/washer fluid pump assembly, remove the three mounting bolts (arrows)

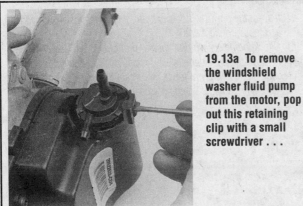

19.13a To remove the windshield washer fluid pump from the motor, pop out this retaining clip with a small screwdriver . . .

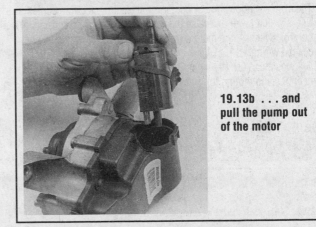

19.13b . . . and pull the pump out of the motor

20 Horn - check and replacement

▶ **Refer to illustration 20.4**

1   Detach the cable from the negative terminal of the battery.

✳✳ CAUTION:

If the vehicle is equipped with a Delco Loc II or Theftlock audio system, make sure you have the correct activation code before disconnecting the battery. See the information at the front of this manual for the radio re-activation procedure.

2   Locate the horn. On some vehicles, it's located underneath the front of the vehicle, generally right behind the bumper. On others, it's mounted to the right inner fenderwell on the right side of the engine compartment.

3   If the horn is underneath, raise the vehicle and place it securely on jackstands and remove the protective panel from the underside of the front left corner of the bumper.

4   Remove the horn mounting bracket bolt (see illustration), lower the horn, unplug the electrical connector and remove the horn.

5   Installation is the reverse of removal.

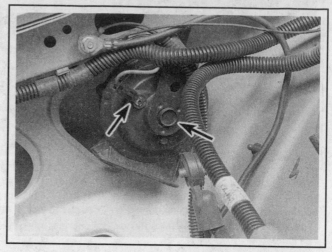

20.4 A typical horn installed in the engine compartment - to replace it, remove the mounting bracket bolt (right arrow) and unplug the connector (left arrow)

21 Cruise control system - description and check

1   The cruise control system maintains vehicle speed with a vacuum actuated servo motor located in the engine compartment, which is connected to the throttle linkage by a cable. The system consists of the servo motor, clutch switch, brake switch, control switches, a relay and associated vacuum hoses. Listed below are some general procedures that may be used to locate common cruise control problems.

2   Locate and check the fuse (see Section 3).

3   Have an assistant operate the brake lights while you check their operation (voltage from the brake light switch deactivates the cruise control).

4   If the brake lights don't come on or don't shut off, correct the problem and retest the cruise control.

5   Visually inspect the wires connected cruise control actuator.

6   Check the vacuum release switch and its hose located at the top of the brake pedal bracket.

7   Check the Vehicle Speed Sensor (see Chapter 6).

22 Power window system - description and check

The power window system operates the electric motors mounted in the doors which lower and raise the windows. The system consists of the control switches, the motors (regulators), glass mechanisms and associated wiring.

Diagnosis can usually be limited to simple checks of the wiring connections and motors for minor faults which can be easily repaired.

These include:

a) *Inspect the power window actuating switches for broken wires and loose connections.*

b) *Check the power window fuse/and or circuit breaker.*

c) *Remove the door panel(s) and check the power window motor wires to see if they're loose or damaged. Inspect the glass mechanisms for damage which could cause binding.*

23 Power door lock system - description and check

The power door lock system operates the door lock actuators mounted in each door. The system consists of the switches, actuators and associated wiring. Diagnosis can usually be limited to simple checks of the wiring connections and actuators for minor faults which can be easily repaired. These include:

a) *Check the system fuse and/or circuit breaker.*

b) *Check the switch wires for damage and loose connections. Check the switches for continuity.*

c) *Remove the door panel(s) and check the actuator wiring connections to see if they're loose or damaged. Inspect the actuator rods (if equipped) to make sure they aren't bent or damaged. Inspect the actuator wiring for damaged or loose connections. The actuator can be checked by applying battery power momentarily. A discernible click indicates that the solenoid is operating properly.*

24 Airbag - general information and precautions

GENERAL INFORMATION

1   Some later models are equipped with a Supplemental Inflatable Restraint (SIR) system, more commonly known as an airbag. This system is designed to protect the driver from serious injury in the event of a head-on or frontal collision. It consists of an airbag module in the center of the steering wheel and, on models so equipped, in the right side of the dashboard, three discriminating (crash) sensors mounted at the front of the vehicle, an arming sensor mounted on the steering column support bracket, an SIR coil assembly mounted under the steering wheel and a diagnostic module located inside the passenger compartment under the dash (1994 and earlier models) or under the right front seat (1995 and later models).

Airbag module and SIR coil

2   The airbag inflator module contains a housing incorporating the cushion (airbag) and inflator unit, mounted in the center of the steering wheel and on models so equipped, in the dashboard on the right side (above the glove box). The inflator assembly is mounted on the back of the housing over a hole through which gas is expelled, inflating the bag almost instantaneously when an electrical signal is sent from the system. The airbag steering column coil assembly is mounted on the steering column under the steering wheel and carries this signal to the module. The coil assembly can transmit an electrical signal regardless of steering wheel position.

Most late models have airbags for both the driver and front-seat passenger (in the instrument panel over the glovebox). On 2001 and later models, optional side-impact airbags are located in the door panels or the rear corners of the front seats. Some 2005 models have an optional "curtain" style airbag system that is mounted at each side of the headliner that is designed to offer side-impact protection to both front and rear-seat occupants.

Sensors

3   The system has four sensors: a forward discriminating sensor mounted on the radiator support, two more discriminating sensors mounted on the frame behind each front bumper bracket, and an arming sensor mounted to the steering column support bracket. These sensors are basically pressure sensitive switches that complete an electrical circuit during an impact of sufficient G force. The arming sensor closes at lower deceleration rates. The electrical signal from these sensors is sent to the Diagnostic Energy Reserve Module (DERM) which then completes the circuit and inflates the airbag.

Diagnostic Energy Reserve Module (DERM)

4   The DERM supplies the current to the airbag system in the event of the collision, even if battery power is cut off. It checks this system every time the vehicle is started, causing the "AIR BAG" light to go on then off, if the system is operating properly. If there is a fault in the system, the light will go on and stay on and the DERM will store fault codes indicating the nature of the fault. If the AIR BAG light goes on and stays on, the vehicle should be taken to a dealer service department or other qualified repair shop immediately for service.

OPERATION

5   For the airbag(s) to deploy, an impact of sufficient force must occur within 30-degrees of the vehicle centerline. When this condition occurs, the circuit to the airbag inflator is closed and the airbag inflates. If the battery is destroyed by the impact, or is too low to power the inflator, a back-up power supply inside the diagnostic/energy reserve module supplies current to the airbag.

Self-diagnosis system

6   A self-diagnosis circuit in the module displays a light when the ignition switch is turned to the On position. If the system is operating normally, the light should go out after seven flashes. If the light doesn't come on, or doesn't go out after seven flashes, or if it comes on while you're driving the vehicle, there's a malfunction in the SIR system. Have it inspected and repaired as soon as possible. Do not attempt to troubleshoot or service the SIR system yourself. Even a small mistake could cause the SIR system to malfunction when you need it.

Servicing components near the SIR system

7   Nevertheless, there are times when you need to remove the steering wheel, radio or service other components on or near the instrument panel or at the front of the vehicle. At these times, you'll be working around components and wiring harnesses for the SIR system. SIR system wiring is easy to identify; they're all covered by a bright yellow conduit. Do not unplug the connectors for the SIR system wiring, except to disable the system. And do not use electrical test equipment on the SIR system wiring. Always disable the SIR system before working near the SIR system components or related wiring.

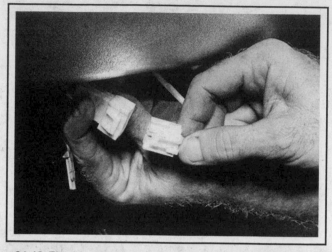

24.10 The driver's side airbag Connector Position Assurance (CPA) connector is located at the base of the steering column; always unlock and unplug it before removing the steering wheel or working in the area of the steering wheel

DISABLING THE SIR SYSTEM

▶ **Refer to illustration 24.10**

8   Turn the steering wheel to the straight ahead position, place the ignition switch key in the Lock position and remove the key. Remove the airbag fuse from the instrument panel fuse block. It's also a good idea to disconnect the cable from the negative terminal of the battery, although this is not actually specified by the manufacturer.

❊ CAUTION:

If the vehicle is equipped with a Delco Loc II audio system, make sure you have the correct activation code before disconnecting the battery. See the information at the front of this manual for the radio re-activation procedure.

24.15b . . . then remove the coil

24.15a To release the SIR coil from the steering shaft, remove this snap-ring . . .

9   If you're working in the vicinity of the steering column, the left side or center of the instrument panel or the console, remove the steering column covers and the left sound insulator panel below the instrument panel.

10   Unplug the yellow Connector Position Assurance (CPA) steering column harness connector (see illustration).

11   If you're working around the right side of the instrument panel, remove the right-side sound insulator panel under the dash, then unplug the Connector Position Assurance connector in the harness to the passenger side airbag module.

ENABLING THE SIR SYSTEM

12   After you've disabled the airbag and performed the necessary service, plug in the yellow CPA connectors. Reinstall the steering column lower trim panel, and the sound insulator panels. Install the airbag fuse. Connect the negative battery terminal.

REMOVING, CENTERING AND INSTALLING THE SIR COIL

▶ **Refer to illustrations 24.15a, 24.15b, 24.16 and 24.18**

13   Anytime some part of the steering system is disassembled for service or replacement, the steering column should be immobilized to ensure that the SIR coil doesn't become uncentered (moved). This can occur, for example, if the steering column is separated from the steering gear, or if the centering spring is pushed down, allowing the hub to rotate while the coil is removed from the steering column. If the coil becomes accidentally uncentered, re-center it as follows before reassembling the steering system.

14   Make sure that the wheels are pointed straight ahead.

15   Remove the coil assembly snap-ring (see illustration) and remove the coil assembly (see illustration).

16   Holding the coil assembly with its bottom side facing up, depress the spring lock (see illustration) and rotate the hub in the direction of

24.16 To center the SIR coil, hold it with its underside facing up, depress the spring lock and rotate the hub in the direction of the arrow on the coil assembly until it stops, then turn it in the opposite direction 2-1/2 turns and release the spring lock

24.18 When properly installed, the airbag coil will be centered with the marks aligned (circle) and the tab fitted between the projections on the top of steering column (arrow)

the arrow until it stops (the arrow is on the back of the coil assembly). The coil ribbon should be wound up snug against the center hub.

17 Rotate the coil hub in the opposite direction 2-1/2 turns, then release the spring lock. The coil is now centered.

18 Install the SIR coil and secure it with the snap-ring. The tab will be at the top and the marks aligned when the coil is properly installed (see illustration).

25 Wiring diagrams - general information

Since it isn't possible to include all wiring diagrams for every year covered by this manual, the following diagrams are those that are typical and most commonly needed.

Prior to troubleshooting any circuits, check the fuse and circuit breakers (if equipped) to make sure they're in good condition. Make sure the battery is properly charged and check the cable connections (see Chapter 1).

When checking a circuit, make sure that all connectors are clean, with no broken or loose terminals. When unplugging a connector, do not pull on the wires. Pull only on the connector housings themselves.

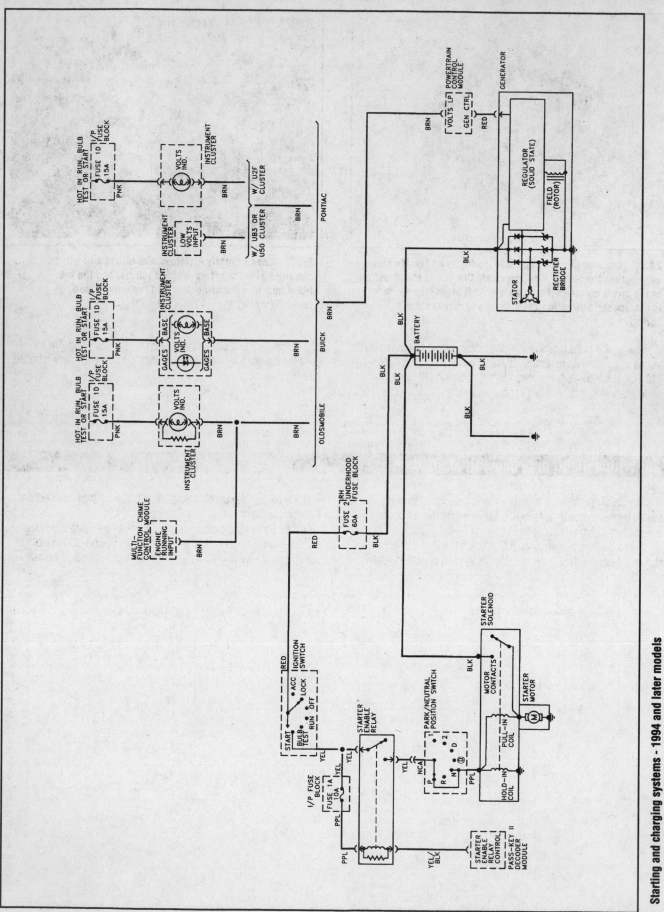

Starting and charging systems - 1994 and later models

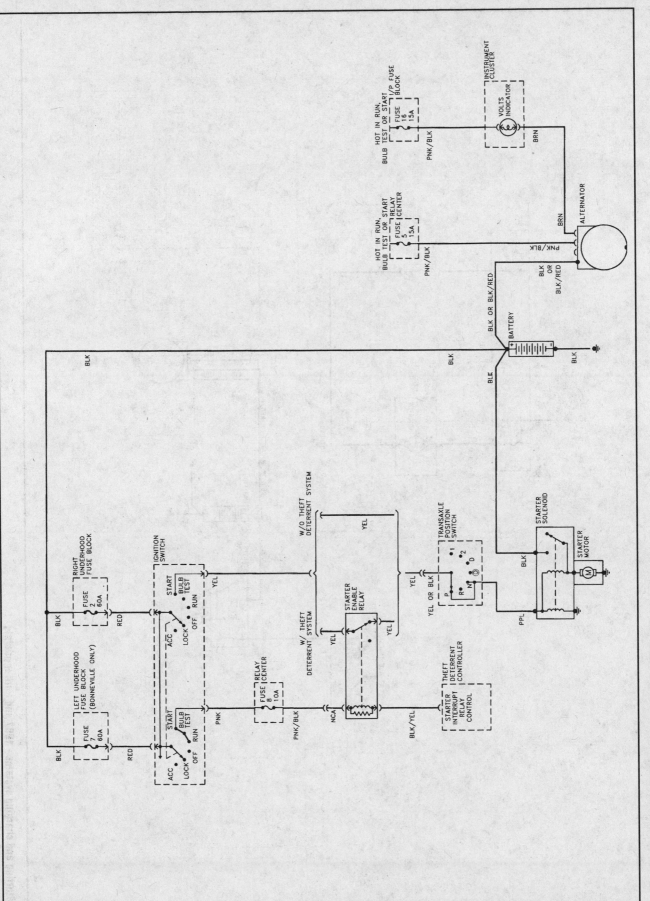

Starting and charging systems - 1988 through 1993 models

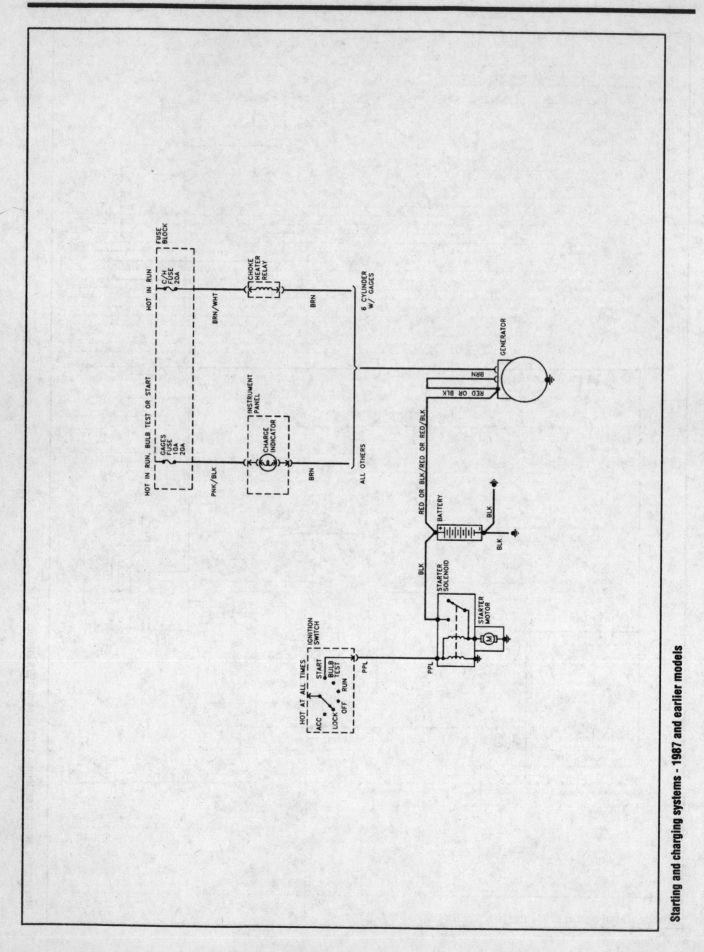

Starting and charging systems - 1987 and earlier models

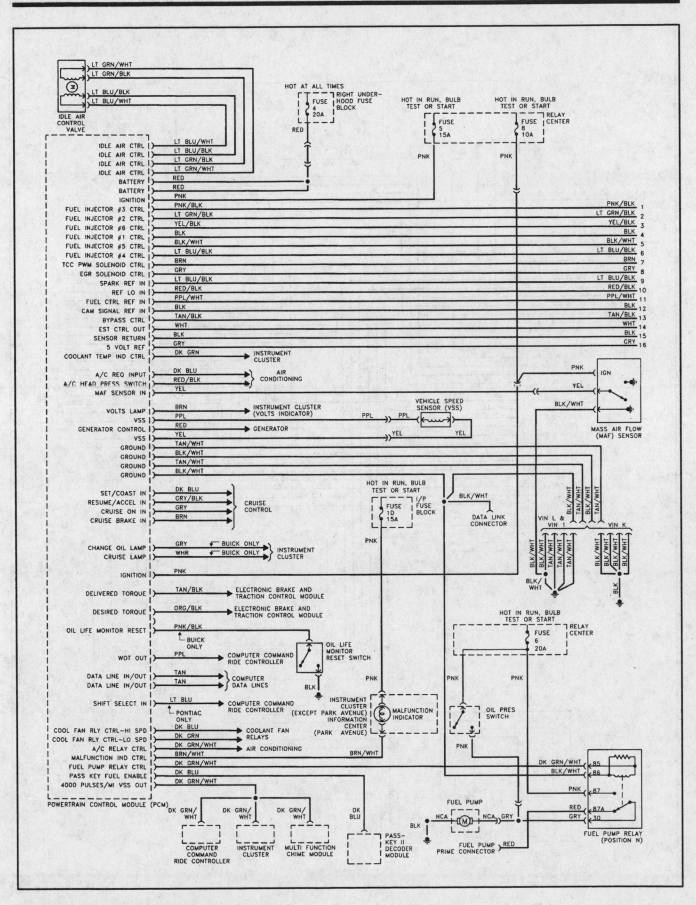

Engine control system - 1994 and later models (1 of 3)

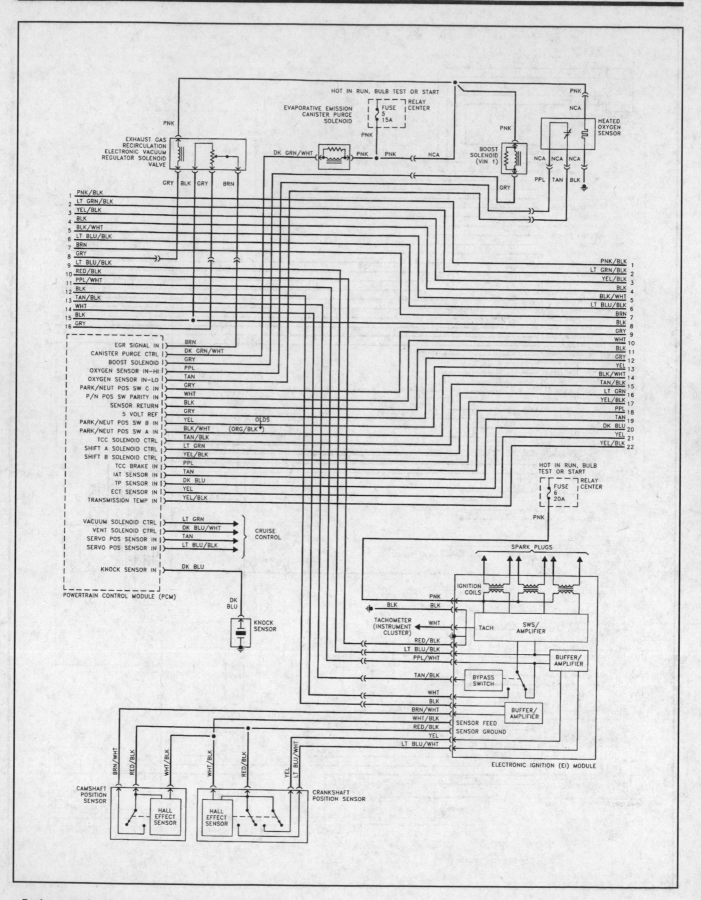

Engine control system - 1994 and later models (2 of 3)

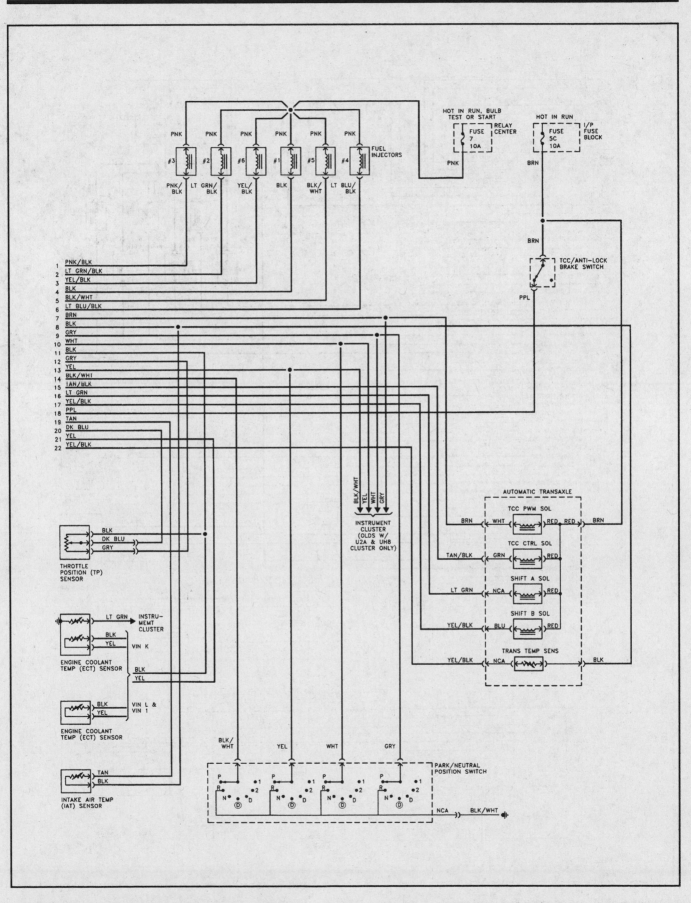

Engine control system - 1994 and later models (3 of 3)

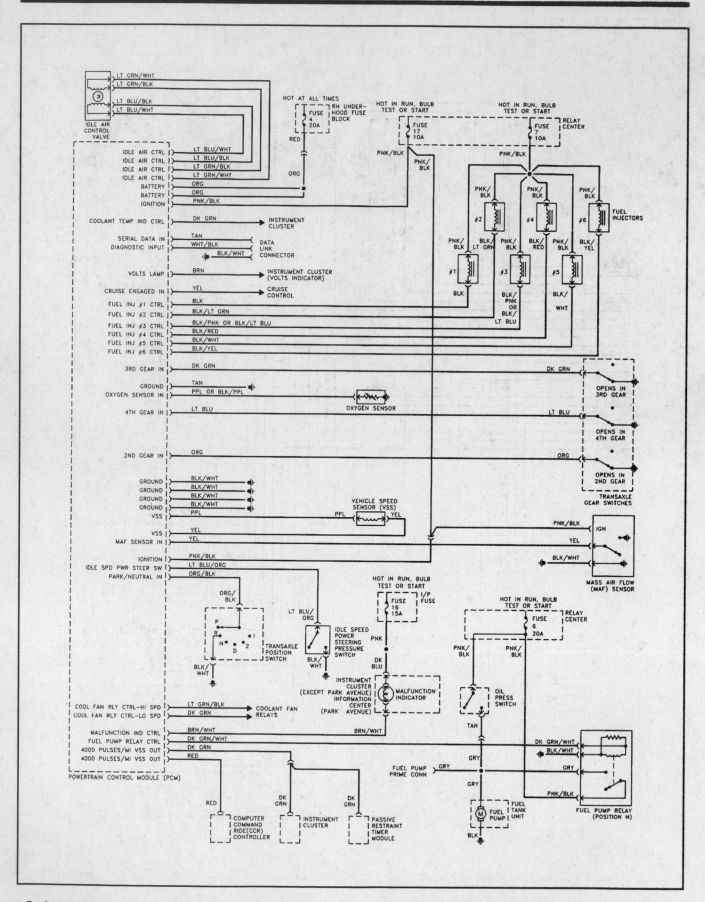

Engine control system - 1988 through 1993 models (1 of 2)

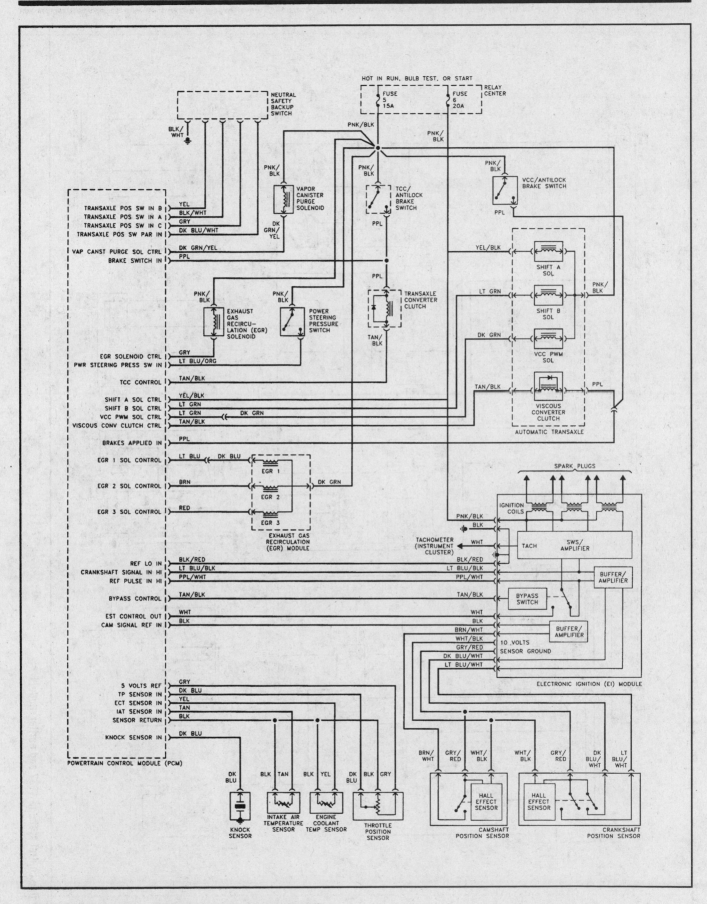

Engine control system - 1988 through 1993 models (2 of 2)

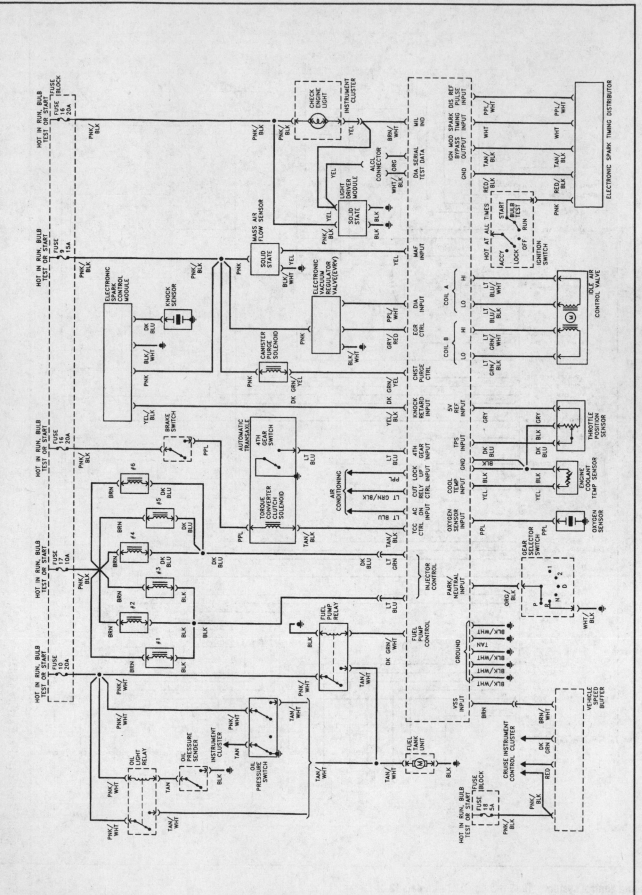

Engine control system - 1987 and earlier models

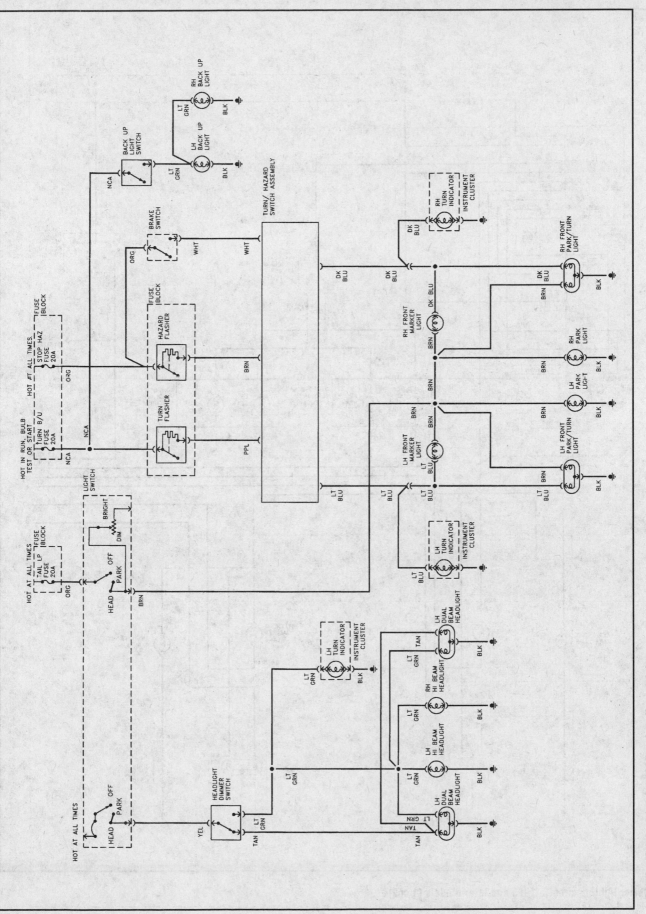

Exterior lighting circuit - 1987 and earlier models

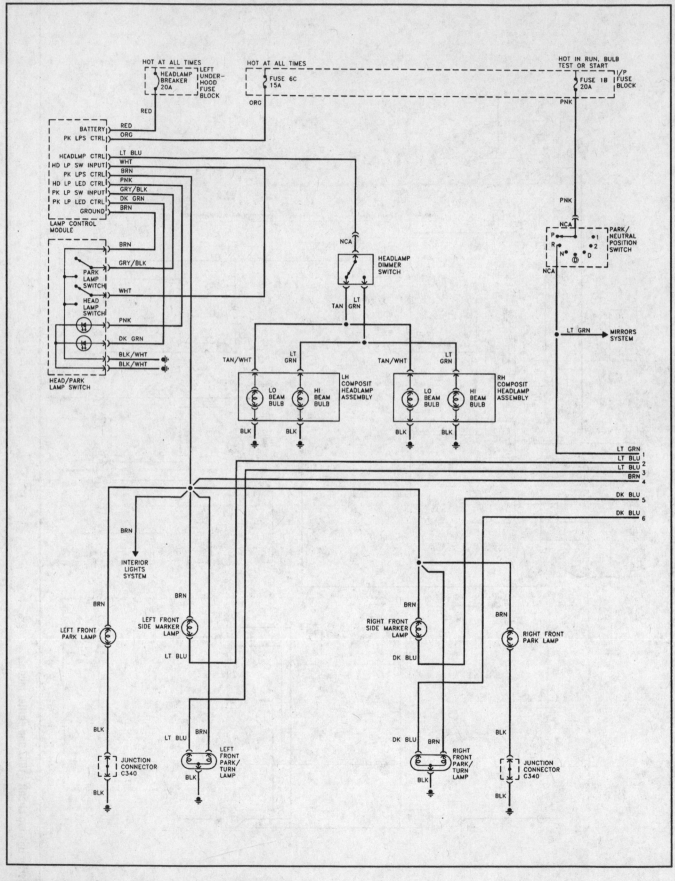

Exterior lighting circuit - 1988 and later models (1 of 2)

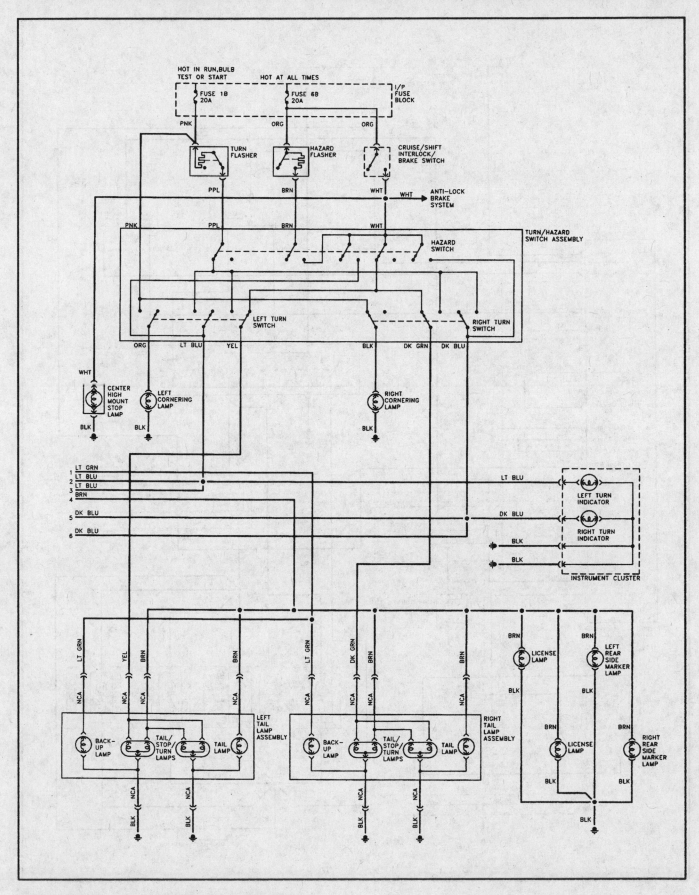

Exterior lighting circuit - 1988 and later models (2 of 2)

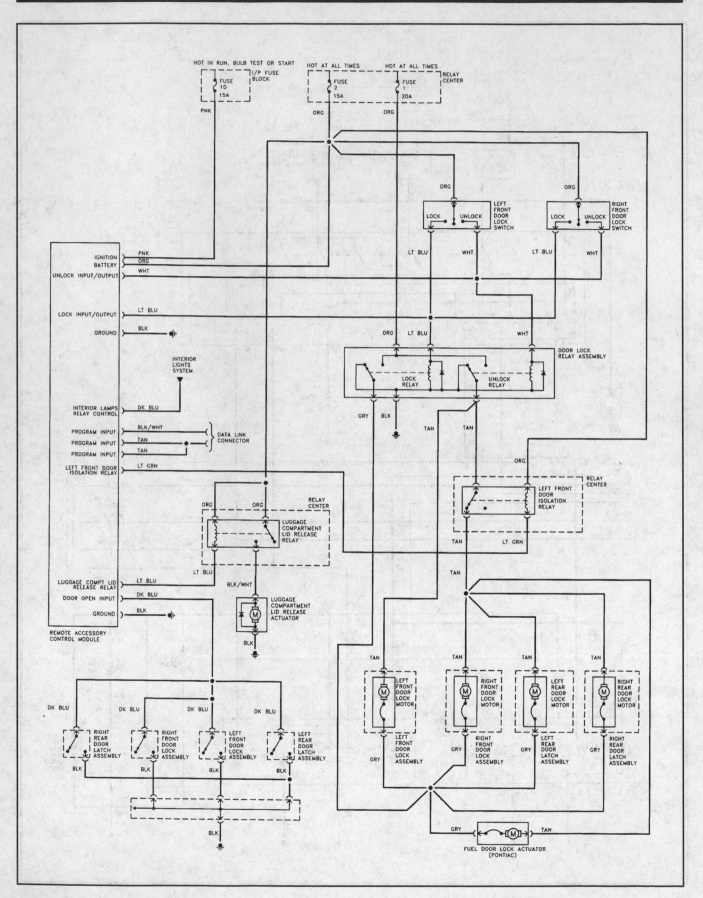

Typical keyless entry system

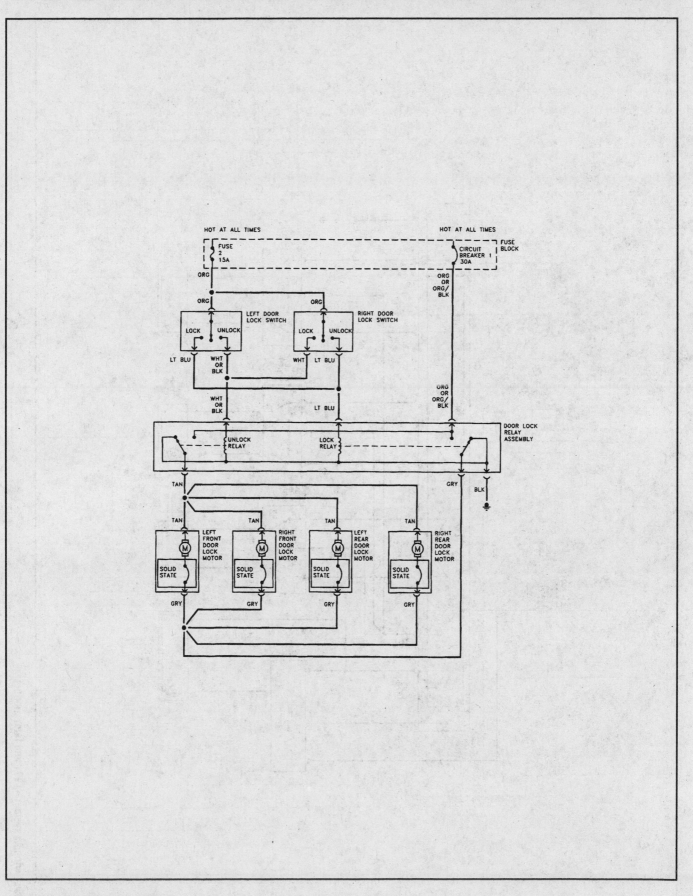

Typical power door lock system

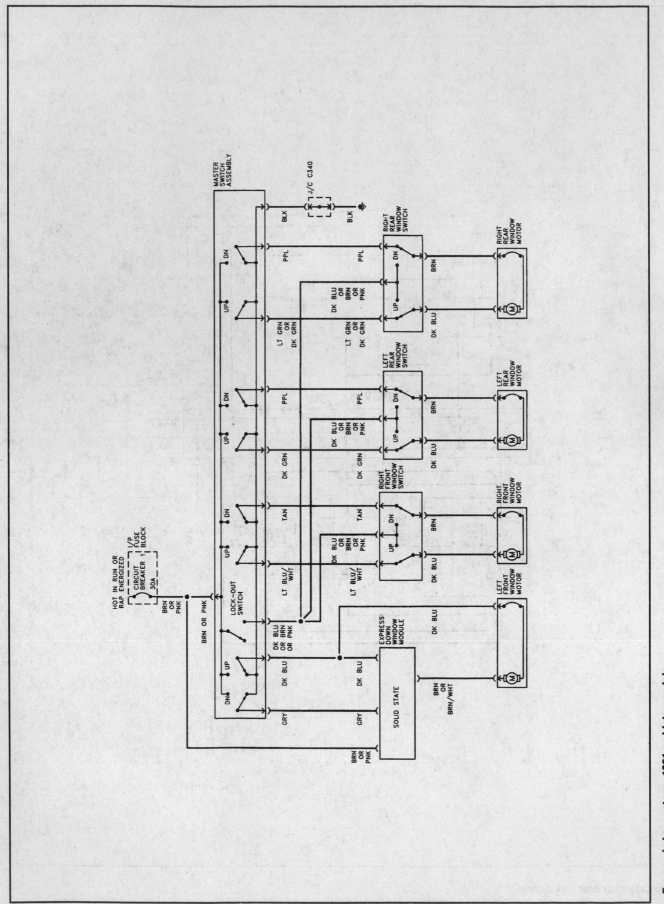

Power window system - 1991 and later models

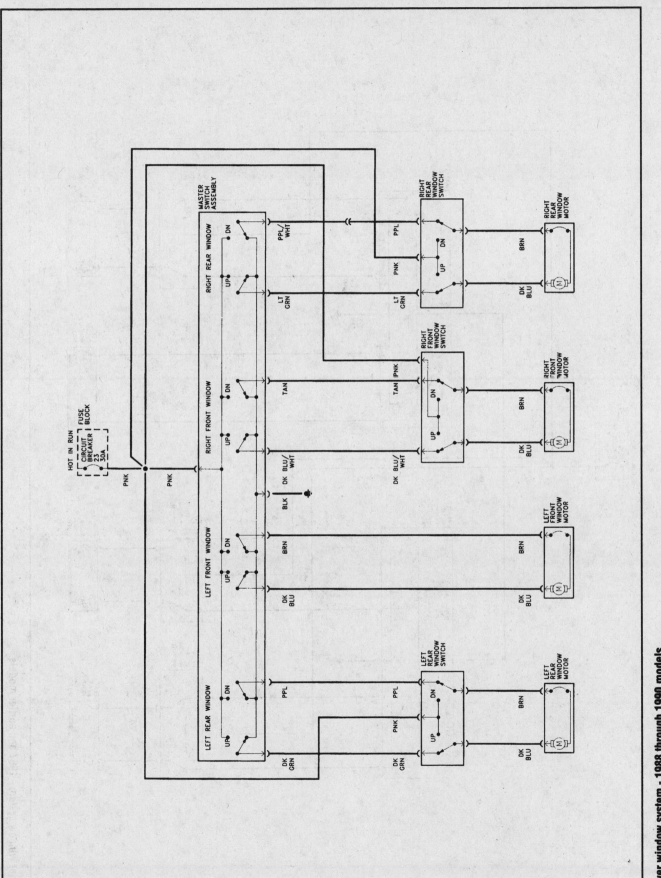

Power window system - 1988 through 1990 models

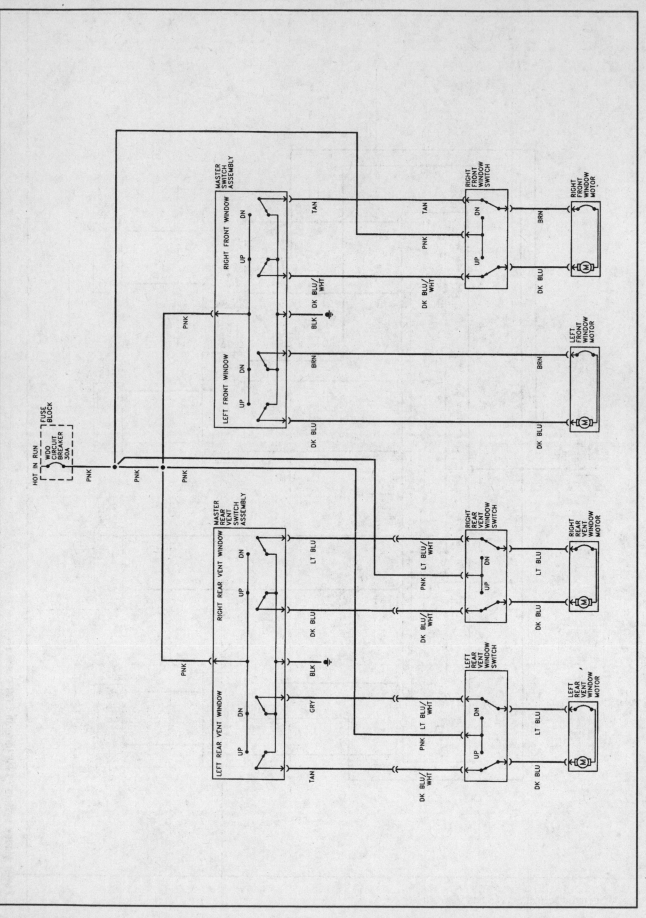

Power window system - 1987 and earlier models

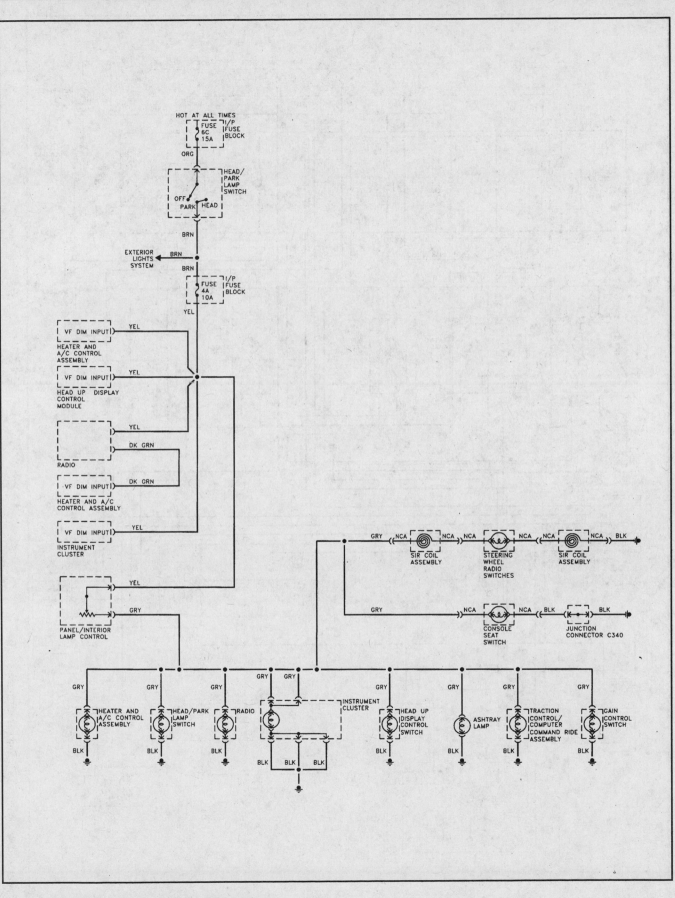

Interior illumination circuit - 1988 and later models

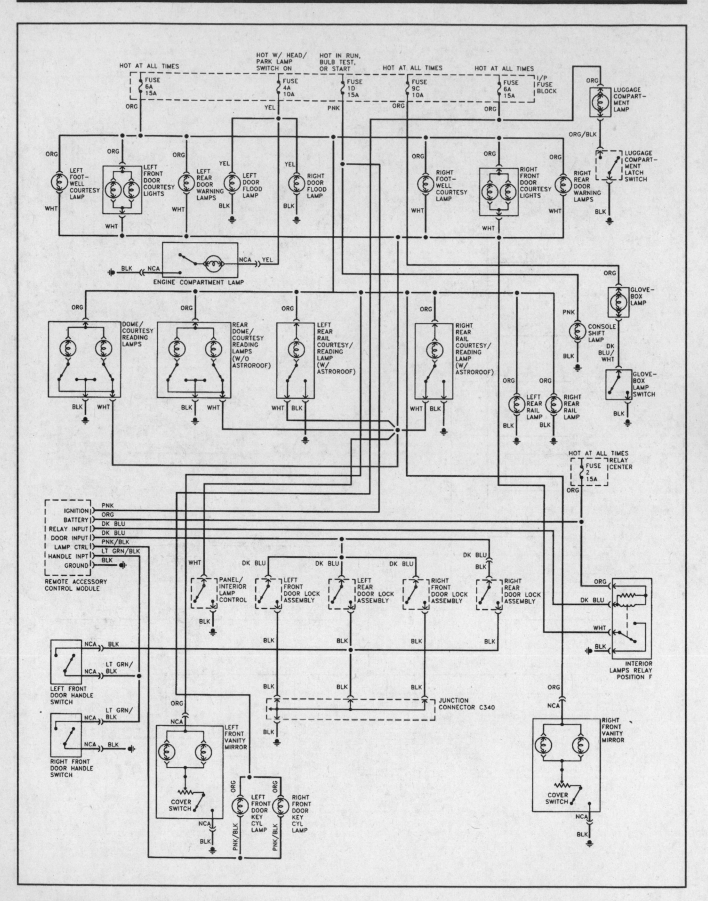

Courtesy light circuit - 1988 and later models

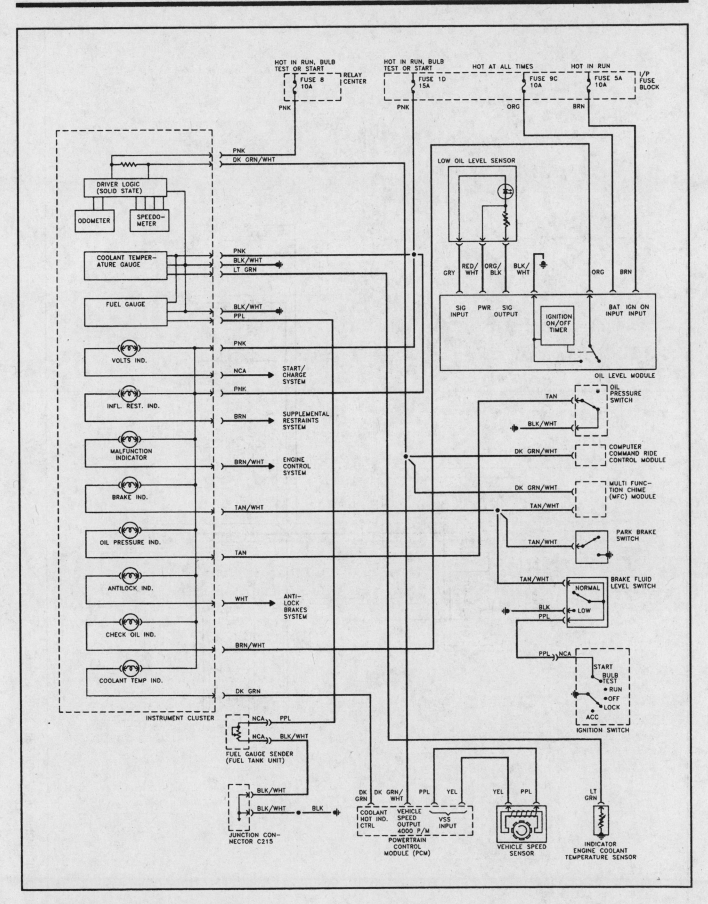

Warning light circuit - 1991 and later models

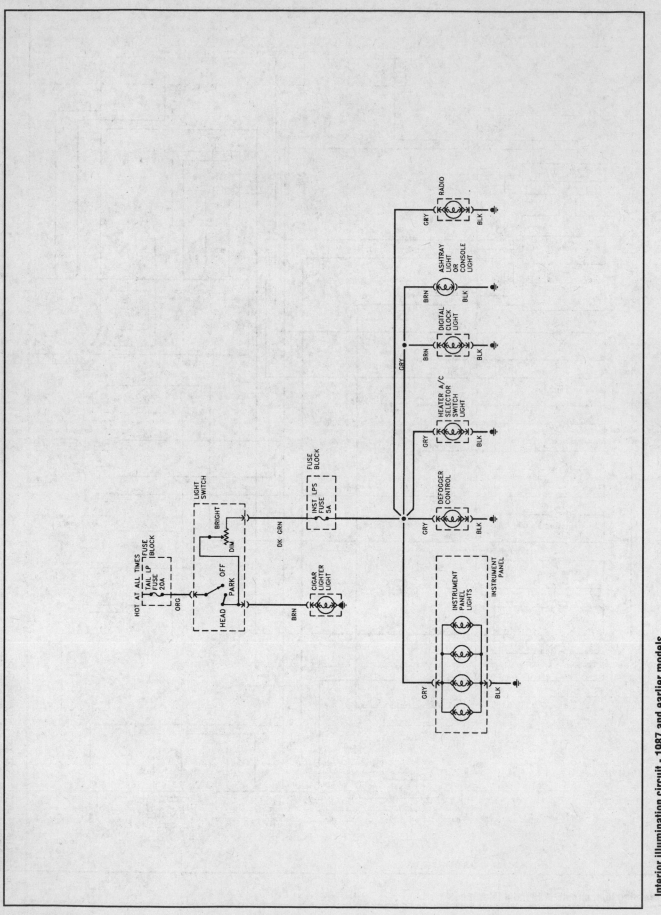

Interior illumination circuit - 1987 and earlier models

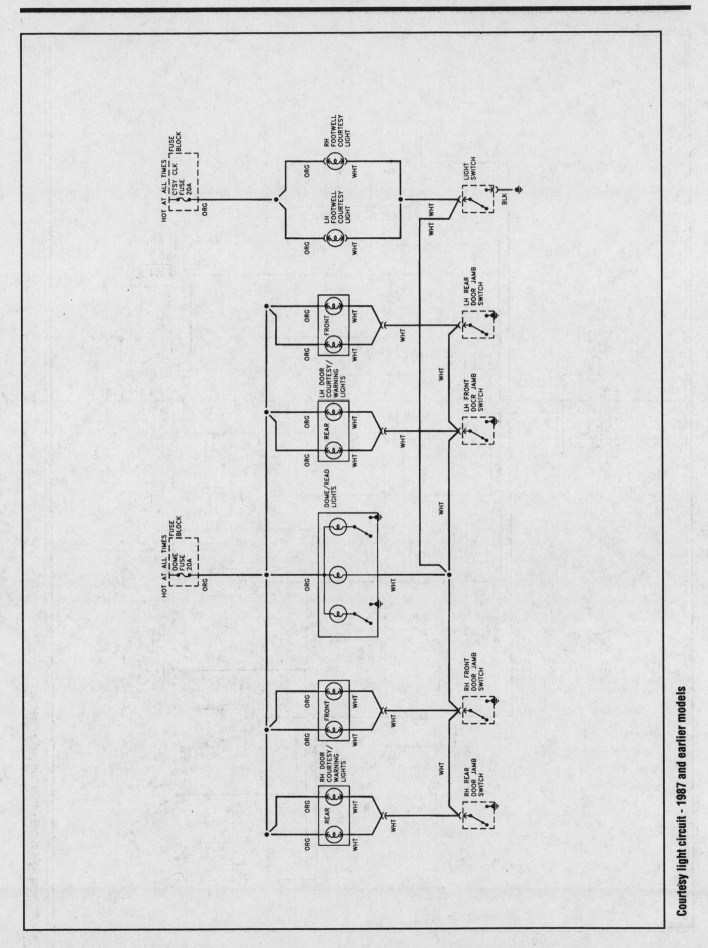

Courtesy light circuit - 1987 and earlier models

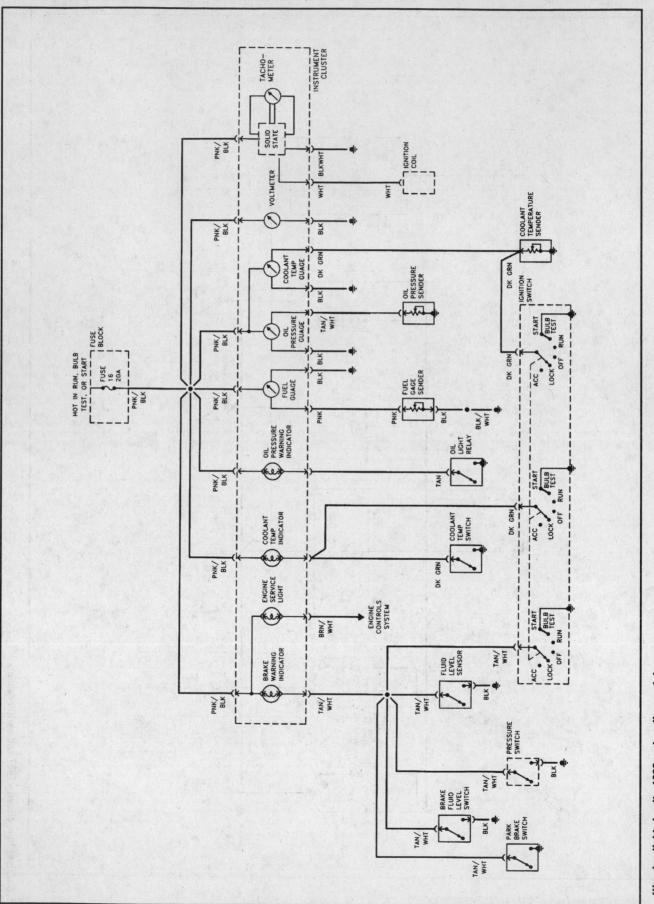

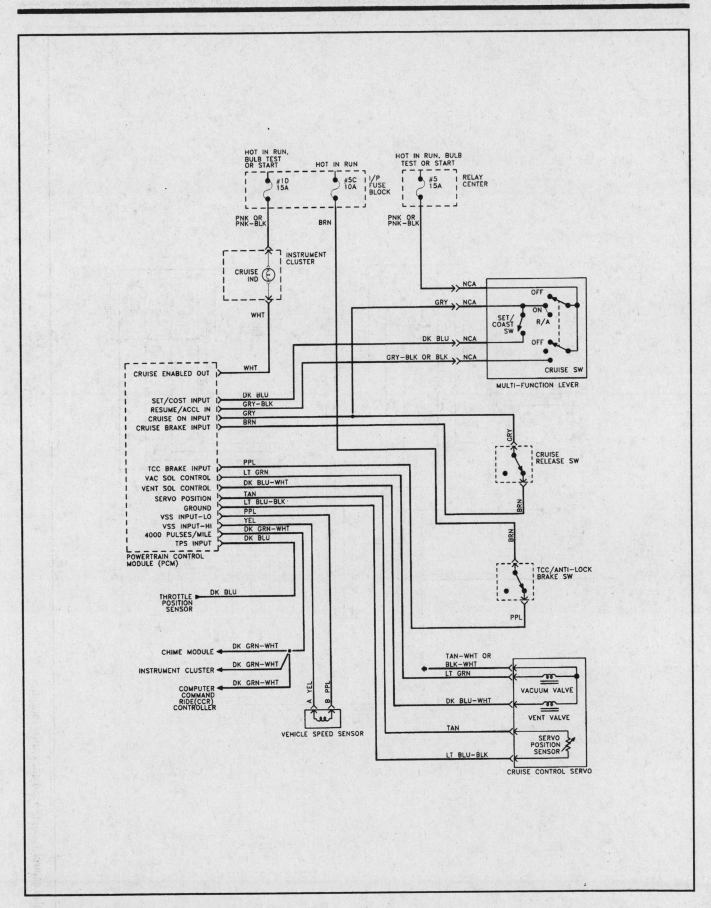

Cruise control system - 1991 and later models

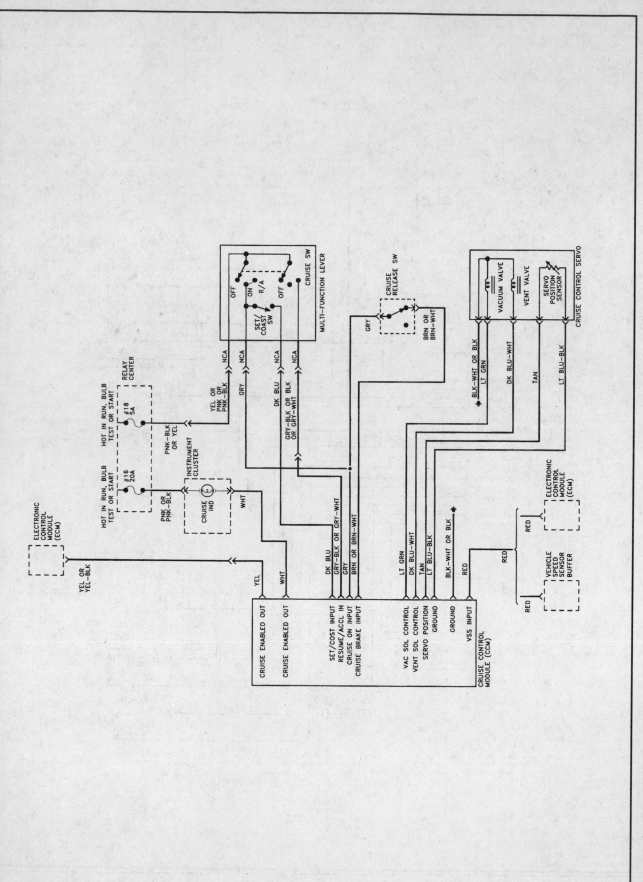

Cruise control system - 1990 and earlier models

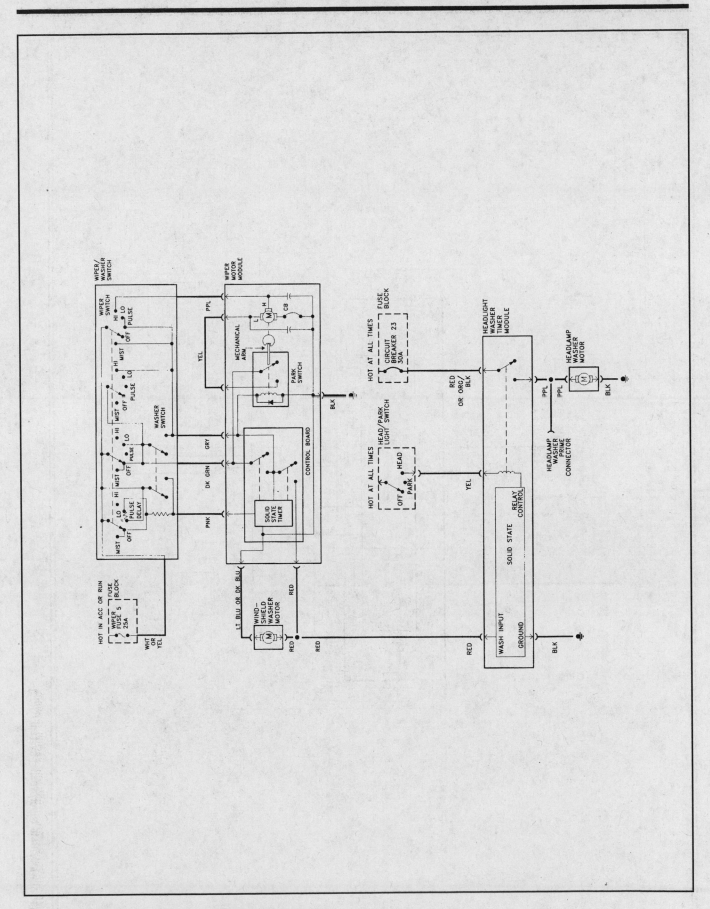

Typical windshield wiper/washer system (Buick/Pontiac)

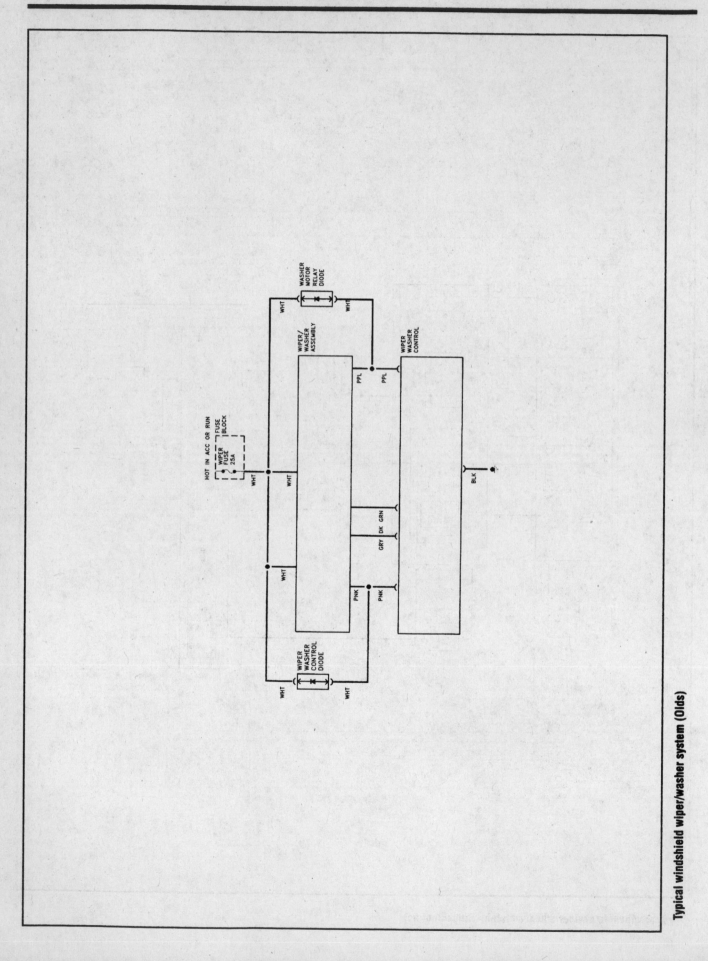

Typical windshield wiper/washer system (Olds)

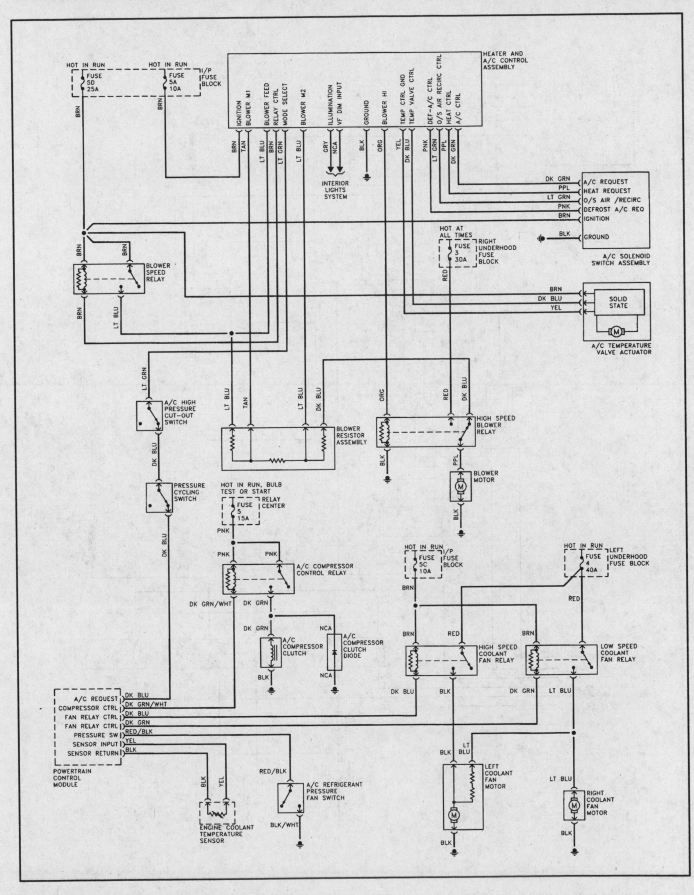

Air conditioning system - 1991 and later models

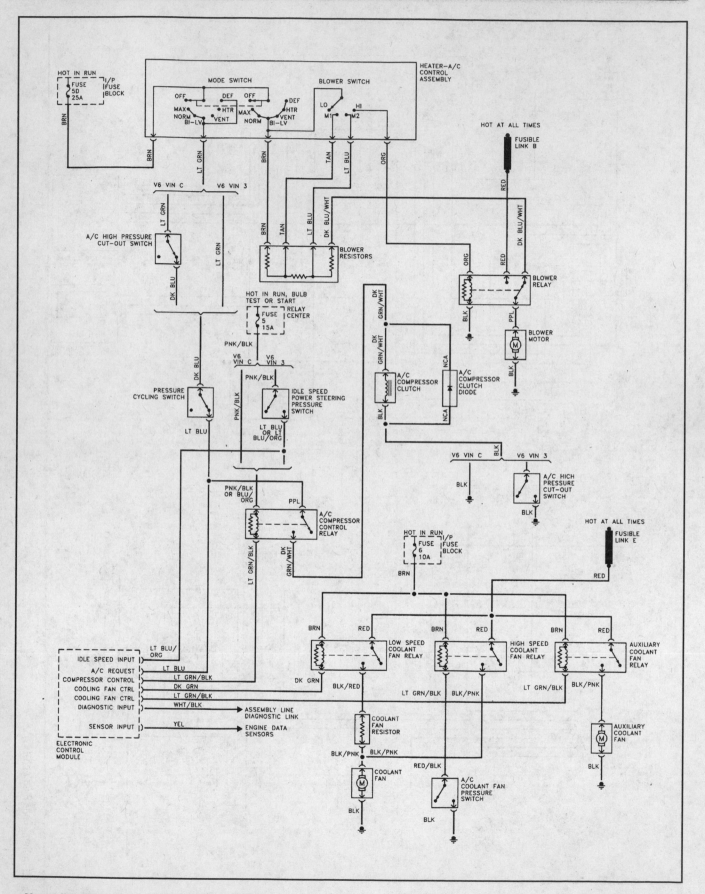

Air conditioning system - 1990 and earlier models

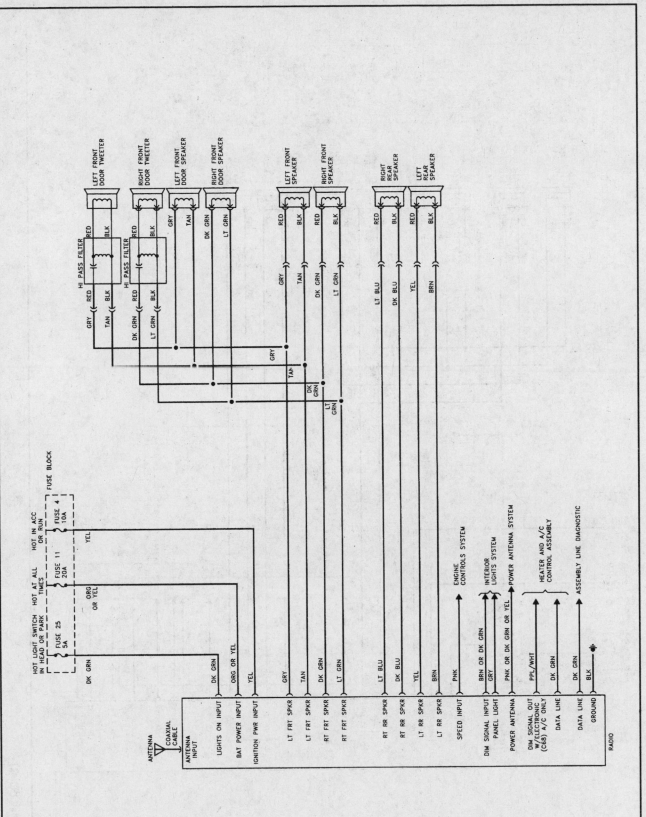

Typical radio circuit (Pontiac/Olds)

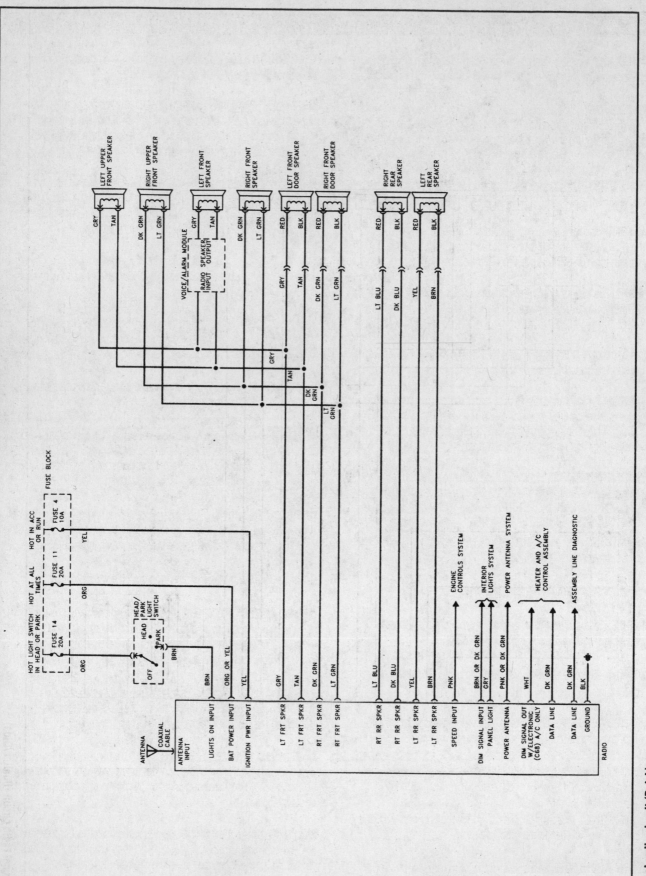

Typical radio circuit (Buick)

GLOSSARY

AIR/FUEL RATIO: The ratio of air-to-gasoline by weight in the fuel mixture drawn into the engine.

AIR INJECTION: One method of reducing harmful exhaust emissions by injecting air into each of the exhaust ports of an engine. The fresh air entering the hot exhaust manifold causes any remaining fuel to be burned before it can exit the tailpipe.

ALTERNATOR: A device used for converting mechanical energy into electrical energy.

AMMETER: An instrument, calibrated in amperes, used to measure the flow of an electrical current in a circuit. Ammeters are always connected in series with the circuit being tested.

AMPERE: The rate of flow of electrical current present when one volt of electrical pressure is applied against one ohm of electrical resistance.

ANALOG COMPUTER: Any microprocessor that uses similar (analogous) electrical signals to make its calculations.

ARMATURE: A laminated, soft iron core wrapped by a wire that converts electrical energy to mechanical energy as in a motor or relay. When rotated in a magnetic field, it changes mechanical energy into electrical energy as in a generator.

ATMOSPHERIC PRESSURE: The pressure on the Earth's surface caused by the weight of the air in the atmosphere. At sea level, this pressure is 14.7 psi at 32°F (101 kPa at 0°C).

ATOMIZATION: The breaking down of a liquid into a fine mist that can be suspended in air.

AXIAL PLAY: Movement parallel to a shaft or bearing bore.

BACKFIRE: The sudden combustion of gases in the intake or exhaust system that results in a loud explosion.

BACKLASH: The clearance or play between two parts, such as meshed gears.

BACKPRESSURE: Restrictions in the exhaust system that slow the exit of exhaust gases from the combustion chamber.

BAKELITE: A heat resistant, plastic insulator material commonly used in printed circuit boards and transistorized components.

BALL BEARING: A bearing made up of hardened inner and outer races between which hardened steel balls roll.

BALLAST RESISTOR: A resistor in the primary ignition circuit that lowers voltage after the engine is started to reduce wear on ignition components.

BEARING: A friction reducing, supportive device usually located between a stationary part and a moving part.

BIMETAL TEMPERATURE SENSOR: Any sensor or switch made of two dissimilar types of metal that bend when heated or cooled due to the different expansion rates of the alloys. These types of sensors usually function as an on/off switch.

BLOWBY: Combustion gases, composed of water vapor and unburned fuel, that leak past the piston rings into the crankcase during normal engine operation. These gases are removed by the PCV system to prevent the buildup of harmful acids in the crankcase.

BRAKE PAD: A brake shoe and lining assembly used with disc brakes.

BRAKE SHOE: The backing for the brake lining. The term is, however, usually applied to the assembly of the brake backing and lining.

BUSHING: A liner, usually removable, for a bearing; an anti-friction liner used in place of a bearing.

CALIPER: A hydraulically activated device in a disc brake system, which is mounted straddling the brake rotor (disc). The caliper contains at least one piston and two brake pads. Hydraulic pressure on the piston(s) forces the pads against the rotor.

CAMSHAFT: A shaft in the engine on which are the lobes (cams) which operate the valves. The camshaft is driven by the crankshaft, via a belt, chain or gears, at one half the crankshaft speed.

CAPACITOR: A device which stores an electrical charge.

CARBON MONOXIDE (CO): A colorless, odorless gas given off as a normal byproduct of combustion. It is poisonous and extremely dangerous in confined areas, building up slowly to toxic levels without warning if adequate ventilation is not available.

CARBURETOR: A device, usually mounted on the intake manifold of an engine, which mixes the air and fuel in the proper proportion to allow even combustion.

CATALYTIC CONVERTER: A device installed in the exhaust system, like a muffler, that converts harmful byproducts of combustion into carbon dioxide and water vapor by means of a heat-producing chemical reaction.

CENTRIFUGAL ADVANCE: A mechanical method of advancing the spark timing by using flyweights in the distributor that react to centrifugal force generated by the distributor shaft rotation.

CHECK VALVE: Any one-way valve installed to permit the flow of air, fuel or vacuum in one direction only.

CHOKE: A device, usually a moveable valve, placed in the intake path of a carburetor to restrict the flow of air.

CIRCUIT: Any unbroken path through which an electrical current can flow. Also used to describe fuel flow in some instances.

CIRCUIT BREAKER: A switch which protects an electrical circuit from overload by opening the circuit when the current flow exceeds a predetermined level. Some circuit breakers must be reset manually, while most reset automatically.

COIL (IGNITION): A transformer in the ignition circuit which steps up the voltage provided to the spark plugs.

COMBINATION MANIFOLD: An assembly which includes both the intake and exhaust manifolds in one casting.

COMBINATION VALVE: A device used in some fuel systems that routes fuel vapors to a charcoal storage canister instead of venting them into the atmosphere. The valve relieves fuel tank pressure and allows fresh air into the tank as the fuel level drops to prevent a vapor lock situation.

COMPRESSION RATIO: The comparison of the total volume of the cylinder and combustion chamber with the piston at BDC and the piston at TDC.

CONDENSER: 1. An electrical device which acts to store an electrical charge, preventing voltage surges. 2. A radiator-like device in the air conditioning system in which refrigerant gas condenses into a liquid, giving off heat.

CONDUCTOR: Any material through which an electrical current can be transmitted easily.

CONTINUITY: Continuous or complete circuit. Can be checked with an ohmmeter.

COUNTERSHAFT: An intermediate shaft which is rotated by a mainshaft and transmits, in turn, that rotation to a working part.

CRANKCASE: The lower part of an engine in which the crankshaft and related parts operate.

CRANKSHAFT: The main driving shaft of an engine which receives reciprocating motion from the pistons and converts it to rotary motion.

CYLINDER: In an engine, the round hole in the engine block in which the piston(s) ride.

CYLINDER BLOCK: The main structural member of an engine in which is found the cylinders, crankshaft and other principal parts.

CYLINDER HEAD: The detachable portion of the engine, usually fastened to the top of the cylinder block and containing all or most of the combustion chambers. On overhead valve engines, it contains the valves and their operating parts. On overhead cam engines, it contains the camshaft as well.

DEAD CENTER: The extreme top or bottom of the piston stroke.

DETONATION: An unwanted explosion of the air/fuel mixture in the combustion chamber caused by excess heat and compression, advanced timing, or an overly lean mixture. Also referred to as "ping".

DIAPHRAGM: A thin, flexible wall separating two cavities, such as in a vacuum advance unit.

DIESELING: A condition in which hot spots in the combustion chamber cause the engine to run on after the key is turned off.

DIFFERENTIAL: A geared assembly which allows the transmission of motion between drive axles, giving one axle the ability to turn faster than the other.

DIODE: An electrical device that will allow current to flow in one direction only.

DISC BRAKE: A hydraulic braking assembly consisting of a brake disc, or rotor, mounted on an axle, and a caliper assembly containing, usually two brake pads which are activated by hydraulic pressure. The pads are forced against the sides of the disc, creating friction which slows the vehicle.

DISTRIBUTOR: A mechanically driven device on an engine which is responsible for electrically firing the spark plug at a predetermined point of the piston stroke.

DOWEL PIN: A pin, inserted in mating holes in two different parts allowing those parts to maintain a fixed relationship.

DRUM BRAKE: A braking system which consists of two brake shoes and one or two wheel cylinders, mounted on a fixed backing plate, and a brake drum, mounted on an axle, which revolves around the assembly.

DWELL: The rate, measured in degrees of shaft rotation, at which an electrical circuit cycles on and off.

ELECTRONIC CONTROL UNIT (ECU): Ignition module, module, amplifier or igniter. See Module for definition.

ELECTRONIC IGNITION: A system in which the timing and firing of the spark plugs is controlled by an electronic control unit, usually called a module. These systems have no points or condenser.

END-PLAY: The measured amount of axial movement in a shaft.

ENGINE: A device that converts heat into mechanical energy.

EXHAUST MANIFOLD: A set of cast passages or pipes which conduct exhaust gases from the engine.

FEELER GAUGE: A blade, usually metal, or precisely predetermined thickness, used to measure the clearance between two parts.

FIRING ORDER: The order in which combustion occurs in the cylinders of an engine. Also the order in which spark is distributed to the plugs by the distributor.

FLOODING: The presence of too much fuel in the intake manifold and combustion chamber which prevents the air/fuel mixture from firing, thereby causing a no-start situation.

FLYWHEEL: A disc shaped part bolted to the rear end of the crankshaft. Around the outer perimeter is affixed the ring gear. The starter drive engages the ring gear, turning the flywheel, which rotates the crankshaft, imparting the initial starting motion to the engine.

FOOT POUND (ft. lbs. or sometimes, ft.lb.): The amount of energy or work needed to raise an item weighing one pound, a distance of one foot.

FUSE: A protective device in a circuit which prevents circuit overload by breaking the circuit when a specific amperage is present. The device is constructed around a strip or wire of a lower amperage rating than the circuit it is designed to protect. When an amperage higher than that stamped on the fuse is present in the circuit, the strip or wire melts, opening the circuit.

GEAR RATIO: The ratio between the number of teeth on meshing gears.

GENERATOR: A device which converts mechanical energy into electrical energy.

HEAT RANGE: The measure of a spark plug's ability to dissipate heat from its firing end. The higher the heat range, the hotter the plug fires.

HUB: The center part of a wheel or gear.

HYDROCARBON (HC): Any chemical compound made up of hydrogen and carbon. A major pollutant formed by the engine as a byproduct of combustion.

HYDROMETER: An instrument used to measure the specific gravity of a solution.

INCH POUND (inch lbs.; sometimes in.lb. or in. lbs.): One twelfth of a foot pound.

INDUCTION: A means of transferring electrical energy in the form of a magnetic field. Principle used in the ignition coil to increase voltage.

INJECTOR: A device which receives metered fuel under relatively low pressure and is activated to inject the fuel into the engine under relatively high pressure at a predetermined time.

INPUT SHAFT: The shaft to which torque is applied, usually carrying the driving gear or gears.

INTAKE MANIFOLD: A casting of passages or pipes used to conduct air or a fuel/air mixture to the cylinders.

JOURNAL: The bearing surface within which a shaft operates.

KEY: A small block usually fitted in a notch between a shaft and a hub to prevent slippage of the two parts.

MANIFOLD: A casting of passages or set of pipes which connect the cylinders to an inlet or outlet source.

MANIFOLD VACUUM: Low pressure in an engine intake manifold formed just below the throttle plates. Manifold vacuum is highest at idle and drops under acceleration.

MASTER CYLINDER: The primary fluid pressurizing device in a hydraulic system. In automotive use, it is found in brake and hydraulic clutch systems and is pedal activated, either directly or, in a power brake system, through the power booster.

MODULE: Electronic control unit, amplifier or igniter of solid state or integrated design which controls the current flow in the ignition primary circuit based on input from the pick-up coil. When the module opens the primary circuit, high secondary voltage is induced in the coil.

NEEDLE BEARING: A bearing which consists of a number (usually a large number) of long, thin rollers.

OHM: (Ω) The unit used to measure the resistance of conductor-to-electrical flow. One ohm is the amount of resistance that limits current flow to one ampere in a circuit with one volt of pressure.

OHMMETER: An instrument used for measuring the resistance, in ohms, in an electrical circuit.

OUTPUT SHAFT: The shaft which transmits torque from a device, such as a transmission.

OVERDRIVE: A gear assembly which produces more shaft revolutions than that transmitted to it.

OVERHEAD CAMSHAFT (OHC): An engine configuration in which the camshaft is mounted on top of the cylinder head and operates the valve either directly or by means of rocker arms.

OVERHEAD VALVE (OHV): An engine configuration in which all of the valves are located in the cylinder head and the camshaft is located in the cylinder block. The camshaft operates the valves via lifters and pushrods.

OXIDES OF NITROGEN (NOx): Chemical compounds of nitrogen produced as a byproduct of combustion. They combine with hydrocarbons to produce smog.

OXYGEN SENSOR: Use with the feedback system to sense the presence of oxygen in the exhaust gas and signal the computer which can reference the voltage signal to an air/fuel ratio.

PINION: The smaller of two meshing gears.

PISTON RING: An open-ended ring with fits into a groove on the outer diameter of the piston. Its chief function is to form a seal between the piston and cylinder wall. Most automotive pistons have three rings: two for compression sealing; one for oil sealing.

PRELOAD: A predetermined load placed on a bearing during assembly or by adjustment.

PRIMARY CIRCUIT: the low voltage side of the ignition system which consists of the ignition switch, ballast resistor or resistance wire, bypass, coil, electronic control unit and pick-up coil as well as the connecting wires and harnesses.

PRESS FIT: The mating of two parts under pressure, due to the inner diameter of one being smaller than the outer diameter of the other, or vice versa; an interference fit.

RACE: The surface on the inner or outer ring of a bearing on which the balls, needles or rollers move.

REGULATOR: A device which maintains the amperage and/or voltage levels of a circuit at predetermined values.

RELAY: A switch which automatically opens and/or closes a circuit.

RESISTANCE: The opposition to the flow of current through a circuit or electrical device, and is measured in ohms. Resistance is equal to the voltage divided by the amperage.

RESISTOR: A device, usually made of wire, which offers a preset amount of resistance in an electrical circuit.

RING GEAR: The name given to a ring-shaped gear attached to a differential case, or affixed to a flywheel or as part of a planetary gear set.

ROLLER BEARING: A bearing made up of hardened inner and outer races between which hardened steel rollers move.

ROTOR: 1. The disc-shaped part of a disc brake assembly, upon which the brake pads bear; also called, brake disc. 2. The device mounted atop the distributor shaft, which passes current to the distributor cap tower contacts.

SECONDARY CIRCUIT: The high voltage side of the ignition system, usually above 20,000 volts. The secondary includes the ignition coil, coil wire, distributor cap and rotor, spark plug wires and spark plugs.

SENDING UNIT: A mechanical, electrical, hydraulic or electromagnetic device which transmits information to a gauge.

SENSOR: Any device designed to measure engine operating conditions or ambient pressures and temperatures. Usually electronic in nature and designed to send a voltage signal to an on-board computer, some sensors may operate as a simple on/off switch or they may provide a variable voltage signal (like a potentiometer) as conditions or measured parameters change.

SHIM: Spacers of precise, predetermined thickness used between parts to establish a proper working relationship.

SLAVE CYLINDER: In automotive use, a device in the hydraulic clutch system which is activated by hydraulic force, disengaging the clutch.

SOLENOID: A coil used to produce a magnetic field, the effect of which is to produce work.

SPARK PLUG: A device screwed into the combustion chamber of a spark ignition engine. The basic construction is a conductive core inside of a ceramic insulator, mounted in an outer conductive base. An electrical charge from the spark plug wire travels along the conductive core and jumps a preset air gap to a grounding point or points at the end of the conductive base. The resultant spark ignites the fuel/air mixture in the combustion chamber.

SPLINES: Ridges machined or cast onto the outer diameter of a shaft or inner diameter of a bore to enable parts to mate without rotation.

TACHOMETER: A device used to measure the rotary speed of an engine, shaft, gear, etc., usually in rotations per minute.

THERMOSTAT: A valve, located in the cooling system of an engine, which is closed when cold and opens gradually in response to engine heating, controlling the temperature of the coolant and rate of coolant flow.

TOP DEAD CENTER (TDC): The point at which the piston reaches the top of its travel on the compression stroke.

TORQUE: The twisting force applied to an object.

TORQUE CONVERTER: A turbine used to transmit power from a driving member to a driven member via hydraulic action, providing changes in drive ratio and torque. In automotive use, it links the driveplate at the rear of the engine to the automatic transmission.

TRANSDUCER: A device used to change a force into an electrical signal.

TRANSISTOR: A semi-conductor component which can be actuated by a small voltage to perform an electrical switching function.

TUNE-UP: A regular maintenance function, usually associated with the replacement and adjustment of parts and components in the electrical and fuel systems of a vehicle for the purpose of attaining optimum performance.

TURBOCHARGER: An exhaust driven pump which compresses intake air and forces it into the combustion chambers at higher than atmospheric pressures. The increased air pressure allows more fuel to be burned and results in increased horsepower being produced.

VACUUM ADVANCE: A device which advances the ignition timing in response to increased engine vacuum.

VACUUM GAUGE: An instrument used to measure the presence of vacuum in a chamber.

VALVE: A device which control the pressure, direction of flow or rate of flow of a liquid or gas.

VALVE CLEARANCE: The measured gap between the end of the valve stem and the rocker arm, cam lobe or follower that activates the valve.

VISCOSITY: The rating of a liquid's internal resistance to flow.

VOLTMETER: An instrument used for measuring electrical force in units called volts. Voltmeters are always connected parallel with the circuit being tested.

WHEEL CYLINDER: Found in the automotive drum brake assembly, it is a device, actuated by hydraulic pressure, which, through internal pistons, pushes the brake shoes outward against the drums.

NOTES

A

MASTER INDEX

Notes